Prebiotic Evolution and Astrobiology

J. Tze-Fei Wong, PhD
Applied Genomics Center
Fok Ying Tung Graduate School
and
Department of Biochemistry
Hong Kong University of Science and Technology
Clear Water Bay, Hong Kong, China

Antonio Lazcano, PhD
Universidad Nacional Autonoma de Mexico
Mexico City, Mexico

LANDES BIOSCIENCE
AUSTIN, TEXAS
USA

PREBIOTIC EVOLUTION AND ASTROBIOLOGY

Landes Bioscience

Please address all inquiries to the Publisher:
Landes Bioscience, 1002 West Avenue, Austin, Texas, USA 78701
Phone: 512.637.6050; FAX: 512.637.6079
www.landesbioscience.com

ISBN: 978-1-58706-330-5

Library of Congress Cataloging-in-Publication Data

Prebiotic evolution and astrobiology / [edited by] J. Tze-Fei Wong, Antonio Lazcano.
    p. ; cm.
  Includes bibliographical references and index.
  ISBN 978-1-58706-330-5
  1. Life--Origin. 2. Exobiology. 3. Biomolecules. 4. RNA. 5. Interstellar matter. I. Wong, J. Tze-Fei (Jeffrey Tze-Fei) II. Lazcano, Antonio.
  [DNLM: 1. Evolution, Molecular. 2. Biogenesis. 3. Exobiology. 4. RNA--genetics. QU 475 P922 2009]
  QH325.P7336 2009
  576.8'3--dc22
                    2009001446

While the authors, editors and publisher believe that drug selection and dosage and the specifications and usage of equipment and devices, as set forth in this book, are in accord with current recommendations and practice at the time of publication, they make no warranty, expressed or implied, with respect to material described in this book. In view of the ongoing research, equipment development, changes in governmental regulations and the rapid accumulation of information relating to the biomedical sciences, the reader is urged to carefully review and evaluate the information provided herein.

DEDICATION

To the memory of Stanley Miller and Leslie Orgel
—Two friendliest giants whose shoulders are the tallest to stand on to peer into the unknown.

JEFFREY TZE-FEI WONG has worked on various aspects of the genetic code for over thirty years, including its birth, its coevolution with the amino acid biosynthetic pathways, its mutation to open up the encoding of unnatural amino acids, and the tracing of its universality to a methanogen root of life. Earlier at the University of Toronto, and in recent years at the Hong Kong University of Science and Technology, he has developed the application of transfer RNAs and aminoacyl-tRNA synthetases as phylogenetic probes. He is the author of *Kinetics of Enzyme Mechanisms,* which is focused on multisubstrate enzyme reactions.

ANTONIO LAZCANO is considered the most distinguished Latin American evolutionist and a world leading authority on the study of the origins and early evolution of life. Born in Mexico, he has worked for over thirty years in a number of problems related to the emergence of life. His books include La Bacteria Prodigiosa (The Miraculous Bacteria), La Chispa de la Vida (The Spark of Life) and El Origen de la Vida (The Origin of Life). He is also the first Latin American scientist that has been Chair of the Gordon Conference on the Origins of Life, and has served on a number of advisory boards including ones for NASA and other international organizations. He was the President of the International Society for the Study of the Origins of Life, being the first Latin American scientist elected to this position. He received an honorary degree from the University of Milano (2008), as well as the Golden Medal for Biological Research (Puebla, 1990), the Medal of the Founders of the Universidad San Francisco de Quito (2005), and the Francesco Redi Medal of the Italian Society of Astrobiology (2008).

CONTENTS

EDITORS

J. Tze-Fei Wong
Applied Genomics Center
Fok Ying Tung Graduate School
and
Department of Biochemistry
Hong Kong University of Science and Technology
Clear Water Bay, Hong Kong, China
Email: bcjtw@ust.hk
Chapters 1,9,14,15

Antonio Lazcano
Universidad Nacional Autonoma de Mexico
Mexico City, Mexico
Email: alar@hp.fciencias.unam.mx

CONTRIBUTORS

Patrice Coll
Universités Paris
Créteil, France
Email: pcoll@lisa.univ-paris12.fr
Chapter 4

Hervé Cottin
Universités Paris
Créteil, France
Email: cottin@lisa.univ-paris12.fr
Chapter 5

Didier Despois
CNRS
Bordeaux, France
Email: despois@obs.u-bordeaux1.fr
Chapter 5

Andrew D. Ellington
Department of Chemistry
    and Biochemistry
Center for Systems
    and Synthetic Biology
University of Texas at Austin
Austin, Texas, USA
Email: andy.ellington@mail.utexas.edu
Chapter 11

Massimo Di Giulio
Laboratory of Molecular Evolution
Institute of Genetics and Biophysics
Naples, Napoli, Italy
Email: digiulio@igb.cnr.it
Chapter 13

Randall A. Hughes
Department of Chemistry
    and Biochemistry
Center for Systems and Synthetic
    Biology
University of Texas at Austin
Austin, Texas, USA
Email: hughes@mail.utexas.edu
Chapter 11

James F. Kasting
Department of Geosciences
Penn State University
University Park, Pennslyvania, USA
Email: kasting@geosc.psu.edu
Chapter 8

Pier Luigi Luisi
Biology Department
University of Rome
Rome, Italy
Email: luisi@mat.ethz.ch
Chapter 12

Pierre-Alain Monnard
Los Alamos National Laboratory
Earth and Environmental Science
    Division
Los Alamos, New Mexico, USA
Email : pmonnard@lanl.gov
Chapter 10

Knud H. Nierhaus
Max-Planck-Institut für Molekulare
    Genetik
Berlin, Germany
Email: nierhaus@molgen.mpg.de
Chapter 2

Markus Pech
Max-Planck-Institut für Molekulare
    Genetik
Berlin, Germany
Email: pech@molgen.mpg.de
Chapter 2

Sandra Pizzarello
Department of Chemistry
    and Biochemistry
Arizona State University
Tempe, Arizona, USA
Email: pizzar@asu.edu
Chapters 6,7

François Raulin
Universités Paris
Créteil, France
Email: raulin@lisa.univ-paris12.fr
Chapter 3

Fabien Stalport
Universités Paris
Créteil, France
Email: stalport@lisa.univ-paris12.fr
Chapter 4

Pasquale Stano
Biology Department
University of Rome
Rome, Italy
Email: stano@uniroma3.it
Chapter 12

PREFACE

With the accelerating pace of genomic analysis and space exploration, the field of prebiotic evolution and astrobiology is poised for a century of unprecedented advances, and there is a need for textbooks for students. The authors of this book, aware of the difficulty of covering the multifaceted subject by any single author, have decided to combine their efforts to provide a suitable book for beginning students from varied disciplines. The book stemmed from a meeting on Basic Questions about the Origin of Life at the Ettore Majorana Foundation and Centre for Scientific Culture in Erice, Sicily in October 2006, where Pier Luigi Luisi laid out a balanced program to approach the subject. The present book basically follows that program.

Chapter 1 addresses, on the foundation of cell biology and molecular biology, the millenia-old questions posed by the essence of life and defining moment of life. Chapter 2 characterizes the minimal cell as the goal of prebiotic evolution. Chapters 3-5 present the realm of astrobiological exploration, centering on Europa, Titan, Enceladus, Mars and the comets, which is highlighted by the historical finding of water on Mars in 2008 as one of the farthest reaching frontiers of science in the 21st Century. Chapters 6 and 7 are focused on meteoric chemistry and the discovery of L-enantiomeric excesses of chiral amino acids on meteorites, which reveals a surprising trait shared by space and biological chemistries. Since primitive Earth was the stage of prebiotic evolution, its conditions were of critical importance to the emergence of life: those conditions are delineated in Chapter 8. The prebiotic syntheses of biomolecules small and large culminating in the appearance of RNA is discussed in Chapters 9 and 10, and Chapter 11 probes the inner workings of the RNA world. Chapter 12 describes the principles of the functional protocell and different models of precellular evolution. The puzzle of split genes of transfer RNA and proteins is analyzed in Chapter 13. The origin, evolution and mutation of the genetic code are discussed in Chapter 14, and Chapter 15 examines the nature and genome of the Last Universal Common Ancestor at the root of life, as well as the formation of different biological domains from this root.

Antonio Lazcano, despite family health concerns over the past months, has devoted a great deal of time and hard work in adjusting and finalizing the balance of the chapters. The many authors have produced most thoughtful, authoritative and readable chapters on their areas so that the student may grasp the background, objective and progress of the major research foci in the field. I am grateful to Hannah Xue, Flora Mat, Ka-Lok Tong, Siu-Kin Ng and Jianhuan Chen for their dedicated research collaboration and assistance in the preparation of this book. Without the understanding and support of my wife Eva, the book would not have been undertaken or completed.

J. Tze-Fei Wong, PhD

CHAPTER 1

Introduction

J. Tze-Fei Wong*

1.1 Old Debate, Young Science

Few human cultures do not ponder the question of the origin of life and try to come up with an answer. The scientific culture is no exception and the search for an explanation has spanned centuries.

Up to the end of the 17th Century, there was no reason to question the occurrence of spontaneous generation as an explanation of departures from like begetting like in biological reproduction, exemplified by the origin of parasitic worms such as the fluke inside the human body. This view was strengthened by the mechanical philosophy of Rene Descartes who reduced natural science to matter and motion, such that agitation of decomposed matter could create an organism. Francesco Redi, however, described experiments proving that maggots were not generated by meat in a state of putrefaction but solely by the flies coming to lay their eggs. Georges Cuvier likewise found spontaneous generation unacceptable and declared, "Life has always arisen from life. We see it being transmitted and never being produced." Finally, spontaneous generation was laid to rest when Louis Pasteur showed that sterilization by heat abolished the appearance of bacterial spoilage.

Yet disproving spontaneous generation of worms and bacteria over days or years is not tantamount to disproving spontaneous generation of life over eons. The origin of life on primitive Earth was clearly spontaneous generation of a different order, a more fitting expression of the philosophy of Descartes with embodiment of matter in atoms and molecules and motion in the kinetic energies they need to enter into chemical evolution. In 1828 F. Wohler had accomplished the synthesis of urea, a bodily constituent, from cyanate and ammonium salt, thus breaking down the conceptual barrier between inorganic and organic compounds. In early 20th Century, Archibald Macallum proposed that life originated from particulate matter: "When we seek to explain the origin of life, we do not require to postulate a highly complex organism...as being the primal parent of all, but rather one which consists of a few molecules only and of such a size that it is beyond the limits of vision with the highest power of the microscope," and the process would go through countless molecular combinations, such that "one giving the right composition resulted in ultramicroscopic particles endowed with the chemical properties of ultramicroscopic organisms."

Thus prebiotic evolution and natural selection were called upon to deliver life. This was followed by W. Lob's synthesis of simple amino acids such as glycine by exposing wet formamide to silent electrical discharge and to ultraviolet light and E.C. Baly's production of formaldehyde and sugar from the action of ultraviolet light on carbon dioxide and water. A. Oparin formulated in 1924 a prebiotic scenario where organic compounds built from inorganics were organized first into coazervate droplets; "as soon as these droplets became separated from the surrounding medium by a more or less definite border, they at once acquired a certain degree of individuality....competition in growth velocity" and at the end "simplest primary organisms have emerged." JBS Haldane could not be more enthusiastic over the potential of prebiotic synthesis of organic compounds that "must have accumulated till the primitive oceans reached the consistency of dilute soup," thus giving birth to the legendary 'primordial soup'.[6-8]

In 1953, Stanley Miller, stimulated by Harold Urey's suggestion of the importance of a reducing atmosphere to prebiotic synthesis, proposed and carried out prebiotic synthesis by passing electric discharge through a simulated primitive Earth atmosphere of methane, ammonia, hydrogen and water. The gases were electrified to produce amino acids and his results electrified the world.[9] Three centuries after the initiation of scientific debate on spontaneous generation, prebiotic evolution has become a frontier fueled by advances in biochemistry, molecular biology, genomics and space exploration. Few areas of scientific enquiry may be expected to yield as many fundamental discoveries in the 21st Century as prebiotic evolution and astrobiology.

1.2 The Astrobiological Dimension

(i) Requirement for Water

Water is constantly looked out for in astrobiological explorations.[10] It is important for a number of reasons. First, water has attractive physical properties. Its anomalous expansion at the freezing point and its high specific heat make it more difficult for a body of water to become frozen solid throughout its entire volume. With a high dielectric constant it is an excellent solvent for charged ions and hydrogen bonding.[11] Without ions and hydrogen bonds, energy generation through proton gradients and familiar types of template-directed replications cannot proceed. Proteins, nucleic acids and similar polymers would depend on water for solvation. Dehydrated, the living cell collapses into an inactive state and often outright death.

Secondly, atmospheric carbon dioxide dissolves in water to form carbonic acid, which weathers silicate rocks (mainly $CaSiO_3$) to form calcium carbonate ($CaCO_3$, or chalk). The process removes carbon dioxide and deposits it as insoluble chalk on the ocean floor. All carbon from Earth's atmosphere would be removed in about 400 million years. The main regeneration process, the 'carbonate

*J. Tze-Fei Wong—Applied Genomics Center, Fok Ying Tung Graduate School and Department of Biochemistry, Hong Kong University of Science and Technology, Clear Water Bay, Hong Kong, China. Email: bcjtw@ust.hk

Prebiotic Evolution and Astrobiology, edited by J. Tze-Fei Wong and Antonio Lazcano. ©2009 Landes Bioscience.

silicate cycle', is brought about by plate tectonics: the chalk on the ocean floor is subducted into the mantle of the Earth, where it is heated up and cracked, releasing carbon dioxide back to the atmosphere via volcanoes. Without this regeneration, which depends on the presence of liquid water, the depletion of free CO_2 on Earth would cause atmospheric synthesis of organic compounds, a rich heterotrophic source of nutrients, as well as any CO_2-dependent autotrophy, to cease.[12]

Thirdly, on Earth the construction of cell membranes depends on the immiscibility of hydrophilic (water liking) substances that dissolve well in water and hydrophobic (water disliking) substances such as oils, fats and petroleum, that like to move away from water. When these two classes of compounds are stirred together, they will not mix but rapidly separate into two phases, as witnessed by the floating of oil slicks on the sea despite churning by giant waves. The immiscibility of these two classes of substances is prerequisite to the construction of cell membranes. Membrane constituents such as phospholipids are amphiphilic, being hydrophilic at the heads of their structures but hydrophobic at the tail end. When suspended in water they line themselves up in a double-layer with the tails from the two layers nestled together, away from the water and maximizing the hydrophobic-hydrophobic contacts between the two layers. The hydrophilic heads of the two layers will face outward, comfortably surrounded by water molecules. These double membranes impart to any cell a compartmental individuality that is essential to the exercise of natural selection at the cellular level. Without water, membranes cannot form. Compounds such as ethanol, dimethyl formamide and dimethyl sulfoxide, even though they possess some of the solvation properties of water, are too miscible with hydrophobic solvents to support the ready formation of membranes. No membranes, no natural selection, how is life going to emerge anywhere? That is why the identification of water in martian soil by NASA's 2008 Phoenix Mars Lander represents such a magnificent breakthrough.

(ii) Habitable Zone

A fundamental concept regarding the suitability of any extraterrestrial site for astrobiological exploration is the Habitable Zone (HZ)—also called the Goldilocks Zone where it not too cold and not too hot. Living matter on Earth consists mostly of the chemical elements HCNO+PSFe. HCNO are elements with low atomic numbers in the periodic table. They can form single (in the case of H) or multiple (in the case of C, N and O) bonds with other atoms and give rise to biopolymers. The construction of biopolymers using much larger atoms would yield macromolecules that are far more limited by steric hindrance with respect to their allowed conformations. Considering what fine tunings of protein, enzyme, nucleic acids and ribozyme conformations are needed by earthly organisms, these lighter HCNO elements would most likely have to form the working core of biomolecules on any planets or moons. However, the multiply bonded molecules made of HCNO typically cannot withstand exceedingly high temperatures. At the other end of the scale, all chemical reactions slow down at low temperatures, decreasing by a factor of two to three for every ten degrees Celsius. This decrease stems from having less energies in the reactant molecules at very low temperatures to cross activation barriers for chemical reactions to occur and applies to all planets (and their moons) in the universe. Therefore it would be difficult to develop life where the temperatures are either too hot or too cold. The range set by the freezing and boiling points of water would narrow the habitable range more specifically to about –5°C to 110°C, the boundaries tolerated by extremophiles on Earth.

Accordingly, based on the premise that habitability requires the presence of liquid water on the planet's surface, the inner edge of the HZ for Earth-like planets of a main sequence star with $CO_2/H_2O/N_2$ atmospheres is determined by loss of water via photolysis and hydrogen escape. The outer edge of the HZ is determined by the formation of CO_2 clouds, which cool a planet's surface by increasing its albedo (reflection of solar/star energy). Conservative estimates for these distances in the Solar System are 0.95 and 1.37 AU, respectively (one astronomical unit, or AU, is equal to 149.6 million km, the average distance of the Earth from the Sun). Because the Sun increases in luminosity as it ages, the HZ evolves outward in time. A conservative estimate for the width of the continuously habitable zone over the history of the Solar System is 0.95 to 1.15 AU. However, other factors can also influence the HZ:[10,12,13]

a. For the Earth, should the Sun weaken unpredictably, CO_2 would condense into CO_2 clouds to increase albedo and lower surface temperature, which in turn would bring more snow and ice and increase albedo further, leading to 'runaway glaciation'. On the other hand, in the event of extreme emission of CO_2 increasing the greenhouse effect, or unpredicted increase in the intensity of solar radiation, surface temperature of the Earth would be elevated, vaporizing more water to the atmosphere and retarding the escape of infrared radiation into space to lead toward a 'runaway greenhouse'.

b. A central star exerts tidal effects on a planet and the tidal braking slows down the axial rotation of the planet. Within a tidal lock radius, the axial rotation becomes synchronized with its orbital rotation. When that happens, one side of the planet always faces the star and it is persistently heated. The other side is in permanent darkness. Water may be completely evaporated through constant heating on the star-lit side even though the temperature of the planet averaged over its two sides is not extreme.

c. Heavier elements such as Fe could furnish reversible reduction-oxidation reactions coupled to energy generation. Therefore a planet likely needs some heavy atoms.

d. The Earth collided with a sizable heavenly body to create the Moon when it was possibly 500 million years old, but since that time there occurred only relatively minor asteroid bombardments. To be potentially habitable over a useful duration of time, a planet has to be located in galactic regions unexposed to frequent large collisions.

e. Since the carbonate silicate cycle is dependent on not only water but also plate tectonics, some plate tectonics movements are necessary. On the other hand, overly unstable plate tectonics will destabilize life.

f. Too small a planet or moon, as is the case of the Moon, will not have strong enough gravity to maintain water or an atmosphere. The surface of too large a planet, like Jupiter, will be covered by a massive ocean of liquid hydrogen.

g. Stellar lifetime on the main sequence decreases with mass. For central stars exceeding 2.2 solar masses, the central hydrogen burning period will last no more than 0.8 billion years, which based on the Earth's experience might be long enough to develop life but not intelligent life.

Given the demanding conditions that must be met by HZ, life is not expected to be a frequent occurrence in the universe. However, the search for extraterrestrial life represents such a unique quest that astrobiological efforts are intensifying, aiming at the most promising intrasolar target sites described in Chapters 3-5. Eventually, as more and more extrasolar planets come to be discovered, the likelihood will increase that some of them too may fall within the HZ of their central stars.

(iii) Panspermia

Primitive Earth was a hospitable site to begin prebiotic evolution, but not necessarily the only conceivable site. In the 19th Century H.E. Richter proposed *lithopanspermia*, advocating that meteorites carried germ cells to Earth from outer space. Early in the 20th Century Svante Arrhenius proposed *radiopanspermia*, suggesting that microbes could have been ejected from planets and scattered in the galaxy. Carried by the radiation pressure of stars, they would arrive on Earth and seed the planet with life. P. Becquerel studied UV interactions with spores and bacteria at very low temperatures in high vacuum and concluded that microbes would be destroyed on the way to Earth.

In recent years, experiments on spacelabs and satellites (LDEF, EURECA, FOTON) identified solar UV radiation as the most destructive factor against unshielded spores, but spores could survive if protected inside rocks against solar UV and galactic cosmic radiation,[7,14] even though secure protection inside rocks might imply problematic release from the rocks as well.

In view of the possible yet unlikely occurrence of Panspermia, it is fitting for scientific investigation to focus on prebiotic evolution on Earth. Besides, the planet where any Panspermia embarked from must have been as hospitable as early Earth to the life forms it sent over. Thus the conditions 'over there' would not be too unlike those on early Earth and much of the understanding gained regarding prebiotic evolution on early Earth is applicable to prebiotic evolution 'over there' as well.

1.3 Pathway to the Cell

(i) Some Basic Questions

The study of prebiotic evolution is defined by basic questions that need to be resolved before the emergence of a living cell can be outlined. Besides the questions examined in other chapters of this book, the present chapter addresses the following questions:

    I. Was catalysis essential to life right from the start?
    II. What were the energy requirements for prebiotic evolution?
    III. Which came first—membranes, replication or metabolism?
    IV. Which came first—heterotrophy or autotrophy?

(ii) Catalysis

Life requires an aqueous medium and moderate temperatures to function, but these are in general not the optimal conditions for organic synthesis. Accordingly, most of the biochemical reactions occurring inside the cell are catalyzed by enzymes, or at earlier times by ribozymes and their analogues. However, an open question is whether prebiotic evolution could have begun without catalysis. The generation time for the uncatalyzed replication of a primitive RNA genome is estimated to be of the order of 4×10^8 years, 10^{10} fold that of a ribozyme-catalyzed one. Thus only two to three uncatalyzed replications could be accommodated in the first 10^9 years of Earth's history. There could be no meaningful biological evolution at this rate. Furthermore, because the multiplication of any RNA genome requires that its physical life time exceeds the time it takes to replicate, the *stability theorem* applies:[15]

$$kLT < 1$$

where k is the rate constant for the hydrolysis of a phosphodiester bond in RNA, L the number of phosphodiester bonds in the genome and T the replication time. For a primitive ribo-organism with three 50-base RNA genes, $L = 147$ and $k = 1.5 \times 10^{-9}$ per minute at pH 7. Therefore T must be less than 8.6 years. So a ribo-organism would face extinction at any time unless it replicated within 8.6 years. To achieve such fast replication, catalysis would be essential right from the beginning. The constraint arising from the stability theorem is equally stringent on any genomes besides RNA genomes, inside or outside the Solar System.

(iii) Energy

The equilibrium position of a chemical reaction is determined by the energetics of the reaction:

$$\Delta G^{o\prime} = -RT\ln K$$

where $\Delta G^{o\prime}$ is the standard free energy change at pH 7, R the gas constant, T the temperature and K the equilibrium constant. When $\Delta G^{o\prime}$ is negative, the reaction is exergonic and its equilibrium position favors the conversion of reactants to reaction products. When $\Delta G^{o\prime}$ is positive, the reaction is endergonic and its equilibrium position does not favor the conversion of reactants to products; instead, it favors the reverse reaction. Any prebiotic or biotic biochemical system depends on many endergonic chemical syntheses (and transports against concentration gradients). An effective means to make an endergonic reaction go is to couple it with an exergonic reaction, so that the combined reaction attains a favorable K, the same way the endergonic process of a car climbing up a hill can be coupled to the exergonic combustion of gasoline. For example, the enzyme hexokinase catalyzes the following reaction:

$$\text{Glucose} + \text{ATP} \rightarrow \text{Glucose-6-phosphate} + \text{ADP}$$

where the exergonic hydrolysis of ATP to ADP (−7.3 kcal/mole) is coupled to the endergonic conversion of glucose to glucose-6-phosphate (+3.3 kcal/mole). Since the energy released by ATP hydrolysis exceeds that needed for glucose-6-phosphate formation, the coupled reaction proceeds well. Likewise, ATP hydrolysis is employed in the body to drive RNA synthesis, DNA synthesis, protein synthesis, biosynthesis of nucleotides, amino acids, polysaccharides, hormone secretion, muscle contraction, nerve conduction etc. The ATP consumed is in turn recharged by glycolytic and mitochondrial phosphorylations. While ATP represents the main energy currency, or energy carrier, of the cell to-day, other high-energy compounds capable of releasing a substantial amount of energy upon hydrolysis may also take the place of ATP, e.g., use of creatine phosphate for energy storage in muscles and of thioesters in fatty acid biosynthesis. Inorganic polyphosphates and thioesters[1,16,17] could be important energy carriers in the prebiotic world prior to the adoption of ATP as standard energy currency of the living world.

For the bacterium *Escherichia coli*, which weighs 2.8×10^{-13} g/cell and contains 4.6×10^6 basepairs of DNA/cell, one round of cell duplication requires 1.7×10^{10} ATP molecules, which translates to a 3,700 ATP molecules per DNA basepair replicated.[15] Since the construction of a DNA double helix consumes only two nucleoside triphosphates per base pair, the cost of operating a living cell is far higher than the construction of a nucleic acid molecule. The minimal genome of a late-stage prebiotic life form contained at least 150 genes (Section 2.7), or about 50,000 basepairs, equivalent to 1% of an *E. coli* genome. Since the efficiency of energy utilization in a prebiotic life form would be lower than that of *E. coli*, one round of prebiotic life form duplication would need at least 1.7×10^8 ATP or thioester or polyphosphate molecules. The high energy bill underlines the importance of energy sources in prebiotic evolution.

Combustion of gasoline is an oxidation and so is the conversion of foodstuff in our bodies to carbon dioxide. In general, reduction-oxidation, or redox, reactions are among the most important exergonic reactions employed to drive energy-carrier production in modern cells and expectedly also in primitive life forms. There were on primitive Earth a variety of primary electron donors and acceptors that could function as reductants and oxidants respectively to provide redox reactions for energy generation:[18]

> Primary electron donors:
> H_2, H_2S, S^0, Fe^{2+}, CH_4, $(NH_4)^+$, CH_2O, photo-electrons
>
> Primary electron acceptors:
> CO_2, CO, S^0, NO, $(SO_4)^{2-}$

Among them, photo-electrons were produced by photosynthesis, CO by atmospheric reaction and NO by lightning. All the others were available from volcanoes or hydrothermal vents. The major electron acceptors were CO_2 and sulfate. A particular constant source of electron donor would be H_2 generated at hydrothermal vents from the conversion of olivine ($Mg_{1.8}Fe_{0.2}SiO_4$) to serpentine ($Mg_{2.7}Fe_{0.3}Si_2O_5[OH]_4$) in the geochemical process of serpentinization.[19] Various redox reactions could contribute to the abiotic synthesis of biomolecules in the environment. Later on, they could also participate in cellular electron transports linked to energy carrier production.

(iv) Membrane, Replication or Metabolism First?

Aside from the prerequisites of catalysis and energy, the three components recognized to be essential to life are membranes needed to provide a discrete structural unit, replicators embodying the informational machinery and metabolism comprising the catalyzed chemical changes. These components must be integrated to yield a living cell. The question is, which of them led the way—did life's emergence follow a Membrane First, Replication First or Metabolism First pathway?[20]

Of the three components, the role of membranes was more straightforward. They might not need a great deal of help from the replicators or metabolites to get started. The amphiphiles supplied by meteorites[21] and aqueous Fischer-Tropsch reactions[22] could spontaneously form membranous vesicles with the ability to grow and divide (Section 12.4). While vesicles could facilitate RNA polymerization (Section 10.4) and maintain metabolite concentrations, their foremost contribution was straightforward: they had to provide discrete housing for the replicators and metabolites to enable natural selection at the level of the individual 'cell'. Prior to moving into the membrane vesicles, the replicators and metabolites could be adsorbed to mineral and clay surfaces, or confined in mineral micro-compartments,[23,24,17] but evolution under those conditions would have to be more communal than cellular in character.

The current debate revolves on the leadership of replication versus metabolism. From the Replication First viewpoint, ribozymes with their complements could self-replicate and through mutations give rise to novel metabolic reactions. In comparison, metabolism was unlikely to be capable of evolution: it is doubtful that the constituents in a metabolic cycle could organize themselves and autocatalytic cycles are generally beset by inherent difficulties.[25,26] So, given the problems of metabolites trying to evolve on their own, replication had to lead metabolism.

From the Metabolism First viewpoint, the discovery that a simple set of low molecular-weight, water soluble organic compounds with low heats of combustion included all the organic acids of the tricarboxylic acid (TCA) cycle is particularly instructive. It suggests that these organic acids might organize themselves into a reductive-TCA cycle. This way, the first steps of prebiotic evolution would be guided by molecular logic.[27] In contrast, if RNA replicators generated their sequences randomly, the resultant sequences were mostly useless. It is estimated that a triple stem-loop structure containing 40-60 nucleotides may offer a reasonable prospect of functioning as a replicase ribozyme. A complete library consisting of just one copy each of the 10^{24} possible 40-mers would weigh about one kilogram. If two or more copies of the sought-for 40-mer are needed, the

requirement increases to a weight of RNA equal to the mass of the Earth.[28] Therefore, left to themselves, the replicators would be totally bogged down by randomness. Under these circumstances, metabolic design based on molecular logic had to take the lead, setting a direction for the replicators.

Unfortunately, both views are correct in their assessments of the fatal weakness of the opposite view, resulting in Robinson's verdict that:

> But significant problems persist with each of the two competing models that have arisen—usually called "genes first" and "metabolism first"—and neither has emerged as a robust and obvious favorite.[29]

If metabolism could not effectively lead replication and replication could not effectively lead metabolism, there was only one possible solution to the dilemma: they must team up right from the start. For any partnership to succeed, each partner should have something to contribute to the other partner. The replicators could contribute to metabolism by coming up with catalysts for novel metabolic reactions. How might the metabolites assist the replicators? Since the most urgent challenge facing the replicators was the need to separate out useful sequences from the sea of useless ones, the metabolites had to provide assistance in this regard. With replicators supplying new catalysts for metabolism and metabolites helping to select useful replicators, together they could quickly develop a strategic blueprint for life based on available building blocks and energy resources, laying down metabolic pathways, cycles and networks consistent with the molecular logic inherent in metabolites with respect to the basic parameters of thermodynamics and water solubility.

(v) Metabolic Selection of Replicators

Two mechanisms that can be used to achieve metabolic selection of replicators are Replicator Induction by Metabolite (REIM) and Replicator Amplification by Stabilization (REAS).

In REIM, the presence of a metabolite induces the synthesis of a replicator. Many RNA molecules are aptamers that bind cognate ligands with specificity. The sequence of an ATP-binding 36-base Aptamer is shown in Figure 1.1A. When this Aptamer binds ATP, the latter is sandwiched between A10 and G11 of the Aptamer by means of base-stacking and hydrogen-bonding to G8 and A12.[30] The Aptamer serves as template for the transcription of its complement, the Anti-Aptamer and vice versa. Once a duplex is formed between Aptamer and Anti-Aptamer, further transcription, being dependent on a partial unfolding of the duplex, is retarded. However, if ATP is added to the duplex, it binds to the Aptamer in competition against the Anti-Aptamer, loosens up the duplex and induces the synthesis of a new strand of Aptamer on the Anti-Aptamer. When this new Aptamer is again pulled away by another ATP molecule, the process repeats itself, leading to the synthesis of multiple copies of the Aptamer (Fig. 1.1B). Such specific induction of functional RNA by metabolite will result in the selective amplification of aptamers and ribozymes that metabolites can bind to over functionless RNA. Upon mutation, the accumulated ribozymes will further generate novel ribozymes that transform the inducer metabolites and their structural analogues to yet newer metabolites, thereby expanding the metabolic pathways and networks.

Previously De Duve proposed that, given the precursors to assemble a wide variety of enzymes in the prereplicative phase of chemical evolution, the binding of a substrate to an enzyme could selectively stabilize that enzyme against lysis and increase its accumulation, upon which the selected enzyme might be further modified and extended to give rise to novel enzyme-catalyzed reactions.[31] The usefulness of the proposed mechanism is supported by the known protection of enzymes against degradation by their substrates, but the

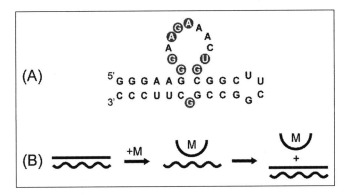

Figure 1.1. Repicator Induction by Metabolite (REIM). A) ATP-binding RNA Aptamer. Conserved residues are shaded.[30] B) Induction by metabolite M. Aptamer is represented by straight line and Anti-Aptamer by wavy line.

difficulty of prereplicative assembly of enzymes of specified sequences might be a limiting factor. Actually, biopolymer stabilization could be even more effective in the replicative phase. From foot-printing experiments, it is known that binding of ligands such as a repressor protein to DNA can protect the contact regions on the DNA. Thus, in REAS, a metabolite can bind to its cognate RNA aptamer or ribozyme and protect it from degradation. Because of the greater susceptibility of RNA than proteins to hydrolysis, such protection can selectively preserve the cognate RNA sequence as other RNA sequences are readily eliminated by hydrolysis. Even more importantly, with ribozymes the effect of stabilization goes beyond catalyst accumulation. On account of the catalyst-cum-replicator nature of ribozymes, stabilization leads to the accumulation of a replicator that can be amplified through replication and also mutate to generate novel ribozymic reactions useful to the expanding metabolism. This combination of amplification and mutation in REAS is illustrated in Figure 1.2, where it potentially might be employed to yield for instance a ribozyme to convert IMP to GMP. Because REAS depends on the degradation of functionless replicators to achieve selection of functional replicators, its effects might be slower than those of REIM. However, because the mechanisms of REIM and REAS are different, they can complement each other in their actions.

Efficient coordination between replication and metabolism is a central requirement in prebiotic evolution. The improbability of the replicators being ever able to evolve their way out of the morass of random sequences unassisted by metabolites suggests that life could not have emerged without selective mechanisms like REIM and REAS.

(vi) Primitive Replicators

Beginning with the discovery of ribozyme activities of RNA, RNA has been a foremost candidate prebiotic replicator, eclipsing earlier considerations of proteins and proteinoids in this capacity. Support for prebiotic RNA replicators includes the following lines of evidence:

I. That ribozymes can perform as replicator-cum-catalysts resolves the question of whether replicators or biocatalysts came first;

II. Ribosomal peptide synthesis is catalyzed by ribozyme;

III. Because RNA replicates through complementary base-pairing and RNA aptamers can bind specific ligands, RNA replicators are well suited to REIM;

IV. Because the phosphodiester bond in RNA is easily hydrolyzed, RNA replicators are well suited to REAS.

As pointed out by Orgel, Lines I and II suffice to establish that an RNA World almost certainly existed prior to the present day DNA-RNA-Protein World.[32] What has to be determined is whether RNA replicators functioned right from the start, or superseded some earlier replicators such as clay platelets,[33] RNA analogues and peptide nucleic acid (PNA) which form a duplex with complementary DNA or RNA.[34] It is not apparent how clay platelets could be subject to metabolic selection by REIM or REAS. PNA, on account of its template-directed replication and more heat-resistant backbone, may work well with REIM and less well with REAS. Another important disadvantage could be the lack of a straightforward prebiotic synthesis of PNA monomers.[32] The lower global flexibility of PNA relative to RNA also means that it might require a larger molecule to display aptamer or biocatalyst activity. Lines III and IV thus favor an early utilization of RNA, or threonucleic acid (TNA), an RNA analogue with a threose-phosphate backbone,[35] as replicators. The discoveries that RNA polymerization can be facilitated by ice[36] and clay[37,38] further support an early entry of RNA.

While a great deal is still unknown about the pathway to the cell, an outline of the core process may be sketched. At an early stage, RNA, TNA or other RNA-like replicators made their clay-assisted appearance, side by side with the first stirrings of mineral and clay-catalyzed metabolic transformations of the available biomolecules in the environment. Interactions between the replicators and metabolites through REIM and REAS gave rise to the selection of metabolically useful replicator-cum-ribozymes, resulting in an expansion of metabolism and biosynthesis. Soon the replicators and metabolites met up with lipoidal membranes spread out on mineral surfaces and became encapsulated in vesicles formed by these membranes. Incessant membrane-replicator-metabolite interactions ensued and the first cell gradually came into focus.

1.4 Heterotrophic Origin

(i) Heterotrophy First vs. Autotrophy First

There are two schools of thought on whether the first cell was a heterotroph feeding on organic compounds available from environmental sources, or an autotroph using only CO_2 and one-carbon compounds from the environment to synthesize all other organic compounds in-house (Fig. 1.3). Prebiotic evolution began with membranes, replicators and metabolites and proceeded to the Last Universal Common Ancestor, or LUCA, from which all extant life descended. Its analysis benefits from both the 'bottom-up' and 'top-down' approaches.[40] The former starts off from building blocks and goes through supramolecular structures and self-sustaining

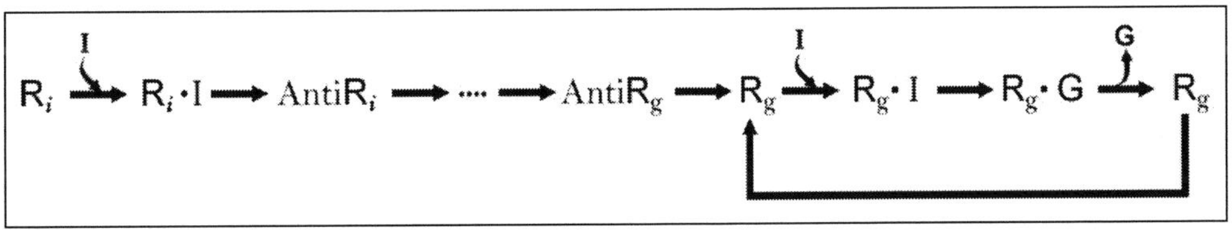

$$R_i \xrightarrow{I} R_i \cdot I \longrightarrow AntiR_i \longrightarrow \cdots\cdots \longrightarrow AntiR_g \longrightarrow R_g \xrightarrow{I} R_g \cdot I \longrightarrow R_g \cdot G \xrightarrow{G} R_g$$

Figure 1.2. Replicator Amplification by Stabilization (REAS). a) Binding of IMP (I) to an aptamer R_i stabilizes R_i; b) R_i directs formation of AntiRi which mutates to AntiRg. c) The latter is transcribed to the ribozyme R_g, which catalyzes conversion of I to GMP (G).

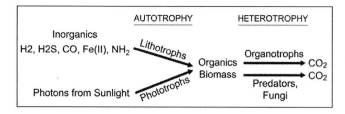

Figure 1.3. Autrophic and heterotrophic organisms (after Nealson and Conrad[39]).

autopoietic organization to reach the cell.[41-43] The latter identifies the nature of LUCA and retraces the trail taken to arrive at LUCA. Both approaches require knowing whether the first living cell was an autotroph or heterotroph.

The experimental abiotic synthesis of Phase 1 amino acids and nucleic acid constituents, along with the finding of meteorites as a rich source of organic compounds, have led to the suggestion of a Heterotrophic First scenario by Lazcano and Miller.[44] However, important discoveries in recent years have demonstrated that a range of organic biomolecules could be derived from the hydrothermal and volcanic sources:

1. Amino acids[45]
2. Lipids[22]
3. Sugars[46]
4. Pyruvic acid[47]
5. α-Hydroxy and α-amino acids[48]
6. Wood-Ljungdahl type pathway for acetyl-thioester formation[49]
7. Fischer-Tropsch synthesis of organic compounds[50]

These discoveries favor the feasibility of an autotrophic origin of life within the hydrothermal-volcanic system and lend support to a *Hot Start Hypothesis* that "life got its start in the scalding mineral rich waters streaming out of deep sea hydrothermal vents".[51] This results in an ongoing Heterotrophy First vs Autotrophy First debate:

> *Proponents of the organic soup theory suggest that life originated through the organization of organic molecules that were produced in the atmosphere by a Miller-Urey type reaction or were delivered to Earth from space......Proponents of the surface metabolism theory, by contrast, contend that metabolism arose autotrophically.... Proponents of both sides have recently delivered hefty criticisms of the other without a compromise in sight.[17]*

The likelihood of Autotrophy First is enhanced by localization of reaction products and diminished by their dilution in the oceans,[52,53] but there is *a priori* indeterminate expectation regarding the extent of dilution. Present day organisms, which include both heterotrophs and autotrophs, are also silent on the metabolic nature of the first cell. Even the identification of an autotrophic LUCA (Section 15.2) cannot settle the debate, for the first cell and

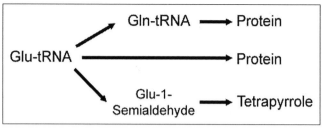

Figure 1.4. Alternate pathways of Glu-tRNA utilization. The middle pathway shows the use of Glu-tRNA for incorporation of Glu into proteins in ribosomal protein synthesis. The top and bottom pathways represent pretran syntheses of two products from Glu-tRNA, viz. Gln-tRNA for incorporating Gln into proteins in ribosomal protein synthesis in many organisms and Glu-1-semialdehyde as precursor of tetrapyrrole in heme biosynthesis.

LUCA might not have the same type of metabolism. Likewise, that 'banded iron' geological formations containing ferric iron became evident only from 2.3 Gya onward only shows that oxygenic photosynthesis was not widespread earlier, without ruling out early photoautotrophy on a small scale. This unsettled debate between autotrophic and heterotrophic origins is unsettling, for it renders the metabolic nature of the first living cell uncertain.

(ii) Triple Convergence

Based on genetic code structure, the coevolution theory of the genetic code proposes that at first only 10 Phase 1 amino acids available from the environment were employed for protein synthesis. Later, another 10 Phase 2 amino acids were produced by the developing amino acid biosynthetic pathways and added to the genetic code,[54,55] some of them such as Gln via pretran synthesis (Fig. 1.4). The theory, now proven by the primordial origin of pretran syntheses (ref. 56 and Section 14.3) indicates that the amino acids employed for protein synthesis were decided by environmental availability, which could be the case only if the cells were heterotrophic.

That the environment only provided half the 20 present day canonical amino acids is supported by the results of atmospheric amino acid synthesis: using high-energy irradiation, Kobayashi et al[57-58] have obtained to date the largest number of abiotically synthesized amino acids and the products include the 10 Phase 1 amino acids and none of the Phase 2 amino acids. It is further supported by the meteoritic amino acids: the largest collection of meteoritic amino acids found so far was discovered by Pizzarello[59] on Antarctica meteorite CR2, comprising again the 10 Phase 1 amino acids and no Phase 2 amino acid (Section 6.2).

The convergence of these three lines of independent evidence stemming from genetic code structure, atmospheric amino acid synthesis and meteoritic amino acids, resulting in complete accord regarding the identities of the amino acids readily available from the prebiotic environment (Table 1.1), provides strong confirmation

---

Table 1.1. Triple convergence of evidence regarding the prebiotic availability of different amino acids

	Gly	Ala	Ser	Asp	Glu	Val	Leu	Ile	Pro	Thr	Phe	Tyr	Arg	His	Trp	Asn	Gln	Lys	Cys	Met
Phase of entry	1	1	1	1	1	1	1	1	1	1	2	2	2	2	2	2	2	2	2	2
Irradiated synthesis	+	+	+	+	+	+	+	+	+	+	0	0	0	0	0	0	0	0	n	n
Meteoritic amino acids	+	+	+	+	+	+	+	+	+	+	0	0	0	0	0	0	0	0	0	0

Classification of Phase 1 and Phase 2 amino acids is based on reference 54. The amino acids produced by atmospheric irradiated synthesis are described in references 57 and 58. The meteoritic amino acids are described in reference 59. Production or presence is indicated by "+" and lack of production or absence by "0". "n" indicates inapplicable on account of the absence of sulfur in the irradiated synthesis.

that environmental availability was a prerequisite to the admission of any amino acid into the earliest genetic code.

If the first cell was an autotroph, it would synthesize in-house all 20 canonical amino acids as in present day blue-green algae. There could be no rational explanation for the triple convergence shown in Table 1.1. On the other hand, if the first cell was a heterotroph, the triple convergence is exactly the outcome expected because the cells, being heterotrophic, had no choice but to utilize only those amino acids available in the environment for protein synthesis, viz. the 10 Phase 1 amino acids. Later on, when the cellular amino acid biosynthetic pathways produced the Phase 2 amino acids, these would be utilized too to increase the chemical versatility of proteins.

On this basis, the first cell was a heterotroph.

(iii) Metabolic Expansion

For a heterotrophic origin, there had to be sufficient organic nutrients in the environment to start off evolution, but not all the metabolites eventually needed would be obtainable from the environment. For example, phosphorylated cofactors like FAD and NADH could not enter into the cell through lipoidal membranes even if they were available in the environment. Metabolic and biosynthetic pathways had to be expanded through such mechanisms as:

a. When the supply of a biomolecule from the environment is exhausted, a biosynthetic pathway for the biomolecule would be developed backwards in *retrograde evolution*. A catalyst is introduced that catalyzes the formation of the biomolecule from an immediate precursor. When that immediate precursor is in turn exhausted, another catalytic step would be added to transform a precursor upstream to the immediate precursor, etc.[60] For instance, Ala, Asp and Glu were Phase 1 amino acids available from the environment. When their supplies in the environment were used up, synthases and transaminases were developed by retrograde evolution to transform pyruvate, oxaloacetate and α-ketoglutarate into Ala, Asp and Glu respectively.

b. In the biosynthesis of a complex pigment such as chlorophyll, for instance, it appears doubtful that the process could begin with chlorophyll from the environment and develop a biosynthetic pathway backwards. Instead, the biosynthetic pathway may grow by *progressive extension* in the forward direction, with each pigment intermediate in the pathway functioning as a biological pigment until it is replaced by a superior pigment derived from it through a forward extension of the pathway. In this manner the biosynthetic pathway for chlorophyll-a would extend from protoporphyrin-9 to Mg vinyl pheoporphyrin-a5 and eventually to chlorophyll-a.[61]

c. The pathways for the formation of some secondary metabolites, alkaloids, antibiotics, etc. might represent *inventive biosynthesis*, where a compound comes to be synthesized as the result of random explorations by catalysts.[54] Since enzymes typically do not have absolute substrate or reaction specificities, they may act on intended or unintended substrates to form unintended side products. Where a side product turns out to be useful, the reaction may be permanently adopted. For example, even though penicillin is useful to *Penicillium*, it is chemically too unstable to have accumulated in the environment to initiate retrograde evolution. The intermediates in its biosynthetic pathway are also not active enough as antibiotics themselves to initiate progressive extension. Thus the metabolic origin of penicillin could be a case of inventive biosynthesis. The pretran syntheses of some Phase 2 amino acids from Phase 1 amino acid-tRNA compounds, which have played important roles in the genetic coding of Phase 2 amino acids (Section 14.4), are examples of inventive biosynthesis. Figure 1.4 shows the pretran syntheses of Gln in many organisms for incorporation into proteins and of Glu-1-semialdehyde[62] to serve as precursor of 5-aminolevulinic acid and heme, from Glu-tRNA. There was no Gln in the prebiotic environment (Section 9.2) and Gln is also not superior to Glu. Likewise Glu-1-semialdehyde is too unstable to accumulate in the environment and Glu cannot replace the function of heme. Accordingly neither retrograde evolution nor progressive extension can be called upon to establish these biosynthetic reactions. Instead, they are most likely the results of inventive biosynthesis.

d. For catalytic reactions involving inorganic ions, a mechanism for metabolic pathway construction that might be applicable to various metallo-enzymes is *enzymatized inorganic catalysis*, whereby a reaction originally catalyzed by inorganic compounds is transformed to a more efficient enzymic (or at an earlier stage, ribozymic) reaction. For example, inorganic iron has a very low catalase acivity splitting hydrogen peroxide into oxygen and water. When the iron is incorporated into a protoporphyrin ring to form heme, the activity is increased one thousand fold. When the heme is attached to the catalase protein, the activity is now one million fold.[61]

1.5 Defining Moment of Life

What Is Life?

If one is to name the longest running riddle, this one must rank high. All humans pass through life once, cherish it and are always fascinated by it. This fascination is precious common ground for humanity, cutting across all barriers of time, geography and language. For prebiotic evolution, solving the riddle is sheer necessity, for it is impossible to know when prebiotic ends and biotic begins without knowing what biotic means. Some definitions of life go back many decades:[63]

FG Hopkins in 1913—A minimum requirement for life is a "dynamic equilibrium in a polyphasic system."

JBS Haldane in 1952—a simple organism such as a bacterial virus contains about 100 bits of negative entropy or information and this is about the amount that would arise spontaneously in 10^9 years in the volume of the primitive ocean.

NW Pirie in 1957—"I argued twenty years ago that a rigid operational definition is not possible. This seems now generally to be accepted."

The 2002 collection of Palyi et al[64] contains a wealth of definitions, e.g.,

Gustaf Arrhenius—The basic ingredients are self-organization, self-replication, evolution through mutation, metabolism and concentrative encapsulation.

Andre Brack—Self-reproduction, mutation and evolution.

David Brin—Energy flows downhill and order, information and manipulative ability rise steeply inside local bundles of space-time.

David Deamer—Semi-permeable boundary structure, a system of polymeric catalysts and a system of polymeric instructions.

Christian de Duve—"Life is what is common to all living things."

Klaus Dose—Membrane, metabolism, control of metabolism, replication and mutability making possible evolution.

Ricardo Guerrero and Lynn Margulis—Matter that makes choices, binds time and breaks gradients.

Romeu Cardoso Guimaraes—Metabolism, growth and reproduction with stability.

Robert Hazen—Metabolism and reproduction with variation, concept of a sequence of discrete emergent steps.

Gerald Joyce—Gives the NASA Working Definition of "LIFE is a self-sustained chemical system capable of undergoing Darwinian evolution" in which self-sustenance is supported by genetically instructed metabolism.

Vladimir Kompanichenko—Organized form of intensified resistance to self-propagating processes of destruction.

Stanley Miller—The origin of life is the origin of evolution. Darwinian evolution requires replication, mutation, selection.

Eors Szathmary—Gives Tibor Ganti's definition based on metabolic network, template macromolecule and encapsulating membrane, all of which are autocatalytic.

Hubert Yockey—Having a genome and a genetic code.

There is deep-seated diversity in the definitions, but also general agreement that life is to be defined based on a combination of constituent attributes rather than any single attribute. The reason is, few of the individual attributes are really unique to life. For instance, self-reproduction/replication is observed with salts crystallizing from supersaturated solutions seeded with a few crystals, there are many geochemical cycles on Earth that are even more ancient than metabolic cycles, all soap bubbles have membranes and evolvability is displayed by ribozymes undergoing in vitro evolution. After all, if evolvability were an exclusive property of the living state, no prebiotic molecular assemblage could have evolved to generate the living state.

Besides having to employ a set of independent attributes for the definition of life based on the actual components of the first living cell, the need to pinpoint the source of the essence or new property that is endowed with purposefulness and unique to life is suggested by:

> *David Abel—"The more we can distil the essence of 'life', the greater the hope of elucidating the lost pathways of abiogenesis."*

> *Hans Kuhn—"Physical objects come into being that behave as if they have a purpose...Let us call physical objects with this fundamentally new property (not present in any ancestral form in a prebiotic universe) living."*

Furthermore, the challenge of addressing the defining moment when nonlife becomes life is posed by:

> *Janet Siefert—"One can begin to postulate on the defining moment or conditions for 'life' that arose here on Earth...I posit that life can be defined as the culmination and the eventual simultaneous occurrence" of the four events of replication, translation, control and cell wall.*

This definition calls attention to the importance of protein catalysts produced from translation, but it might be too restrictive in precluding any possibility of the translation-less ribo-organisms of RNA World being living creatures. Translation cannot be an independent attribute. Instead, in order for natural selection to be effective, the phenotypes expressed through the catalysts must be determined unambiguously by the genotypes embodied in the replicators, which may choose to act as catalysts themselves as in the case of ribozymes, or to arrange the production of surrogate catalysts as in one-gene-one-enzyme coding. Candidate replicators

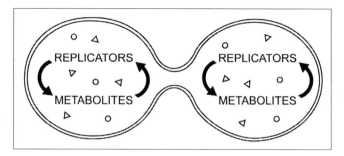

Figure 1.5. The defining moment of life, with thioesters (triangles) and ATP (circles) (Section 9.6) energizing the first cell division. The arrows depict coordination.

that know how to replicate but cannot contribute catalysis, e.g., replicating salt crystals, will remain irrelevant to life. Therefore the replicative function in life's emergence on Earth or elsewhere has to be fulfilled by replicators that are inherently not only mutable in order to make evolution possible, but also catalysis-competent by virtue of either their own catalytic ability (e.g., ribozymes), or that of catalysts specified by them through translation (e.g., enzymes). In view of this, translation in Siefert's simultaneous occurrence might be replaced by metabolism, in accord with the definitions of Arrhenius, Dose, Ganti and others.

Membranes, metabolism and replication, three constituent components that have come together from independent prebiotic origins, make the cell functional because, and only because, their partnership is a fully integrated one. The integration, painstakingly built with an evolved tool kit of catalysis, genetic encoding, encapsulation, REIM, REAS, energy coupling, transport gradients, signal transduction etc., opens up an organizational state of molecules with a higher level of complexity than pathways and autocatalytic cycles. The capability of the integrated whole so radically exceeds the individual capabilities of the three component parts that the towering excess amounts to a new property, an essence of life, undreamed of in the nonliving world. The extreme adaptability of this organizational state, acquired through heredity, mutability and evolution, equips it with the purpose-oriented behavior characteristic of the living. In time, this impressive adaptability would permeate from the cellular to the multicellular, neural and societal levels. On this basis, the defining moment of life is the moment when successful integration is signaled. Since self-reproduction is fundamental to life and prerequisite to evolution, it makes a suitable signal. Therefore the defining moment of life may be identified as follows:

> Life is the integration of membranes, metabolism and mutable, catalysis-competent replication into a single entity. The defining moment of life is the moment of successful integration signaled by the self-reproduction of such an entity.

Accordingly, the birth of the living cell took place at its first division. This historic event is shown in Figure 1.5.

References

Further Readings

Highly readable books to provide an appreciation of the basic issues of prebiotic evolution and the diversity of scientific approaches to the subject:

1. De Duve C. Blueprint for a Cell. Neil Patterson Publishers, 1991.
2. Lahav N. Biogenesis. Theories of Life's Origin. Oxford University Press, 1999.
3. Luisi PL. Emergence of Life. From chemical origins to synthetic biology. Cambridge University Press, 2006.
4. Morowitz HJ. Beginnings of Cellular Life. Yale University Press, 1984.
5. Palyi G, Zucchi C, Caglioti L, eds. Fundamentals of Life. Elsevier, 2002.

Specific References

6. Fairley J. The spontaneous generation controversy from Descartes to Oparin. John Hopkins Univ Press 1977.

7. Raulin-Cerceau F. Historical review of the origin of life and astrobiology. In: Seckbach J, ed. Origins, Genesis, Evolution and Diversity of Life. Kluwer Academic Publishers, 2004:17-33.

8. Oparin AI (translated by Morgulis S). The Origin of Life. MacMillan 1938:109-195.

9. Bada JL, Lazcano A. Prebiotic soup—revisiting the Miller experiment. Science 2003; 300:745-746.

10. Franck S, von Bloh W, Bounama C et al. Habitable zones in extrasolar planetary systems. In: Horneck G, Baumstark-Khan C, eds. Astrobiology. The Quest for the Conditions of Life. Springer-Verlag, 2001:47-55.

11. Henderson LJ. The Fitness of the Environment: an Inquiry into the Biological Significance of the Properties of Matter. Boston: MacMillan, 1913.

12. Ulmschneider P. Intelligent Life in the Universe. Springer Verlag, 2003:51-71.

13. Kasting JF, Whitmire DP, Reynolds RT. Habitable zones around main sequence stars. Icarus 1993; 101:108-128.

14. Horneck G, Mileikowsky C, Melosh HJ et al. Viable transfer of microorganisms in the solar system and beyond. In: Horneck G, Baumstark-Khan C, eds. Astrobiology. The Quest for the Conditions of Life. Springer-Verlag, 2001:57-76.

15. Wong JT, Xue H. Self-perfecting evolution of heteropolymer building blocks and sequences as the basis of life. In: Palyi G, Zucchi C, Caglioti L, eds. Fundamentals of Life. Elsevier, 2002:473-494.

16. Schwartz AW. Phosphorus in prebiotic chemistry. Phil Trans R Soc B 2006; 361:1743-1749.

17. Russell MJ, Martin W. The rocky roots of the acetyl-CoA pathway. Trends in Biochem Sci 2004; 29:358-363.

18. Canfield DE, Rosing MT, Bjerrum C. Early anaerobic metabolisms. Phil Trans R Soc B 2006; 361:1819-1836.

19. Zierenberg RA, Adams MWW, Arp AJ. Life in extreme environments: hydrothermal vents. Proc Natl Acad Sci USA 2000; 97:12961-12962.

20. Oro J, Lazcano A. A holistic precellular organization model. In: Ponnamperuma C, Eirich FR, eds. Prebiological Self-Organization of Matter. A. Deepak, 1990:11-33.

21. Deamer DW. Prebiotic amphiphilic compounds, In: Seckbach J, eds. Origins, Genesis, Evolution and Diversity of Life. Kluwer Academic Publishers, 2004:77-89.

22. Rushdi AI, Simonet BRT. Lipid formation by aqueous Fischer-Tropsch type synthesis over a temperature range of 100-400°C. Orig Life Evol Biosph 2001; 31:103-118.

23. Wachterhauser G. Before enzymes and templates: theory of surface metabolism. Microbiol Rev 1988; 52:452-484.

24. Wachterhauser G. From volcanic origins of chemoautotrophic life to Bacteria, Archaea and Eukarya. Phil Trans R Soc B 2006; 361:1787-1808.

25. Orgel LE. Self-organizing biochemical cycles. Proc Natl Acad Sci USA 2000; 97:12503-12507.

26. Orgel LE. The implausibility of metabolic cycles on the prebiotic Earth. PLoS 2008; 6:e18.

27. Morowitz HJ, Kostelnik JD, Yang J et al. The origin of intermediary metabolism. Proc Natl Acad Sci USA 2000; 97:7704-7708.

28. Joyce GF, Orgel LE. Prospects for understanding the origin of the RNA World. In: Gesteland RF, Cech TR, Atkins JF, eds. The RNA World, 2nd ed. Cold Spring Harbor Laboratory Press, 1999:49-77.

29. Robinson R. Jump-starting a cellular world: investigating the origin of life, from soup to networks. PLoS Biology 2005; 3:e396.

30. Puglisi JD, Williamson JR. RNA interaction with small ligands and peptides. In: Gesteland RF, Cech TR, Atkins JF, eds. The RNA World, 2nd ed. Cold Spring Harbor Laboratory Press, 1999:403-425.

31. De Duve C. Selection of differential molecular survival: a possible mechanism of early chemical evolution. Proc Natl Acad Sci USA 1987; 84:8253-8256.

32. Orgel LE. Prebiotic chemistry and the origin of the RNA World. Crit Rev Biochem Mol Biol 2004; 39:99-123.

33. Cairns-Smith AG. The Life Puzzle. University Toronto Press, 1971:129-136.

34. Sen S, Nilsson L. MD simulations of homomorphous PNA, DNA and RNA single strands: characterization and comparison of conformations and dynamics. J Am Chem Soc 2001; 123:7414-7422.

35. Schoning K, Scholz P, Guntha S et al. Chemical etiology of nucleic acid structure: the alpha-threofuranosyl-(3′→2′) oligonucleotide system. Science 2000; 290:1347-1351.

36. Monard P-A, Kanavarioti A, Deamer DW. Eutectic phase polymerization of activated ribonucleotide mixtures yields quasi-equimolar incorporation of purine and pyrimidine nucleobases. J Am Chem Soc 2003; 125:13734-13740.

37. Ferris JP, Hill AR, Liu R et al. Synthesis of long prebiotic oligomers on mineral surfaces. Nature 1996; 381:59-61.

38. Ferris JP. Montmorillonite-catalysed formation of RNA oligomers: the possible role of catalysis in the origin of life. Phil Trans R Soc B 2006; 361:1777-1786.

39. Nealson KH, Conrad PG. Life: past, present and future. Phil Trans R Soc B 1999; 354:1923-1939.

40. Jortner J. Conditions for the emergence of life on the early Earth: summary and reflections. Phil Trans R Soc B 2006; 361:1877-1891.

41. Maturana H, Varela F. Autopoiesis and Cognition: The Realization of the Living. Springer, 1980.

42. Margulis L, Guerrero R. Origins of life to evolution of microbial communities: a minimalist approach. In: Ponnamperuma C, Eirich FR, eds. Prebiological Self-Organization of Matter. A. Deepak, 1990:261-278.

43. Lehn JM. Toward self-organization and complex matter. Science 2002; 295:2400-2403.

44. Lazcano A, Miller SL. On the origin of metabolic pathways. J Mol Evol 1999; 49:424-431.

45. Hennet RJC, Holm NG, Engel MH. Abiotic synthesis of amino acids under hydrothermal conditions and the origin of life: a perpetual phenomenon? Naturwissenschaften 1992; 79:361-365.

46. Zubay G. Studies on the lead-catalysed synthesis of aldopentoses. Orig Life Evol Biosph 1998; 28:13-26.

47. Cody GD, Boctor NZ, Filley TR et al. Primordial cabonylated iron-sulfur compounds and the synthesis of pyruvate. Science 2000; 289:1337-1340.

48. Huber C, Wachterhauser G. α-Hydroxy and α-amino acids under possible hadean, volcanic origin-of-life conditions. Science 2006; 314:630-632.

49. Martin W, Russell MJ. On the origin of biochemistry at an alkaline hydrothermal vent. Phil Trans R Soc B 2007; 362:1887-1925.

50. McCollom TM, Seewald JS. Abiotic synthesis of organic compounds in deep-sea hydrothermal environments. Chem Rev 2007; 107:382-401.

51. Zimmer C. How and where did life on Earth arise? Science 2005; 309:89.

52. Bada JL, Fegley Jr B, Miller SL et al. Debating evidence for the origin of life on Earth. Science 2007; 315:937-938.

53. Wachtershauser G, Huber C. Response to "Debating evidence for the origin of life on Earth". Science 2007; 315:938-939.

54. Wong JT. Coevolution of the genetic code and amino acid biosynthesis. Trends Biochem Sci 1981; 6:33-35.

55. Wong JT. Coevolution theory of the genetic code at age thirty. BioEssays 2005; 27:416-425.

56. Wong JT. Question 6: Coevolution theory of the genetic code: a proven theory. Orig Life Evol Biosph 2007; 37:403-408.

57. Kobayashi K, Tsuchiya M, Oshima T et al. Abiotic synthesis of amino acids and imidazole by proton irradiation of simulated primitive Earth atmospheres. Orig Life Evol Biosph 1990; 20:99-109.

58. Kobayashi K, Kaneko T, Saito T et al. Amino acid formation in gas mixtures by high energy particle irradiation. Orig Life Evol Biosph 1998; 28:155-165.

59. Pizzarello S. Meteorites and the chemistry that preceded life's origin. In: Wong J T-F, Lazcano A, eds. Prebiotic Evolution and Astrobiology. Austin: Landes Bioscience, 2008.

60. Horowitz NH. On the evolution of biochemical syntheses. Proc Natl Acad Sci USA 1945; 31:153-157.

61. Granick S. Speculations on the origins and evolution of photosynthesis. Annals NY Acad Sci 1957; 69:292-308.

62. Randau L, Schauer S, Ambrogelly A et al. tRNA recognition by glutamy-tRNA reductase. J Biol Chem 2004; 279:34931-34937.

63. Pirie NW. Some assumptions underlying discussion on the origins of life. Annals NY Acad Sci 1957; 69:369-376.

64. Palyi G, Zucchi C, Caglioti L. Short definitions of life. In: Palyi G, Zucchi C, Caglioti L, eds. Fundamentals of Life. Elsevier, 2002:15-55.

CHAPTER 2

The Minimal Cell

Markus Pech and Knud H. Nierhaus*

2.1 Introduction

In this chapter we consider the beginning of life under the aspect of common and minimal principles governing the existence of a cell, in particular we will take into account the question of "the minimal gene set" that is necessary for life.

All living organisms are made of cells. A single cell is an entity, separated from other cells by a membrane or a cell wall, filled with a variety of chemical materials and subcellular structures and equals more a gel than an aqueous solution (see Table 2.1). The major characteristics that set living cells apart from nonliving chemical systems are (i) self-feeding or nutrition, (ii) replication or growth, (iii) differentiation, (iv) chemical signalling and (v) evolution.

2.2 Chemical Components of a Cell

An analysis of the chemical composition of the living cell shows that it is quite different from the chemical composition of the Earth. The cell therefore is not a random assortment of chemical elements but instead a selective chemical system composed of a restricted set of elements, primarily of H, C, N, O, P and S (see Fig. 2.1), which make up more than 99% of its weight.[1] Remarkably, the most abundant molecule of the cell is not special at all—water accounts for about 70% of the weight of the living cells and therefore most intracellular reactions occur in an aqueous environment. In contrast all cellular structures and biological molecules contain an abundance of carbon. Most important, carbon is able to combine not only with many other elements but also with itself, thus forming chains and rings and thereby generating large chemical structures of considerable diversity and complexity with no obvious upper limit to their size.

The main cellular components are chemical structures called macromolecules, which are built up of individual building blocks connected in specific ways. Such a single building block is called a monomer, represented in general by small organic compounds that are grouped in classes according to their chemical properties. In cell biochemistry there are four important sets of monomers, which built up the cellular components: sugars, the monomeric constituent of polysaccharides; fatty acids, the monomeric units of lipids; nucleotides, the basic units of the nucleic acids (DNA and RNA); and amino acids, the monomeric constituent of proteins. Nucleic acids and proteins can be considered as informational macromolecules because the sequence of monomeric units within them is highly specific and carries biological information and the means to process this information. In contrast, the sequence of monomers in polysaccharides and lipids is usually highly repetitive and the sequence itself is generally of less functional importance. However, both the nature and the sequence of the monomeric units of any macromolecule are important in distinguishing it chemically from related macromolecules.

2.3 Prokaryotic and Eukaryotic Cells

In the phylogeny of the living world we can distinct two major types of cells called prokaryote and eukaryote, which are structurally very different (Fig. 2.2). The prokaryotes are divided into two kingdoms of life: the bacteria and the archaea. One of the most characteristic features is the lacking of a "true" cell nucleus (prokaryote: from old Greek pro- before + karyon nut or kernel), as well as any other membrane-bound organelles. The genomic DNA, in most instances a single circular chromosome, is localised in an irregularly shaped region within the cell called nucleoid. The other major components within the prokaryotic cell, other than water, include ribosomes, macromolecules, small organic molecules (mainly precursors of macromolecules) and various inorganic ions. The outer barrier of the cell is the cytoplasmic membrane, a thin and flexible layer, and a cell wall which is composed of peptidoglycan in bacteria. In contrast eukaryotes (from old Greek eu- good or true + karyon nut or kernel) have a nucleus containing their genomic DNA and protected by a double membrane. Besides this basic feature we find in the cytoplasm, in addition to ribosomes and macromolecules, in general several membrane-bound organelles and compartments such as endoplasmic reticulum, Golgi apparatus and mitochondria, as well as vesicles like peroxysomes and lysosomes. Furthermore, the eukaryotic cell comprises a cytoskeleton, a series of internal structures that structurally support the cell and help organize and move its internal components. In comparison to prokaryotes where the cells of different species are almost similar, the set of cellular components may vary in eukaryotes depending on the organism or the tissue. Plant cells comprise in addition a large central vacuole, as well as chloroplasts, in tissues involved in photosynthesis. The cell walls in plants are made up of cellulose and protein and in many cases of lignin, in contrast to the cell walls of fungi, which are made up of chitin. The cells of animals lack a rigid cell wall and have as outer barrier only a cell membrane.

2.4 The Beginning of Life on Earth and the Genetic Code

Sequence analysis of genomes of species from all three domains of life showed that primarily proteins related with energy processes and protein synthesis, transcription and replication can be considered to

*Corresponding Author: Knud H. Nierhaus—Max-Planck-Institut für Molekulare Genetik, Ihnestrasse 73, AG Ribosomen, 14195 Berlin, Germany. Email: nierhaus@molgen.mpg.de

Prebiotic Evolution and Astrobiology, edited by J. Tze-Fei Wong and Antonio Lazcano. ©2009 Landes Bioscience.

Table 2.1. Approximate molecular composition of a bacterial cell. Adapted from reference 23.

Component	Percent of Total Weight	Approx. Number of Molecules Per Cell	Number of Different Kinds
Water	70%	4×10^{10}	1
DNA	1%	1-4	1
RNA	6%	5×10^5	3,000
Protein	15%	1×10^6	3,000
Nucleotides and precursors	0.4%	1.2×10^7	200
Amino acids and precursors	0.4%	3×10^7	50
Carbohydrates (polysaccharides and precursors)	3%	2.5×10^8	200
Lipids	2%	2.5×10^7	50
Ions	1%	2.5×10^8	20
Waste products and intermediates	0.2%	1.5×10^7	200

be generally common and conserved in the three domains, in spite of differences in detail.[2]

Here, the undisputed steps in the beginning of life on Earth are compiled, before a retrograde approach is presented outlining a possible minimal set of components required for protein synthesis, based on our knowledge of the modern translational apparatus of the bacterium *Escherichia coli*, since evidence suggests that the bacterial domain is most deeply rooted in the universal evolutionary tree (see however Section 15.2). Figure 2.3 shows a universal small subunit ribosomal RNA tree with some recently established numbers of age, which are in overall agreement with the evidence from comparisons

of ribosomal RNA, ribosomal proteins and membrane composition. Because protein synthesis is an energy demanding process, we further continue that consideration by addressing one of the main problems of early life, namely avoiding wasteful energy loss.

Describing the evolutionary time at which a feature of live appeared is, in many cases, more like a flower arrangement, rather than a sound determination. However, a few solid data do exist. One cornerstone for our estimate of early life is seen in the stromatolites, which represent ancient cells. The oldest "stromatolites" reported are from rocks of Isua Supracrustal Belt, Greenland, dated at 3,750 million years (Ma) ago, but they have been questioned as to whether

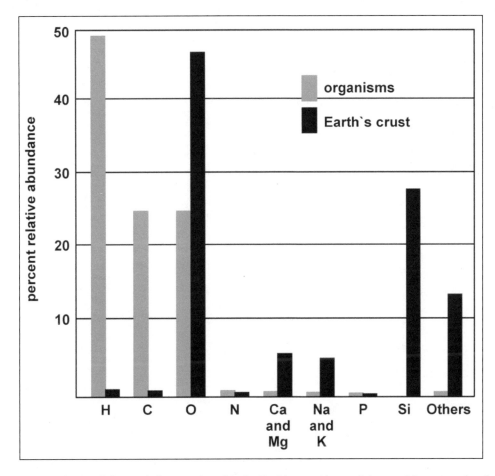

Figure 2.1. The relative abundance of chemical elements found in the Earth's crust (the nonliving world) compared to that of living organisms. The relative abundance is expressed as a percentage of the total number of atoms present. Adapted from reference 25.

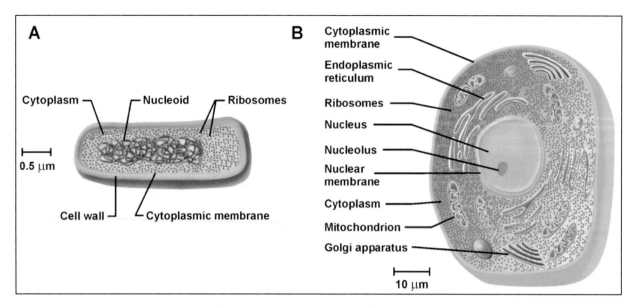

Figure 2.2. Comparison of (A) a prokaryotic cell and (B) a eukaryotic animal cell. For explanations see text. Adapted from reference 26.

they can really be considered as the first imprint of life. More solid data exist about formations that are 3,500 Ma old containing remnants of ancient cells: (i) One is the Pilbara region of Western Australia with an age of 3,430 Ma; a recent report supports the suggestion that these Pilbara-Craton structures might be of biotic origin.[3] (ii) Another of about the same age with evidence of microbial biomarkers is the pillow lava from the Baberton Greenstone Belt in South Africa.[4] Such ancient cells must have genes (from RNA?) and a translational apparatus, i.e., the genetic code has an age of 3.2 to 3.6 billion years. This estimate has been backed up in an elegant study of Eigen and coworkers,[5] where tRNA sequences from various organisms were used to conclude that the genetic code has an age of $3,300 \pm 300$ Ma. It follows that chemical evolution, the development of the genetic code and the existence of the "RNA world" must be squeezed into a time span of 400 to 800 Ma, corresponding to the time gap between the formation of the first rocks (4,000 Ma) and the appearance of the first cells (3,600 to 3,200 Ma).

Another important landmark is the observation, in many iron deposits around the Earth at the geological layer of about 2 billion years ago (2,000 Ma), of Fe[III] (ferric state) precipitates indicating the appearance of the oxidizing power of O_2, a product of photosynthesis. The earliest Fe[III] deposits are found in the Hamersley iron formation in Western Australia.[6] At deeper layers, Fe[II] (ferrous state) deposits are usually present. It follows that cyanobacterial photosynthesis developed before 2,000 Ma. Appearance of the pollutant O_2 in the atmosphere was a major threat to early life, since every cell contained and still contains a reducing milieu—a relic of the origin of life when the atmosphere was reducing. The consequence was obviously a massive extinction. Only a few cells survived due to a membrane composition that prevented the passage of oxygen into the cell. Over the longer run the cells eventually turned the appearance of atmospheric oxygen into a major advantage by inventing respiration.

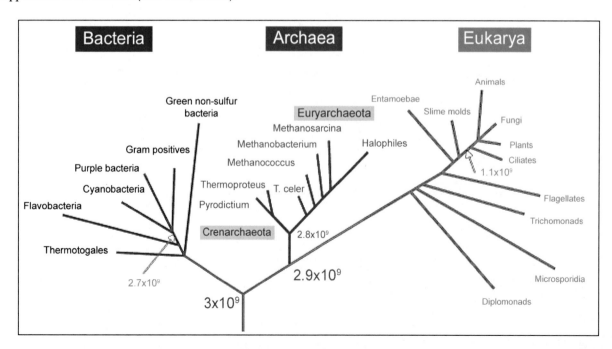

Figure 2.3. Small subunit ribosomal RNA phylogenetic tree of the three living domains Bacteria, Archaea and Eukarya. The numbers depicting dates in years ago are according to references 6,27; see also reference 28.

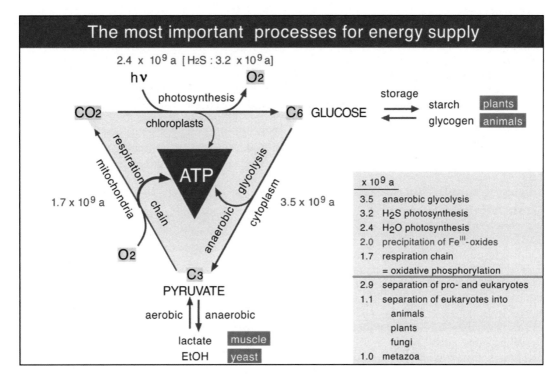

Figure 2.4. The three essential energy (ATP) production lines are shown together with their rough dates of origin in red. Cellular ATP production started with anaerobic glycolysis about 3.500 Ma years ago followed by photosynthesis and finally by respiration. The last step provided the energy required for the development of highly organized multicellular life. The inset (grey) gives some additional data.

Eukaryotic cells developed mitochondria by engulfing respiring bacteria, specifically an α-proteobacterial ancestor,[7] leading to a dramatic improvement in energy production taking the form of ATP, the energy currency of life. Respiration means an 18-fold improvement in the energy efficiency of ATP formation compared to the most ancient energy-production pathway, viz. anaerobic glycolysis. We can define three important stages of evolution relating to ATP production (Fig. 2.4): (i) Before respiration (>2,600 Ma ago), the major energy production pathway was anaerobic glycolysis, where C6 sugars (glucose, fructose) were broken down to C3 carbohydrates (pyruvate). It is thought that the sugars were taken from the environment, formed by high temperatures and electrical discharges through the ancient atmosphere according to the Stanley Miller experiments mimicking the ancient atmospheric composition and physical factors (flashes, temperature[8]). (ii) At 2,600-2,400 Ma ago photosynthesis was invented, abolishing the dependence of life on exogenous energy-rich compounds, such as C6 sugars. (iii) <2,000 Ma ago some of the enzymes developed for

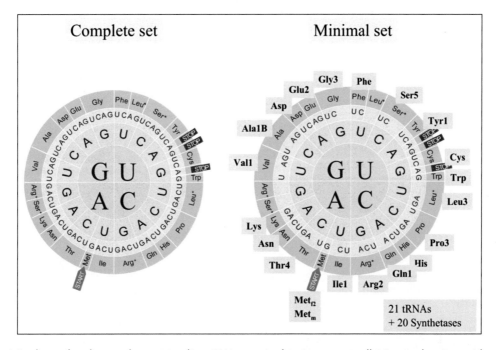

Figure 2.5. A minimal set of codons and corresponding tRNAs required to incorporate all 20 natural amino acids with a constrained genetic code lexicon. Left panel, standard genetic lexicon in the form of the codon sun. Right panel, the constrained codon sun. For explanations see text.

photosynthesis were modified slightly and re-used to catalyze the corresponding reaction for respiration. Indeed, the essential subunits of the cytochrome b_6f-complex in chloroplasts are paralogues of those of the bc1-complex in mitochondria and the bc-complex in bacteria, pumping H^+ out of the chloroplast stroma, the mitochondrial matrix or the cytoplasm, respectively.[9] Invention of respiration was a major achievement in life development, since now the cell had paradisiacal potential for energy production, increased by the aforementioned factor of 18. Thus, the way was now paved for the development of multicellular organisms and the principle of labour division among various tissues.

2.5 Possibilities of Simplifying the Existing Bacterial Translational Apparatus

To present a possible minimal set of components required for protein synthesis it is necessary to introduce the key players, which are involved in this fundamental and essential process. The genomic DNA contains the instructions (genes) used in the development and functioning of all living organisms. Protein-encoding genes will be first transcribed by an RNA polymerase to produce a complementary messenger RNA (mRNA). In the second step of protein synthesis (translation) the mRNA is decoded to produce a specific polypeptide according to the mRNA sequence. Therein involved are several components: first of all the ribosome, the 'factory' of protein synthesis. The ribosome is a huge complex composed of proteins and RNA (rRNA), where the amino acids are assembled into a polypeptide chain. The other important components are transfer RNAs (tRNAs transport amino acids to the ribosome), as well as initiation factors (IF), elongation factors (EF) and termination factors (RF).

In early times of life, translation of mRNAs started just at the 5'-end with a sense codon, later with a special initiation codon AUG, but without a 5'-untranslated region and a Shine-Dalgarno sequence. The Shine-Dalgarno sequence is complementary to the 3'-end of the 16S ribosomal RNA and should ensure by base pairing the effective formation of the ribosome-mRNA complex. In contrast to general belief, on average only 55% bacterial transcripts with a 5'-untranslated sequence contain a Shine-Dalgarno sequence.[10] Nondissociated 70S ribosomes can initiate translation using AUG codons located at or near the 5'-end, preferentially starting at the first up to the third nucleotide position of an mRNA.[11] In fact, the commonly used translation of poly(U) into poly(Phe) uses this type of ancient initiation without even an AUG codon at the 5'-end. In order to describe the potential of minimizing a modern bacterial translational apparatus, we will assume that the translation of all transcripts will start at or near the 5'-end of an mRNA, or when organized in a polycistronic mRNA, the intercistronic regions will be short as in mRNAs coding for ribosomal proteins, where the 70S-type of initiation is valid for all cistrons except the first one (R. Gupta and K. H. Nierhaus, unpublished). The canonical initiation with 30S subunits will be called 30S-type initiation (see refs. 12,13).

(i) Number of tRNAs

The genetic code consists of 64 codons, three of which are stop codons (Fig. 2.5, left panel). The 61 sense codons are read by 42 different tRNAs in *E. coli*. Note that two tRNAs are considered as different when their anticodons are different; tRNAs containing the same anticodon but different sequences outside the anticodon are considered to be the same kind of species, or isoacceptors. If we take the most often used codon for a distinct amino acid, then a minimal set of 21 tRNAs will suffice to translate all proteins—20 tRNAs to decode codons for the 20 amino acids during the elongation phase (elongator tRNAs) plus the initiator tRNA , which decodes the

initiator codon AUG at the beginning of the coding region of an mRNA. However, the codon usage in mRNAs would be restricted to those codons that can be deciphered by the minimal set of 20 elongator tRNAs.

One anticodon can usually read two and sometimes three codons, differing in the third codon position, for example,

the codons   5'-UU**C/U**   can be read by
the anticodon   3'-AAG.

This "fuzzy" reading at the third position of a codon is called the wobbling feature of anticodons. When we take into account this wobbling feature of tRNAs, the codon lexicon of a genome to be translated by the minimal set of 20 elongator tRNAs reduces to the constrained system shown in Figure 2.5, right panel.

(ii) Reduction of the Ribosome

Not all the molecules composing an *E. coli* ribosome are essential. About one third of the ribosomal proteins can be deleted one at a time with the cell remaining viable. However, in many cases the strains lacking a ribosomal protein are very sick, as indicated by decreased growth rates by up to a factor of 7 to 10, or conditional lethality, e.g., cold or temperature sensitivity (ref. 14 and see Table 2.2). Consideration of a minimal ribosome should target first these proteins, although it is unknown whether all these proteins can be omitted simultaneously without threatening viability. Interestingly, the shortest known rRNAs are those in mitochondrial ribosomes from the higher eukaryotes, which are ~30% shorter than the *E. coli* rRNAs (whereas the mitochondrial rRNAs of e.g., yeast are comparable in length to *E. coli* rRNAs). These mitochondrial ribosomes compensate for the lack of rRNA sequences by the presence of additional proteins,[15] so that the overall size of a mitochondrial ribosome is even larger than the corresponding one from *E. coli*. Therefore, based on recent ribosomal structures, it does not seem very likely that a ribosome could be derived where both the number of ribosomal proteins is reduced as in the Dabbs study[14] and the rRNA shortened as in the mitochondrial rRNAs.

Table 2.2. Ribosomal proteins missing in mutants of *Escherichia Coli*

Protein	Mutant Designation	Phenotype
L1	RD19, MV17/10	-
L11	AM68, AM76, AM77	-
L15	AM16-98	cs
L19	AM149	-
L24	AM290	ts
L27	AM125	cs
L28	AM81, AM108	cs
L29	AM111	-
L30	AM10	-
L33	AM90, AM108	cs
S1	VTS03	-
S6	AM80	- unpublished
S9	AM83	cs
S13	AM109-113	- unpublished
S17	AM111	ts
S20	VT514	ts

cs, cold sensitive; ts, temperature sensitive. Data taken from reference 24.

Table 2.3. Dispensable factors for protein synthesis in a minimal cell

Initiation Factors	Elongation Factors	Termination Factors
IF1, IF2 and ~~IF3~~	EF-Tu (EF1), EF-G (EF2) and EF-Ts; ~~EF4 (LepA)~~	RF1, ~~RF2~~ and ~~RF3 + RRF~~

(iii) Reducing Ribosomal Factors

Table 2.3 lists the translational factors active in a bacterial system. For initiation, three factors are required; IF1, IF2 and IF3. Since initiation, translation and termination can be performed with 70S ribosomes,[16] the anti-association factor IF3 is not needed in the proposed minimal system.

The elongation phase knows four factors, the classical EF-Tu, EF-G and EF-Ts and the recently detected EF4 (Lep A; ref. 17). EF4 is essential at growth conditions at pH ≤6.5 and/or high salts (K+, Mg2+) or low temperature and is not needed in the presence of well controlled media (Z. Karim and K. H. Nierhaus, 2007, submitted).

In the termination phase there are four active factors (plus EF-G). RF1 and RF2 decode the stop codons (UAG + UAA) and (UGA + UAA), respectively. RF2 can be omitted when the stop codons employed are UAG and UAA, since some species have lost RF2 and use UGA to encode the amino-acid tryptophan. Another termination factor, RF3, fine-tunes the termination process by improving the release of RF1 and RF2 from the ribosome. However, this factor is not essential in E. coli and even missing in 72 of the 191 genomes analysed recently.[18] The ribosome recycling factor (RRF) is essential for the dissociation of the 70S ribosomes following translation termination, so that the 30S subunits can re-enter canonical 30S-type initiation. By using the 70S-type of initiation exclusively, as we suggest, RRF would not be required (see also ref. 16).

Therefore, of the 11 factors essential for translation in bacterial systems, only six are needed for the minimal system suggested here (Table 2.3).

2.6 Optimized Energy Consumption for Protein Synthesis

The synthesis of new proteins is, as we mentioned above, an extremely energy demanding process. Adding up the energy required for (i) the synthesis of a codon, which is the nucleic acid information unit for an amino acid, (ii) the charging reaction of a tRNA by its synthetase with the correct (cognate) amino acid and (iii) the subsequent incorporation of this amino acid into the nascent peptide chain, a total of 10 energy-rich bonds need to be sacrificed. In this context an energy-rich bond means an acidic-anhydride bond of adjacent phosphate residues of ATP or GTP, each with an energy content of about $\Delta G'^0 = -6$ kcal/mol. This enormous energy requirement explains why the cell has developed intricate systems for controlling energy consumption. This is certainly the Achilles' heel of modern in vitro translation systems, where usually no more of 5% of the energy is used for actual protein synthesis, with the rest being wasted by uncontrolled and useless energy drains. In striking contrast, a bacterial cell can channel up to 70% of its energy consumption into the synthesis of proteins.

In the following section we will outline the major points one has to follow in order to avoid the waste of energy and to direct most of the energy—as in a living cell—into useful work, in this case protein synthesis. We will focus on at least two major reasons for energy losses that can be identified in in vitro systems: (i) the metabolic energy drain and (ii) idle ribosomes and an excess of elongation factor EF-G.

(i) Preventing Metabolic Energy Drain

The standard way to regenerate energy (ATP, GTP) used in protein synthesis is to add phosphoenolpyruvate (PEP) in mM concentrations and the enzyme pyruvate kinase that transfers the phosphate group from PEP to AMP and ADP yielding ADP and ATP, respectively (Fig. 2.6A). Likewise, GMP and GDP are also substrates for this enzyme. PEP contains a phosphoester bond, the energy richest bond of all biological samples with an energy content of $\Delta G'^0 = -12$ kcal/mol. But this system allows only the synthesis of limited amounts of proteins on account of the inhibitory effect resulting from the generation of orthophosphate. Orthophosphate by-products not only reduce the pH, but also bind Mg2+ ions, thus reducing the free Mg2+ concentration and severely affecting the structure of all ribonucleoproteins in the system, especially the ribosome. The two positive valences of Mg2+ shield the negatively charged phosphate groups of the rRNA and maintain the high density of the ribosome. Thus reducing the free Mg2+ concentration sets the repelling phosphate groups free and expands the ribosome, impairing its efficiency.

One way to get rid of this unwanted effect is to recycle the liberated phosphate by addition of pyruvate oxidase, together with some prosthetic groups, such as TPP and FAD (thiamine pyrophosphate and flavin dinucleotide, respectively). This system generates acetylphosphate, a standard cellular regeneration system, by using molecular oxygen and H_2O (Fig. 2.6B) and this improvement in conditions leads to a definite prolongation in protein synthesis in vivo.[19]

However, this system is still far from ideal. The need of molecular oxygen is the drawback, since oxygen uptake is limiting and becomes worse when larger reaction volumes are employed owing to decreased relative surface area. Jewett and Swartz recently reported a convincing solution to this problem by combining the classical PEP/pyruvate kinase system with the synthesis of acetylphosphate via acetyl-CoA, they efficiently recycle orthophosphate (Pi) during the synthesis of acetylphosphate (Fig. 2.6C). This solution represents a breakthrough: for the first time several mg of proteins can be synthesized per ml, making the in vitro system a serious alternative for the synthesis of proteins for structural and functional analysis.

(ii) Preventing Idle Ribosomes in the Presence of an Excess of EF-G

We already mentioned that four proteins participate in translation during the elongation phase. Two of them, EF-Tu and EF-G, are components of the central activity of the ribosome, the elongation cycle, where in a cycle of reactions the nascent chain is extended by one amino acid (aa). EF-Tu brings the aa in the form of aa-tRNA to the decoding centre of the ribosome and EF-G translocates the tRNAs on the ribosome by one codon length (reviewed in ref. 20). These two factors, together with EF-Ts and EF4, belong to the superfamily of GTP-binding (G) proteins, which is an important class of regulatory proteins. They bind GDP or GTP and undergo a conformational change depending on the presence of which nucleotide is bound. In the GTP conformer (the "on"-state) they bind to their target and promote a distinct reaction of the target, with the energy being paid by the binding energy generated during the binding step of the G protein. The target then signals that the reaction has been successfully accomplished and triggers the hydrolytic

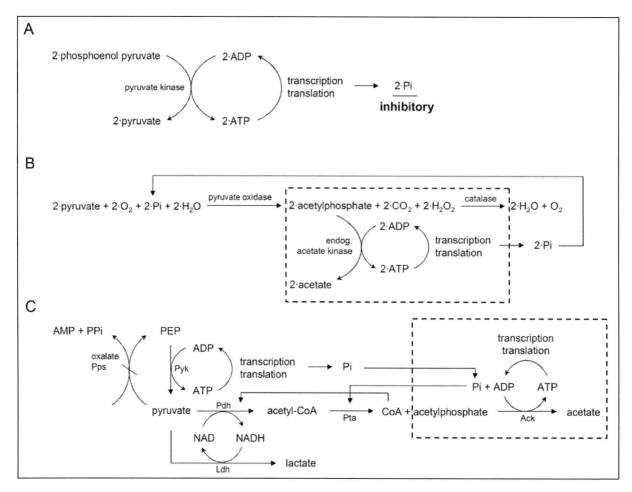

Figure 2.6. Various ATP regeneration systems for in vitro protein expression. A) Conventional scheme for energy regeneration using phosphoenol puryvate (PEP) and pyruvate kinase. The generated orthophosphate (Pi) inhibits protein synthesis. B) Improved energy regeneration using pyruvate, pyruvate oxidase and endogenous acetate kinase (according to ref. 19) C) Near in vivo system based on phospho*trans*-acetylase (Pta) to produce acetylphosphate. The boxes in B and C show a common set of reactions for the regeneration of ATP. The PEP synthase (Pps) is blocked by the presence of oxalate. Ack, acetate kinase; Ldh, lactate dehydrogenase; Pdh, pyruvate dehydrogenase; Pyk, pyruvate kinase. (Modified from ref. 29).

reaction "GTP → GDP + Pi" in the corresponding enzymatic centre of the G protein. Upon release of the inorganic phosphate, the G protein switches into the GDP conformer ("off"-state) and, losing its affinity for the target, dissociates from it.

Usually a G protein recognizes a specific configureuration in the target before promoting or even triggering a specific reaction. This is also true for EF-G; it recognizes specifically the ribosome state before translocation and dissociates from the ribosome following GTP hydrolysis and Pi release. However, EF-G is the only G protein involved in translation that also interacts effectively with empty or idle ribosomes, probably because empty ribosomes seem to have "conformational memory" of the two main states before and after translocation.[21] Thus empty ribosomes can trigger EF-G dependent GTP hydrolysis with a high turnover, which is not seen with EF-Tu. It follows that in a minimal system with poor energy resources, both ribosomes and EF-G need to be present in limiting amounts. This condition is attained when the molar ratio of EF-G:ribosomes is about 0.2 to 0.3, i.e., 2-3 EF-G molecules per 10 ribosomes.

2.7. Conclusion

In this chapter we have tried to estimate the minimal set of genes needed for a minimal living cell. A brave estimate of the minimal components for the translational apparatus today comprises no more than 200 genes, of which more than 120 are associated with the translational apparatus, encoding about 40 genes for ribosomal proteins, two rRNAs (omitting the 5S rRNA), 21 tRNAs, 20 synthetases, six factors and at least 20 tRNA modifying enzymes. In addition a minimum of 30 genes are needed for both the generation of household energy and the synthesis of at least some of the amino acids (note: since some of the amino acids were formed in the Stanley Miller type experiments mimicking the atmosphere and the physical environment of more than 3 billion years ago,[8] they could be taken up from the primordial soup by the earliest cells and thus did not need to be synthesized).

Even if we consider a "limping" life form, we end up with a total of at least 150 genes. Interestingly, this is in a reasonable agreement with the most restricted life form known today, namely the bacterium *Carsonella ruddii* with 182 genes.[22] This bacterium is an obligatory endosymbiont living in phloem sap-feeding psyllids ("jumping plant lice"), in specialized cells (bacteriocytes) made by these insects. The species belongs to the γ-proteobacteria as does *Escherichia coli*, is transmitted by vertical transmission through the host generations (no exogenous infections!), contains a full translation system and produces some amino acids for the host. We therefore think that it is not possible to construct a minimal cell "getting-by" with significantly less than 150 genes.

References

Specific References

1. Neidhardt FC. Chemical composition of Escherichia coli. In: Neidhardt FC et al, eds. Escherichia coli and Salmonella typhimurium: Cellular and Molecular Biology. Washington, D.C: American Society for Microbiology, 1987:3-6.
2. Andrade MA, OC, Sander C et al. Functional classes in the three domains of life. J Mol Evol 1999; 49:551-557.
3. Allwood AC et al. Stromatolite reef from the Early Archaean era of Australia. Nature 2006; 441:714-718.
4. Banerjee NR et al. Preservation of ~3.4-3.5 Ga microbial biomarkers in pillow lavas and hyaloclastites from the Barberton Greenstone Belt, South Africa. Earth Planet Sci Lett 2006; 241:707-722.
5. Eigen M et al. How old is the genetic code? Statistical geometry of tRNA provides an answer. Science 1989; 244:673-678.
6. Knoll AH. A new molecular window on early life. Science 1999; 285:1025-1026.
7. Gray MW, Burger G, Lang BF. The origin and early evolution of mitochondria. Genome Biol 2001; 2:reviews1018.1-5.
8. Miller SL. A production of amino acids under possible primitive Earth conditions. Science 1953; 117:528-529.
9. Alberts B et al. Chapter 14. Energy conversion: mitochondria and chloroplasts, in molecular biology of the cell, New York: Garland Science, 2002.
10. Chang BS, Halgamuge, Tang SL. Analysis of SD sequences in completed microbial genomes: Non-SD-led genes are as common as SD-led genes. Gene 2006; 373:90-99.
11. Moll I et al. Translation initiation with 70S ribosomes: An alternative pathway for leaderless mRNAs. Nucleic Acids Res 2004; 32:3354-3363.
12. Gualerzi CO et al. Initiation factors in the early events of mRNA translation in bacteria. Cold Spring Harb Symp Quant Biol 2001; 66:363-76.
13. Antoun A et al. How initiation factors maximize the accuracy of tRNA selection in initiation of bacterial protein synthesis. Mol Cell 2006; 23:183-193.
14. Dabbs ER. Mutant studies on the prokaryotic ribosome. Hardesty B and Kramer G, eds. Structure, Function and Genetics of Ribosomes. New York: Springer-Verlag, 1986:733-748.
15. O'Brien TW et al. The translation system of mammalian mitochondria. Biochem Biophys Acta 1990; 1050:174-178.
16. Umekage S, Ueda T. Spermidine inhibits transient and stable ribosome subunit dissociation. FEBS Lett 2006; 580:1222-1226.
17. Qin Y et al. The highly conserved LepA is a ribosomal elongation factor that back-translocates the ribosome. Cell 2006; 127:721-733.
18. Margus T, Remm M, Tenson T. Phylogenetic distribution of translational GTPases in bacteria. BMC Genomics 2007; 8:15.
19. Kim DM, Swartz JR. Prolonging cell-free protein synthesis by selective reagent additions. Biotechnol Prog 2000; 16:385-90.
20. Wilson DN, Nierhaus KH. The ribosome through the looking glass. Angew Chem Int Ed Engl 2003; 42:3464-3486.
21. Mesters JR et al. Synergism between the GTPase activities of EF-Tu.GTP and EF-G.GTP on empty ribosomes. Elongation factors as stimulators of the ribosomal oscillation between two conformations. J Mol Biol 1994; 242:644-654.
22. Nakabachi A, YA, Toh H et al. The 160-kilobase genome of the bacterial endosymbiont carsonella. Science 2006; 314:267.
23. Luria S, Gould S, Singer S. A view of life. Menlo Park, California: Benjamin/Cummings Pub. Co. 1981.
24. Dabbs ER. Mutants lacking individual ribosomal proteins as a tool to investigate ribosomal properties. Biochimie 1991; 73:639-645.
25. Alberts B et al. Chapter 2. Small molecules, energy and biosynthesis, in molecular biology of the cell. New York: Garland Science, 2002.
26. Madigan M, Martinko J, Parker J. Brock biology of microorganisms. 9th ed. London: Prentice Hall International, 2000.
27. Brocks JJ, LG, Buick R et al. Archean molecular fossils and the early rise of eukaryotes. Science 1999; 285:1033-1036.
28. Woese C. Interpreting the universal phylogenetic tree. Proc Nat Acad Sci USA 2000; 97:8392-8396.
29. Jewett MC, SJ. Mimicking the escherichia coli cytoplasmic environment activates long-lived and efficient cell-free protein synthesis. Biotechnol and Bioengin 2004; 86:19-26.

CHAPTER 3

Planetary Astrobiology—The Outer Solar System

François Raulin*

3.1 Planetary Astrobiology, Habitability and the Case of the Solar System

Astrobiology, the study of Life in the Universe, is more precisely the study of the origins, evolution, distribution and destiny of life in the whole universe. This very wide and interdisciplinary domain thus includes not only the search for extraterrestrial life, but also the study of the origin and evolution of life on our planet.[1-3] Although obvious, it is essential to remember that the Earth is so far the only place we know where life is present and that terrestrial life is the only example we have. Searching for a second example requires some assumptions about what is life, what an extraterrestrial living system is made of and how it can be detected.[4]

What is life is a difficult question: how to define a whole, of which we only know one element? We are thus forced to base most of our approaches on this one element and to take the terrestrial life as a reference. Thus it is often considered that life is the property of a living system, exchanging matter and energy with its environment to keep a high level of information and complexity, to be capable of re-producing itself, to adapt to the outside constraints and to evolve by natural selection. On Earth, Life emerged after a period of chemical evolution that involved—through complex and numerous processes of prebiotic chemistry—liquid water, organic matter and energy. The resulting living systems are built on carbon chemistry and their development and evolution requires liquid water and organic and inorganic nutrients. The availability of these ingredients seems a sine qua non for the emergence and evolution of Life. Starting from the terrestrial example, astrobiologists search for extraterrestrial places were similar conditions are or were present.

In fact, one can distinguish different categories of planetary bodies of prime interest for astrobiology. There are bodies where a complex organic chemistry is going on. The study of the chemical processes and structures involved in this chemistry is crucial for understanding the general processes of complexification of matter in the universe, which is essential in the evolutionary steps to life. In that domain the study of the organic chemistry in comets and meteorites is of paramount importance, since their organic content has probably directly participated in the prebiotic chemistry on Earth.

There are also planetary bodies which show today some similarities with our planet before the emergence of life. The study of such environments is also of tremendous importance since most of the conditions which were present on the primitive Earth have disappeared today, erased by geological processes and by life itself.

Now, if we want to understand the processes which allowed the origin of life on Earth and check our ideas and concepts we need to place them in a realistic environment: the availability today of planetary bodies showing analogies with the early Earth is a unique opportunity. In that domain, Titan, the largest satellite of Saturn is a precious target.

And, finally, there are extraterrestrial planetary bodies where life, either extinct or extant, may be present. Those places are characterized by past conditions compatible with the development of complex prebiotic processes over a period long enough for the emergence of life (or conditions compatible with the importation of living systems from other places), followed by conditions compatible with habitability. One of the main parameter which drives the habitability of a planetary body is the presence of liquid water. Mars, like the Earth, very likely had large bodies of liquid water on its surface for long period of times—several hundreds of millions of years—in its early history. This makes the red planet the most attractive body in the solar system for searching for traces of extraterrestrial bio-signatures. Indeed, if Life was—or is still—present on Mars, those traces may be reachable today in the close sub-surface, since the Mars environment, in spite of a drastic evolution of its atmosphere, has probably kept part of these traces owing to the lack of strong tectonic activity.

But there are other places in the solar system where liquid water is probably present. This is the case of three out of the four Galilean satellites of Jupiter: Ganimede, Callisto and Europa (Fig. 3.1). This is also the case of Titan, the largest satellite of Saturn and, more recently evidenced, that of Enceladus, a smaller satellite of the same giant planet. Although we have so far no direct evidence of these internal oceans, the most interesting cases are those of Europa and Enceladus, since if they exist, the internal liquid water bodies may be in contact with rocky materials, facilitating redox reactions that provide chemical energy to sustain prebiotic processes as well as energy for living systems. In the following we will describe the case of the most important object of the outer solar system for astrobiology.[5,6]

3.2 The Jovian System: The Case of Europa

(i) Europa's Ocean

With a diameter of 3120 km, Europa is one of the four largest satellites of Jupiter, the so called "Galilean" satellites which were discovered by Galileo Galilei in 1610. With the "grand tour" of the Voyager mission in the outer solar system, a tremendous amount of new information has been obtained on the giant planets and their

*François Raulin—LISA, Universités Paris 12 et Paris 7, CNRS, 61 Av. du Général de Gaulle, 94010, Créteil, France. Email: raulin@lisa.univ-paris12.fr

Prebiotic Evolution and Astrobiology, edited by J. Tze-Fei Wong and Antonio Lazcano. ©2009 Landes Bioscience.

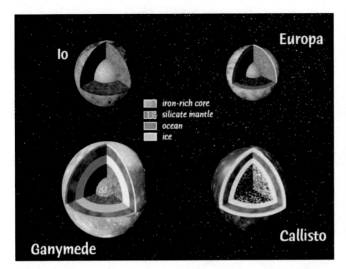

Figure 3.1. Possible internal structure of the four Galilean satellites (adapted from NASA/JPL).

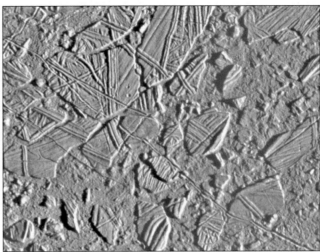

Figure 3.3. High resolution image of Europa's surface (area of about 34 x 42 km) taken by the Galileo mission in 1997, from a distance of 53490 km. It shows crustal plates broken apart and shifted, resembling features seen on polar ice seas on Earth. Credit: NASA/JPL.

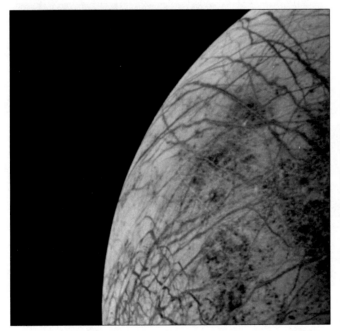

Figure 3.2. Image of Europa taken during Voyager 2 closest approach, in July 1979. The surface is young and shows complex array of streaks indicating that the crust has been fractured and filled by materials released from the interior. Credit: NASA/JPL.

Figure 3.4. Artist view of a cryobot exploring Europa's ocean. Credit: NASA/JPL.

satellites (Fig. 3.2). In particular, from the data collected during the flyby of Jupiter by the two Voyager spacecrafts in 1979, models of the internal structure of the Galilean satellites started to be developed, some suggesting the possible presence of an internal ocean in several of them, particularly in Europa (Fig. 3.1). However, most of our current knowledge on this satellite comes from the Galileo mission which explored the Jupiter system from 1995 to 2003. The high resolution images of Europa's surface showed chaotic terrains, poorly craterized, suggesting resurfacing processes and with many deep fractures indicative of a relatively fluid sub-surface (Fig. 3.3). These observations re-enforced the assumption of the presence of an internal ocean and suggested that it may be not very far from the surface. The mean density (compared to water) of Europa, is about 3, suggesting that it is made of silicates (density of 3-3.5) and water ice (density of ~0.9). The Galileo mission detected a strong induc-

tive magnetic field in Europa. It can be explained by the presence either of an iron core, or of a water ocean more than 10 km thick and with a conductivity similar to that of the terrestrial oceans.[7] So far the available data do not allow discriminating between the two possibilities, but the presence of an internal water ocean that may be located (relatively) close to the surface crust of water ice seems likely.

Over the past thirty years, many papers have been published on this subject (see for instance refs. 7, 8 for a review). The subsurface ocean could be about 100 km thick, covered by an ice shell of about 15 km and with a water temperature of 4°C.[9] More recently Hand and Chyba[10] have re-examined the possible properties of Europa's ocean on the basis of the Galileo data. Some of the main conclusions of their work indicate the possibility that the European ocean is a near-saturation magnesium sulphate aqueous solution, with an ice shell thickness of 4 km or less.

From Prebiotic Chemistry to Life on Europa?

With the possible presence of liquid water in its (not too far) subsurface, Europa is a target of great interest for the astrobiologist community. The other conditions (availability of organic matter and energy) necessary for the emergence of Life may have also been present on/in Europa in its early history. As it very likely occurred on Earth, many organics may have been imported to Europa from meteorites. Furthermore, Galilean satellites may be largely made of material from carbonaceous meteorites, such as CII chondrites. Those include about 2.5% C and 13.5% of H_2O. Since Europa may contain more than 7% of water in mass, if the assumption that it is made of CII chondrites is valid, it may also contain about 1% by mass of carbon atoms,[11] half of which is insoluble complex organic matter. The latter may produce, under hydrolysis by heterogeneous processes, organics of biological interest. The water soluble organic fraction may represent about 0.05% of the mass of Europa and include many compounds of biological interest.

Assuming that most of the water is in the liquid phase, one can estimate a concentration of dissolved organic carbon in Europa's oceans as high as about 1% by mass. Such a concentration of organics, depending on their nature, may be sufficient for efficient prebiotic evolution. For instance, a concentration of 0.1 M (equivalent to 0.3% by mass) of HCN in aqueous solution is large enough to induce the polymerization of HCN toward its tetramer and higher oligomers, source of amino acids, purines and pyrimidine bases upon hydrolysis. Low temperatures reduce the rate constants of prebiotic chemical reactions, but they may also increase the concentrations of reacting organics through eutectic effects and thus increase the rates of reaction. In addition, high pressure conditions such as those found in subsurface oceans may also induce chemical condensation reactions that are essential for the formation of biological macromolecules starting from their monomers, such as polypeptides and polynucleotides from their building blocks of amino acids and nucleotides.

In addition, hydrothermal vents may be present on the floor of the oceans, providing favourable locations for increases in chemical complexity, thanks to the heterogeneous processes that can occur at the interface between the hot gases and the liquid and solid phases. These processes can be favoured by the potential catalytic properties of the mineral phases and by the high thermal gradients, which protect the products from thermal degradation. However, recent work on the salinity of the ocean by Hand and Chyba[10] suggests a high concentration of salts, which could substantially reduce the oligomerization processes essential to the prebiotic emergence of replicating macromolecules. Thus the present conditions in the ocean may not be very favourable to the origin of life, but this does not exclude that the latter might have happened much earlier when the conditions in the ocean could have been different.

(ii) Europa's Habitability

If Life had emerged in Europa's ocean, one should wonder about the habitability of this environment. First, the high energy radiations in the vicinity of Jupiter may look like a major problem for the persistence of Life. But the ice crust of several kilometres appears to be a good protection shell. The temperature of the ocean—likely to be in the 0°C range or even warmer[9]—should not be a problem for the development of life. The potential high salt concentration, large possible pH range and high pressure (100 MPa maximum) is not an obstacle if we take into account the existence of extremophiles living on Earth under more extreme conditions. Thus the presence of living systems in Europa's ocean cannot be ruled out.[11]

How extensive is the putative Europa's biota? Reynolds et al[12] examined the different possible sources of energy available to bio-systems on Europa. Taking into account only the solar energy availability, they estimated a possible biomass density of 2×10^{-6} g m^{-2}. This value is quite small compared to the lower limit of the biomass for the

Table 3.1. Examples of terrestrial organisms of interest for Europa (adapted from Oro et al, ref. 11)

Methanogenic Archaebacteria
Heterotrophic (fermentation): organics + H_2 => CH_4 + biosynthetic products
organics = methanol, methylamine, formate, acetate, ..
Autotrophic (CO_2 reduction): CO_2 + H_2 or $Fe°$ => CH_4 + biosynthetic products

Thermophilic Archaebacteria
Heterotrophic: (fermentation): yeast extract => CO_2 + biosynthetic products
(S respiration): organics + S => H_2S + CO_2 + biosynthetic products
Autotrophic (S reduction): CO_2 + H_2 + S => H_2S + biosynthetic products
organics = alcohols, sugars, formate, acetate, ..

Photosynthetic Bacteria
Photoheterotrophic (anaerobic): organics + light => biosynthetic products
Photoautotrophic (anaerobic): CO_2 + $2H_2S$ + light => S_2 + H_2O + (HCHO)
Photoautotrophic (oxygenic): CO_2 + H_2O + light => O_2 + (HCHO)

Earth, which is about 1000 to 10000 g m^{-2}. Nevertheless, although very limited, it corresponds to a nonnegligible biota. In addition, these calculations are only based on solar energy; thus the current value of Europa's possible biomass may be higher if the other energy sources present on the Galilean satellite, in particular heat delivery by radiogenic heat flow, are important. Indeed, Chyba and Phillips[8] estimated a total steady state biomass for Europa of about 10^{13}-10^{15} g, assuming additional production of hydrogen and oxygen, in the ice and the oceans through the decay of radioactive ^{40}K and a biomass turnover time of about 1000 years. Averaged over the surface of the planetary body, this gives a biomass density of 0.3-30 g m^{-2}.

What kind of life can we expect? Biological evolution to eukaryotic life cannot be excluded but prokaryotic anaerobic life is more probable. Archaebacteria-like organisms seem the most likely biota on Europa.[13] Terrestrial archaebacteria are indeed good examples of what can be expected. There is a large number of possible examples with different metabolic activities[11] as shown on Table 3.1.

(iii) Astrobiological Exploration of Europa

How could we detect such life on Europa? A powerful approach would be a "Hydrobot/cryobot" mission with a cryobot melter to go through the several kilometres thick ice crust and to release a submersible in the (still hypothetical) Europa oceans (Fig. 3.4). The submersible could be equipped to search for hypothetical hydrothermal vents and for hypothetical micro-organisms in their vicinity. However, there are many crucial problems associated with such a mission, specially its cost and many technological difficulties. In particular it requires a very difficult adaptation of technology used on Earth to space application. The terrestrial example of Lake Vostok in Antarctica[14] gives a clear illustration of the problem. After many years of drilling, the top of this subsurface lake, located below 3 km of water ice, is almost reached. However, scientists are now hesitating to go into it, because of the crucial problem of biological contamination. This problem would be even more crucial for an extraterrestrial liquid water body like Europa's ocean, because of its much lower biomass density. Indeed, the biological activity in the hypothetical Europa oceans may be quite limited and difficult to detect and the importance of potential biological contamination much higher.

Consequently, an orbiter mission seems—for the moment— much more reasonable. This would have been possible with the

NASA-JIMO (Jupiter Icy Moons) mission which was planning to offer a detailed mapping of Europa and to provide data allowing confirmation of the existence of internal oceans. This mission has been recently cancelled because of a loss of funding. Other missions to Europa are under study in the frame of the ESA Cosmic Vision announcement of opportunities. These missions could allow confirmation of the presence of an internal Europa ocean and a search for traces of biological activity by looking for molecular signatures of metabolic activity such as CH_4, H_2S, HCHO, etc. that could be released at the surface of the satellite.

3.3 The Saturn System: The Case of Titan and Enceladus

(i) Some Historical Background and General Data[6]

Discovered in 1655 by the Dutch astronomer Christiaan Huygens, the largest moon of Saturn was much later named "Titan" by John Herschel in the nineteenth century. Observations of limb darkening around the disk of Titan by the Spanish astronomer José Comas-Sola in 1908 suggested the presence of an appreciable atmosphere around Titan. This hypothesis was taken seriously into consideration by the British astronomer James Jeans who calculated that a planetary body like Titan can keep a substantial part of its atmosphere (most of the constituents with a molecular mass exceeding about 16 Dalton) in spite of escape processes due to its relatively small size (diameter 5150 km), if its temperature is low enough (less than ~100 K). The presence of an atmosphere around Titan was definitively confirmed in 1944, with the detection the absorption bands of gaseous methane (CH_4) by the American astronomer Gerard Kuiper. He thus demonstrated that this atmosphere includes methane as one of its constituents and even succeeded in deriving from his observations that the partial pressure of methane at Titan's surface should be about 0.1 bar (100 hPa). The detection of methane in Titan's atmosphere was then confirmed by many observations from ground based telescopes and by the several space missions that flew by Saturn.

With a radius of 2575 km, Titan (Fig. 3.5) is the second largest satellite of the Solar System after Ganymede (R = 2631 km). Like Saturn, its period of rotation around the Sun is about 30 years and its mean distance to the Sun is about 9.5 astronomical units (AU) corresponding to a received solar flux of about 1% of the flux at the level of the Earth orbit. If considered as a black body, the mean temperature of Titan is about 82 K. Titan is in fact the only satellite of the solar system to have a noticeable atmosphere, as well as the only planetary body with an atmosphere close to that of the Earth. Indeed, Titan's atmosphere is mainly made of dinitrogen (N_2) and its surface pressure is 1.5 bar (1500 hPa). As mentioned above, the atmosphere also includes a few percent of methane. The surface temperature is about 94 K, higher by about 12 K than the black body temperature because of a noticeable greenhouse effect (20 K, mainly produced by atmospheric methane and dihydrogen), in spite of the anti-green house effect (~ −8 K) produced by the haze particles. In fact, owing to its low surface temperature Titan's atmosphere near the surface is about 5 times denser than the Earth's atmosphere.

Titan spins around Saturn with a period of 16 Earth-days, with synchronous rotation. Thus the Titan solid surface rotates slowly; however its atmosphere presents a super-rotation caused by strong zonal winds. With a distance from Saturn equal to about 20 Saturnian radii, Titan is far enough from the giant planet to avoid interactions with the rings but still close enough to allow interaction of its atmosphere with the electrons of the magnetosphere of Saturn which thus play a role in its chemical evolution.

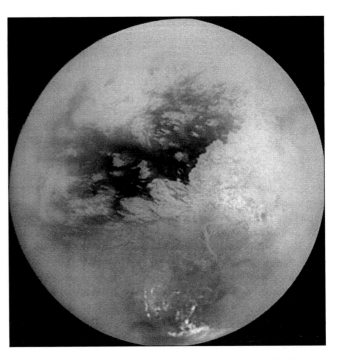

Figure 3.5. Mosaic of Titan's surface from images taken by the narrow angle camera of the Cassini spacecraft. The large bright white area is named Xanadu. The resolution is about 1.3 km/pixel. Credit: NASA/JPL/Space Science Institute.

(ii) Theoretical Modelings

Titan's orbit is slightly eccentric (e = 0.029): this suggests the absence of a shallow ocean on Titan's surface which would tend to reduce the eccentricity value, transforming the orbit into a circular one. Now, the mean density of Titan is 1.88 g cm⁻³. This low value (even lower than Europa) indicates an internal structure made of low density materials—ices, like water ice, of a mean density of approximately 1 g cm⁻³ mixed with high density materials, such as rocks and silicates, of a mean density of about 3 g cm⁻³.

Titan originated about 4.6 billion years ago from the accretion of small planetesimals present in Saturn's sub-nebula. The heat generated by gravity processes and radioactive decay melted the icy components making the rocks sink toward the centre of the forming Titan. The resulting initial internal structure was a central core made of undifferentiated silicate, mixing rocks and ices, covered by a silicate outer core and a liquid mantle of water-ammonia mixture, in contact with a dense atmosphere (Fig. 3.6). The origin and composition of this primordial atmosphere are still under debate. Recent models of the evolution of the Saturn sub-nebula indicate that the planetesimals which formed Titan included CH_4 and NH_3 in the form of hydrates (and not CO or N_2 which are not efficiently trapped as hydrates). Then CH_4 and NH_3 were progressively released to the atmosphere. NH_3 was converted into N_2 by photochemical processes in the upper regions of the atmosphere or by impact shock catalyzed dissociations. At that time Titan's surface was warm enough to maintain an aqueous environment for some (up to 100) millions of years. As its energies became less abundant, it cooled down and liquid water started freezing out of the H_2O-NH_3 solution. An ice I crust (Ice I is the form of all natural snow and ice on Earth) formed, the thickness of which rapidly increased to 30 km in about 70 million years, covering an internal ocean initially in contact with a rocky bottom. The high pressure at this bottom induced the crystallization of water ice VI (high pressure water ice) and ammonia hydrates, enriching the ocean in

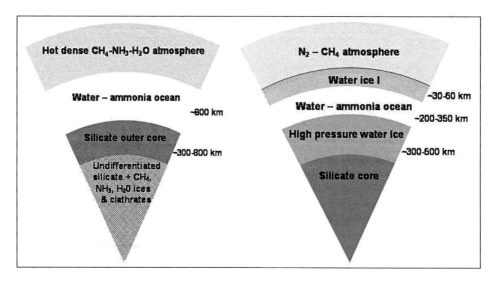

Figure 3.6. Internal structure of Titan; left: primordial Titan; right present Titan (adapted from Icarus, Vol. 146, Fortes[24]; Copyright 2000 with permission from Elsevier).

ammonia. The thickness and composition of this ocean depend on the heat flow and convective transport; it could be about 200 km thick, containing up to 15 wt % of NH_3.

Concerning the evolution and present state of the atmosphere, several kinetic models, mainly photochemical models, have been published. Their main output is the vertical concentration profile of major and trace species. Because of the lack of detailed observations and data on the eddy diffusion coefficient prior to Cassini-Huygens, these profiles were relatively poorly constrained. However, these models provided a preliminary understanding of the chemical processes involved in this complex atmospheric chemistry. They assumed that the chemical scheme starts with the dissociation of CH_4 and N_2 by the solar UV photons and the electrons from Saturn's magnetosphere. The resulting primary species, formed in the high atmosphere, yield simple hydrocarbons and nitrogen-containing molecules, particularly C_2H_2 and HCN. Once they are formed, theses two compounds diffuse to the lower atmospheric regions where they can still absorb the available UV photons (in the near UV) because their UV spectra extend to that region. Although less energetic than the Mid-UV, these photons can produce photo-dissociation. The resulting radicals produce more complex hydrocarbons and nitriles. In the chemical processes polyynes, such as C_4H_2 and C_6H_2 and cyanopolyynes, like HC_5N, may play an important role in the formation of high molecular weight products. Additional CH_4 dissociation probably occurs in the low stratosphere through photo-catalytic processes involving C_2H_2 and polyynes. The end product of this chemistry would be a macromolecular organic compound mainly made of C, H and N. However, as described below, the measurements reformed by the Cassini-Huygens mission[15,16] on Titan's ionosphere suggest that organic processes in this high atmospheric zone must also contribute to the formation of this macromolecular matter. But detailed models of this ion chemistry coupled to photochemical models still need to be developed.

(iii) Laboratory Experiments

Another approach to study these processes is to carry out laboratory experiments. There are experiments specifically designed for determining some of the parameters involved in the evolution of the planetary body, such as rate constants, thermodynamic data, or spectroscopic data. There are also experiments simulating partly or globally the evolution of the considered environment. Such so-called "simulation experiments" can integrate the many different physical and chemical processes involved. One of the very first simulation

experiments carried out in the laboratory was the historical experiment of Stanley Miller to mimic the chemical evolution processes in the primitive Earth environment. In the absence of any direct observational data of such an environment, this was indeed a unique tool. A similar approach has been used more recently by several groups to study the evolution and organic chemistry of Titan's atmosphere using first closed reactors and more recently open flow reactors, with a plasma discharge as energy source and a flow mixture of dinitrogen and methane (Fig. 3.7).

In Titan's atmosphere the main energy sources are solar UV radiation and mid-energy electrons coming from Saturn's magnetosphere. The most abundant solar photons in the mid-UV, compatible with methane photo-dissociation ($\lambda < \sim 150$ nm), are the Lyman α (121.6 nm) photons. This radiation can easily be simulated in the laboratory by using monochromatic lamps. But Lyman α photons do not allow the photo-dissociation of N_2, which requires photons of much shorter wavelength ($\lambda < \sim 100$ nm). This very far UV is difficult to obtain in the laboratory with irradiation systems compatible with closed or flow reactors. For that reason, most of the laboratory experiments simulating the chemical evolution of Titan's atmosphere, which have been carried out so far on N_2—CH_4 initial gas mixtures using an electron impact energy source, mimic Titan's atmospheric electrons impinging from Saturn magnetosphere. Another approach, compatible with a short wavelength UV source, is to use an initial gas mixture with containing an N-compound that can be photo-dissociated by UV photons of longer wavelengths and is also present in Titan's atmosphere—although at trace levels. Examples are the molecules HCN and HC_3N.

From the many experiments which have been carried out since the early 1980s to simulate Titan's atmospheric chemistry in the laboratory, more than 200 different organic molecules have been detected in the gas phase. They are mainly hydrocarbons and nitriles. In those experiments all the gaseous organic species observed in Titan's atmosphere, as well as many others not (yet) observed, have been detected, in particular polyynes such as C_6H_2, C_8H_2 and cyanopolyyne HC_4-CN. These compounds are also included in photochemical models of Titan's atmosphere, where they might play a key role in the chemical schemes allowing the transition from gas phase products to aerosols. Of further astrobiological interest is the formation of organic compounds with asymmetric carbon and of O-containing compounds when CO (present in Titan's atmosphere at the 100 ppm level) is present, the most abundant O-organics being oxirane (also named ethylene oxide) $(CH_2)_2O$.

Simulation experiments also produce solid organics, usually named tholins, a generic name invented by Carl Sagan in the late 1970s.[17] These *"Titan tholins"* are supposed to be laboratory analogues of Titan's aerosols. They have been extensively studied since the first work by Sagan and Khare more than 20 years ago. These laboratory analogues show very different properties depending on the experimental conditions. For instance, the average C/N ratio of the product may vary between values of less than 1 and more than 11. The molecular composition of the Titan tholins is still poorly known, but it is well established that they are made of macromolecules of largely irregular structures. IR and UV spectroscopy and analysis by pyrolysis-GC-MS techniques show the presence of aliphatic and aromatic hydrocarbon groups, of CN, NH_2 and C = NH groups. Their hydrolysis, even at neutral pH, releases amino-acids; it is also of astrobiological importance to mention that some (terrestrial) microorganisms can use them as nutrients.

(iv) Observations: The Cassini-Huygens Case

Of course, none of the two approaches described above, namely modeling and simulation experiments, can be developed without any observational data such as those provided by the NASA-ESA Cassini-Huygens mission which combines remote sensing and in-situ observations.[15,16] The Cassini spacecraft, carrying the Huygens Titan atmospheric probe, was launched on October 15, 1997. After a seven-year interplanetary trajectory, it reached Saturn in 2004, with Saturn orbit insertion on July 1, 2004, allowing Cassini to become a new Saturn satellite. At the end of 2004, after two Titan encounters, it released the Huygens probe on the third orbit around Saturn on December 25, 2004. Huygens penetrated Titan's atmosphere on January 14, 2005. Huygens landed after a descent of about 2.5 hours and continued to function for three hours on the surface. Cassini recovered surface data for more than 1 hour and 10 minutes, before it left the field of view of the landed probe.

The exploration of Titan is one of the main objectives of the Cassini-Huygens mission. The four year nominal mission (mid 2004 to mid 2008, currently extended) includes 74 orbits around Saturn with 44 close Titan fly-bys. Several of the twelve instruments of the orbiter (Table 3.2A) and most of the six instruments of the probe (Table 3.2B) provide data of astrobiological interest. On Cassini, the Ion and Neutral Mass Spectrometer (INMS), the Composite Infrared Spectrometer (CIRS) and the Ultraviolet Imaging Spectrograph (UVIS) determine the chemical composition of different zones of Titan's atmosphere. They are able in particular to detect many organics including new species and to determine their vertical concentration profile. The Cassini Radar, the cameras of the Imaging Science Subsystem (ISS) and the Visual and IR Mapping Spectrometer (VIMS) can map Titan's surface through the haze layers and provide information on the morphology, geology and chemical composition of the surface.

On the Huygens probe,[15,18] the GC-MS instrument has performed a detailed chemical analysis of the atmosphere during the descent of the probe and also after the probe has landed.[19] The Aerosol Collector and Pyrolyzer (ACP) experiment managed to collect Titan's aerosols during the descent of the probe and to heat them at different temperatures up to high temperatures for pyrolyzing the refractory part of the collected particles. The gases produced were transferred to the GC-MS instrument for molecular analysis. This was the first direct in-situ molecular and elemental analysis of Titan's hazes.[20] The Huygens Atmospheric Structure Instrument (HASI) determined, in particular, the vertical profiles of pressure and temperature. The Descent Imager/Spectral Radiometer (DISR) measured the radiation budget of the atmosphere, investigated the

cloud structure and took images of the surface (Fig. 3.8). The Surface Science Package (SSP) provided information on the physical state and chemical composition of the surface.

The data from the Cassini orbiter give a global mapping of Titan with observations of potential temporal and spatial variations. The data of the Huygens entry probe give very detailed information of one particular location of Titan. Although the Cassini mission is far from ended, the synergy provided by the complementarities of the orbiter and descent probe data has already revealed a new vision of Titan with three main astrobiological aspects: (i) many similarities with the Earth, including the primitive Earth; (ii) a complex organic chemistry; and (iii) potential habitability.[21]

(v) Main Astrobiological Aspects of Titan

Titan's dense atmosphere, which extends up to about 1500 km, is mainly composed of N_2 like the atmosphere of the Earth. The other main constituents are CH_4 (~1.6-2% in the stratosphere as measured by the CIRS on Cassini and the GC-MS on Huygens) and H_2 (~0.1%). Although Titan is much colder, with a troposphere (~94-~70 K), a tropopause (70.4 K) and a stratosphere (~70-175 K), its atmosphere as recently evidenced by Cassini-Huygens presents a similar complex structure and includes, like the Earth, a mesosphere and a thermosphere. These similarities are linked to the presence of greenhouse gases and anti-greenhouse elements in both atmospheres. On Titan CH_4 and H_2 are equivalent to terrestrial condensable H_2O and noncondensable CO_2. In addition, the haze particles and clouds in Titan's atmosphere play an anti-greenhouse effect similar to that of the terrestrial atmospheric aerosols and clouds.

Methane on Titan seems to play the role of water on Earth with a complex cycle not yet fully understood. Cassini ISS has detected surface features near the South Pole that might be a lake; radar has discovered the presence of lakes and seas in the north polar region, likely to be of liquid methane and ethane (Fig. 3.9). Moreover, the DISR instrument on Huygens has provided pictures of Titan's surface that clearly show dentritic structures resembling terrestrial fluvial net (Fig. 3.8). These features, free of crater impacts, are in a relatively young terrain where liquid flowed recently. As shown by the Huygens GC-MS data, the CH_4 abundance in the low troposphere above the landing site reaches the saturation level at an altitude of approximately 8 km, allowing the possible formation of clouds. The potential resulting rain may be the cause of the observed fluvial-like net. In addition, GC-MS data also suggest the presence of condensed methane on the landing site. Other observations from the Cassini instruments clearly show the presence of various surface features of different origins indicative of volcanic, tectonic, sedimentological and meteorological processes similar to those that are found on Earth (Fig. 3.10).

INMS and GC-MS have detected the presence of argon (Ar) in Titan's atmosphere. Like in the Earth atmosphere, the most abundant isotope is ^{40}Ar, which should come from the radioactive decay of ^{40}K. Its stratospheric mole fraction, measured by GC-MS, is about 4×10^{-5}. The abundance of primordial argon (^{36}Ar) is about 200 times smaller. The other primordial noble gases have a mixing ratio of less than 10 ppb. This strongly suggests that Titan's atmosphere is of a secondary origin formed by the degassing of trapped gases. Now, N_2 could not have been efficiently trapped in the icy planetesimals from which Titan formed, but NH_3 could. Thus it is believed that Titan's primordial atmosphere was initially made of NH_3 which was then progressively transformed into N_2 by photolysis and/or impact driven chemical processes. The $^{14}N/^{15}N$ ratio measured by GC-MS in the stratosphere (= 183) is 1.5 times smaller than that of primordial nitrogen. This indicates that the atmosphere was probably lost several times since its formation with the simultaneous

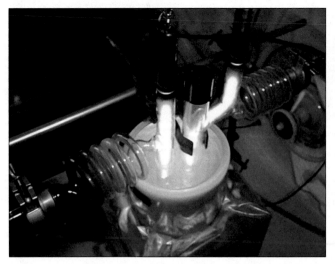

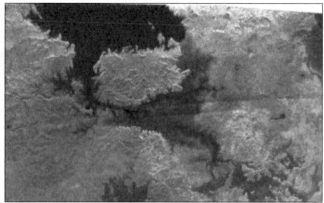

Figure 3.9. Cassini Radar image of Titan's surface at high latitude regions obtained in February 2007, showing the presence of a large lake (several 100 km, with a ~100 km size island. Credit: NASA/JPL.

Figure 3.7. Experimental device for laboratory simulation of Titan's atmosphere chemistry at LISA (Laboratoire Interuniversitaire des Systèmes Atmosphériques, Universités Paris 12 et Paris 7). A gas mixture of N_2 (98%) and CH_4 (2%) flows continuously flows through an open reactor and is maintained at low pressure (~1 hPa) and low Temperature (~150 K) (Credit : P. Coll).

Figure 3.10. Top: The see sand dunes seen by Cassini on Titan surface (top) look like Namibian sand dunes on Earth (bottom). Credit: upper photo NASA/JPL—lower photo; NASA/JSC.

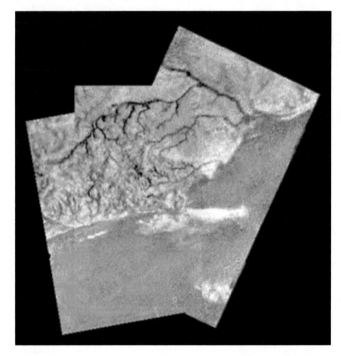

Figure 3.8. Image of Titan's surface taken by the DISR instrument on Huygens at 6.5 km altitude and showing channel networks, highlands and dark-bright interface. Credit: ESA/NASA/JPL/University of Arizona.

transformation of methane into organics, suggesting the presence of large deposits of organics on Titan's surface.

(vi) An Active Organic Chemistry

Analogies can also be made between the active organic chemistry which is currently going on on Titan and the prebiotic chemistry on primitive Earth. Several organics have already been detected in Titan's atmosphere, including compounds usually considered as key molecules in terrestrial prebiotic chemistry, such as hydrogen cyanide (HCN), cyanoacetylene (HC_3N) and cyanogen (C_2N_2). The detected organics in the stratosphere[21] are hydrocarbons (both with saturated and unsaturated chains) and N-organic compounds, mainly nitriles, as expected from laboratory simulation experiments. Since the arrival of Cassini in the Saturn system, the presence of water and benzene has been unambiguously confirmed by the CIRS instrument.

Direct analysis of the ionosphere by the INMS instrument during the low altitude Cassini fly-bys of Titan shows the presence of many organic species at detectable levels at very high altitudes (1100-1300 km). The mass range of this instrument is limited to a maximum mass of 100 Daltons. However, extrapolation of the data strongly suggests that high molecular weight species up to several thousand Daltons may be present in the ionosphere.[22] These new data open a fully new vision of the organic processes occurring in Titan's atmosphere, with a strong implication of the ionospheric chemistry in the formation of complex organic compounds in Titan's environment, which was not envisaged before (Fig. 3.11).

On the contrary, the GC-MS instrument on board Huygens[19] did not detect many organics in the low atmosphere. This is probably due to their condensation on the aerosols. These particles, for which no direct data on chemical composition were available before, have been analyzed by the ACP instrument.[20] The data show that the aerosols are made of refractory organics that release HCN and NH_3 during pyrolysis. Moreover, the nature of the pyrolysates indicates the potential presence of nitrile, amino and/or imino groups.

Table 3.2A. *Instruments on the Cassini spacecraft, InterDisciplinary Programs (IDP), the leading scientists and the potential for astrobiological return of their investigation*

Instrument or IDP	P.I., T.L. or IDS[*]	Country	Astrobiological Return
Optical Remote Sensing Instruments			
Composite infrared spectrometer (CIRS)	V. Kunde, M. Flasar	USA	+++
Imaging science subsystem (ISS)	C. Porco	USA	+++
Ultraviolet imaging spectrograph (UVIS)	L. Esposito	USA	++
VIS/IR mapping spectrometer (VIMS)	R. Brown	USA	++
Fields Particles and Waves Instruments			
Cassini plasma spectrometer	D. Young	USA	+
Cosmic dust analysis	E. Grün	Germany	+
Ion and neutral mass spectrometer	H. Waite	USA	+++
Magnetometer	D. Southwood, M. Dougherty	USA	NA[#]
Magnetospheric imaging instrument	S. Krimigis	USA	NA
Radio and plasma wave spectrometer	D. Gurnett	USA	NA
Microwave Remote Sensing			
Cassini radar	C. Elachi	USA	+++
Radio science subsystem	A. Kliore	USA	++
Interdisciplinary Program			
Magnetosphere and plasma	M. Blanc	France	+
Rings and dust	J.N. Cuzzi	USA	+
Magnetosphere and plasma	T.I. Gombosi	USA	+
Atmospheres	T. Owen	USA	+++
Satellites and asteroids	L.A. Soderblom	USA	+
Aeronomy and solar wind interaction	D.F. Strobel	USA	++

[*]P.I. = Principal Investigator; T.L. = Team Leader; IDS = InterDisciplinary Scientist. [#]NA = not applicable.

This result strongly supports the hypothesis that the aerosols have a molecular composition very close to that of the laboratory tholins. These particles are probably made of a refractory organic nucleus, covered with condensed volatile compounds, with a mean diameter of the order of 1 μm (Fig. 3.12).

After sedimentation the aerosols should accumulate on the surface to form a deposit of complex refractory organics and frozen volatiles. The DISR data show the presence of water ice but no clear evidence—so far—of tholins. On the other hand, analysis of the atmosphere near the surface by GC-MS subsequent to Huygens'

Table 3.2B. *Instruments on the Huygens entry probe, InterDisciplinary Programs (IDP), the leading scientists and the potential for astrobiological return of their investigation*

Instrument or IDP	P.I. or IDS[*]	Country	Astrobiological Return
Gas chromatograph-mass spectrometer (GC-MS)	H. Niemann	USA	+++
Aerosol collector and pyrolyser	G. Israël	France	+++
Huygens atmospheric structure instrument	M. Fulchignoni	Italy	++
Descent imager/spectral radiometer (DISR)	M. Tomasko	USA	+++
Doppler wind experiment	M. Bird	Germany	+
Surface science package (SSP)	J. Zarnecki	UK	+++
Interdisciplinary Program			
Aeronomy	D. Gautier	France	++
Atmosphere/surface interactions	J.I. Lunine	USA	++
Chemistry and exobiology	F. Raulin	France	+++

[*]P.I. = Principal Investigator, IDS = InterDisciplinary Scientist.

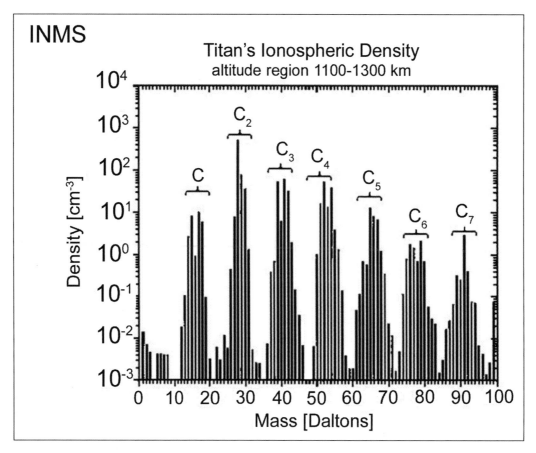

Figure 3.11. Averaged mass spectrum of Titan's ionosphere taken by the Cassini-INMS instrument near 1,200 km altitude. The spectrum shows signature of organic compounds including up to 7 (carbon/nitrogen) atoms. Image Credit: NASA/JPL/University of Michigan.

landing indicates the likely presence of many organics, including N-organics, C3 and C4 hydrocarbons and benzene, vaporized by the heating of the probe. This is in agreement with the hypothesis that the surface is rich in condensed volatile organics and that most of the organic compounds are in the condensed phase in the low atmosphere.

With this picture of Titan's organic chemistry, the chemical evolution of the main atmospheric constituents—dinitrogen and methane—produces mainly ethane which accumulates on the surface or the near sub-surface, eventually dissolving in methane-ethane lakes and seas and complex refractory organics which accumulate on the surface together with condensed volatile organic compounds such as HCN and benzene. In spite of the low temperature, contrary to what was often said, Titan is not a congealed Earth: the chemical system is not frozen. Titan is an evolving planetary body and so is its chemistry. Once sedimented on Titan's surface, the aerosols and their complex organic content may follow a chemical evolution of astrobiological interest. Laboratory experiments show that Titan tholins can release many compounds of biological interest, such as amino acids, once in contact with liquid water. This seems possible even with water ice. Those processes could be particularly favourable in zones of Titan's surface where cryovolcanism is occurring. Thus one can envision the possible presence of such compounds on Titan's surface or near subsurface. In situ measurement of Titan's surface would thus offer a unique opportunity to study by a ground truth approach some of the many processes that could be involved in prebiotic chemistry, including isotopic and enantiomeric fractionations.

In fact, even the presence of liquid water on Titan's surface is a possibility in spite of the low surface temperature. Cometary impacts can melt the surface water ice and provide, during periods as long as about 1000 years, conditions of terrestrial-like prebiotic syntheses with localized liquid water bodies.[23] The hypothetical internal water-ammonia ocean mentioned above is also a place where an active prebiotic chemistry can occur, like in the case of Europa. These prebiotic processes are less likely now, since Titan's ocean is probably not in contact any more with the silicate layer, but they might have occurred in the early history of Titan, with the possible presence of hydrothermal activity allowing a CHON (Carbon, Hydrogen, Oxygen and Nitrogen) prebiotic chemistry over long periods of time (several million years). Thus even the presence of life in this hidden ocean today cannot be excluded.

(vii) Habitability of Titan

Indeed, although Titan's surface is too cold and the energy available in its environments is not abundant enough to provide conditions of habitability, the subsurface ocean may be suitable for life. As suggested by Fortes[24] the possible mean temperature of the ocean (260 K and even 300°K in the vicinity of cryovolcanic hotspots), is not an obstacle to the development of living systems. The same is true for the pressure and pH parameters. At a depth of 200 km, the expected pressure in Titan's ocean is about 500 MPa (5 kbar), not incompatible with life as shown by terrestrial examples. The pH (aqueous solution with up to ~15% by weight of NH_3) would be ~11.5: on Earth, bacteria can grow at pH 12. Even the limited energy resources do not exclude the sustaining of life.

Fortes has estimated that an energy flux of about 5×10^8 W may be available in Titan's ocean for bioactivities. In the terrestrial biosphere this value corresponds to the production of about 4×10^{11} mol of ATP per year and about 2×10^{13} g of biomass per year. Assuming an average turnover for the living systems of the order of a year, the biomass density would be 1 g/m^2, which is almost in the range of

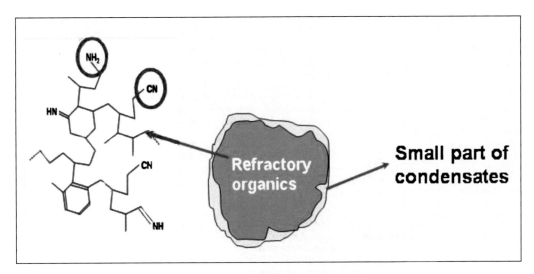

Figure 3.12, left. Model of the chemical composition of Titan's aerosols, derived from the Huygens ACP data. Credit:—F. Raulin, in G. Horneck and P. Rettberg Eds Complete Course in Astrobiology, page 232, 2007. Copyright Wiley-VCH Verlag GmbH and Co. KGaA. Reproduced with permission.

Figure 3.13, right. As seen by the Cassini camera, large plumes are ejecting fine particles in the south polar region of Enceladus (left image), like gigantic terrestrial geysers. In the same region, the CIRS instrument detected warmer area (89-91 K, compared to the surrounding temperature of 74-81 K) and correlated to the "tiger-stripes" looking fractures (right image). Credit: Left: NASA/JPL/Space Science Institute; Right: NASA/JPL/GSFC/Space Science Institute.

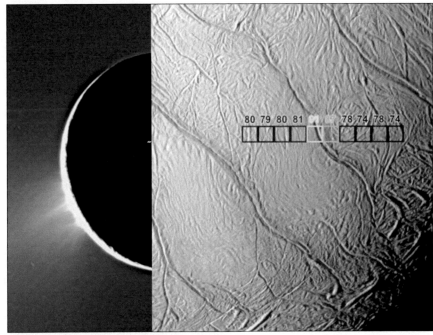

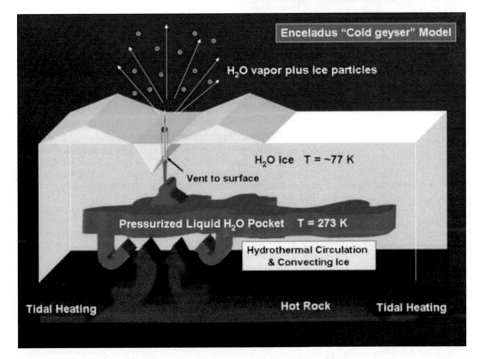

Figure 3.14, left. A model of Enceladus plumes formation, implying the presence of subsurface reservoirs of pressurized liquid water, in contact with hot rocky materials. Credit: NASA/JPL/Space Science Institute.

Europa's ocean. Cassini-Huygens data do not evidence any sign of biological activity on Titan's surface. In particular, the measurement of the carbon isotopic ratio in methane strongly suggests a non biological origin. However, these data do not exclude the possibility of the presence of a limited biota in the sub-surface ocean of Titan.

Enceladus Surprises

Of course the Cassini-Huygens mission is not over yet and could give still provide surprising discoveries. This was already the case with Enceladus. Before the Cassini-Huygens mission, very little was known about this small satellite of Saturn (its radius is only 250 km). Since the observations of Enceladus by the Cassini orbiter instruments, this astonishing satellite is becoming a new astrobiological planetary target.

The images taken during the close fly by of the satellite by the Cassini orbiter show huge plumes of icy particles, continuously ejected from the south polar regions. Moreover, its surface is covered with fractures, looking like tiger stripes and warmer than the surrounding areas, which seems to be in contact with the internal part (Fig. 3.13). This suggests the presence of pressurized liquid water inside the satellite, which may be partially in contact with the hot rock (Fig. 3.14). Dinitrogen, N_2 and/or CO and several hydrocarbons (methane, propane and acetylene) have been detected by Cassini-INMS in the plume. N_2 may be produced by the thermal decomposition of ammonia, eventually catalyzed by the presence of minerals. Thus ammonia could be present in the internal liquid water bodies together with organics, offering the possibility of an evolved prebiotic chemistry. Indeed those environmental conditions are favourable for the syntheses of various organic compounds, including many of the molecules that are the building blocks of life!

3.4 Conclusions: From Earth to beyond the Solar System

Without anticipating too much the content of the following chapters, there are several places in the solar system of interest for astrobiology, in addition to the Earth. Mars remains the most interesting one for discovering traces of past or even present life, but there are also several location of great interest in the outer solar system for looking at prebiotic processes or searching for bio-signatures: Europa, Titan and even Enceladus are among those important targets. However, from what we know today about the habitable bodies, it seems very likely that the Earth is the only place in the solar system where macroscopic life is present.

Now, the discovery of extrasolar planets opens new possibilities for finding extraterrestrial life with a higher degree of evolution. Within a decade about 300 exoplanets have already been detected with quite limited tools, allowing practically only the detection of giant planets of very large Earth like planets. Unexpected giant planets have been discovered, orbiting so close to their star that their temperature is very high. Some of these exoplanets may be covered by a global and hot liquid water ocean, compatible with a developed aquatic life. Some telluric-like exoplanets have already been detected, but they are larger (about twice) than the Earth and seem to be located in nonhabitable zones. Nevertheless, soon we should be able to detect exoplanets with size and general conditions much closer to those of planet Earth.

But how can we search for life on these exoplanets? The detection of atmospheric bio-signatures is a possibility, but it requires spectroscopic tools which will probably not be available before one or even several decades, with ambitious space missions such as Darwin. However, one approach is already available, even if very speculative: the SETI (Search for Extra-Terrestrial Intelligence) approach. Even

if the probability for life to evolve toward intelligent life and technologically advanced civilizations is very low (which we do not know), it cannot be excluded that what happened on Earth also happened on another habitable planet outside the solar system, especially if the number of those planets is very high (which we do not know either). There is a very strong time constraint linked to the age and the life time of such civilization. For instance, the technologically "advanced" terrestrial civilization is only detectable through its radio signals from stars located at less than about 100 light years from our sun. Nevertheless, in spite of its highly speculative assumptions and constraints, searching for hypothetical signals (in the radio range, but also in other spectral domains) remains so far the only tool available to detect life activity on extrasolar planets.

References

Further Readings

1. Gilmour I, Sephton MA (eds). An introduction to astrobiology. Cambridge: USA. Cambridge University Press, 2004:
2. Horneck G, Reetberg P (eds). Complete course in astrobiology. Germany: Wiley-VCH Pub., Weinheim, 2007:
3. Pudritz RE, Higgs P, Stone J (eds). Planetary systems and the origins of life. Cambridge Astrobiology Series 2007; in press.
4. Schulze Makuch D, Irwin LN. Life in the universe. Expectation and constraints. Springer, Berlin 2004.
5. Encrenaz T, Kallenbach R, Owen TC et al. Theouter planets and their moons. Space sciences series of ISSI. Springer, Berlin 2005.
6. Lorenz R, Mitton J. Lifting Titan's veil- exploring the giant moon of Saturn. Cambridge University Press, Cambridge, UK 2002.

Specific References

7. Sotin C, Prieur D. Jupiter's moon Europa: geology and habitability. In ref. 2 2007; 253-271.
8. Chyba CF, Phillips CB. Europa as an abode of life. Origin Life Evol Biosph 2002; 32:47-68.
9. Melosh HJ, Ekholm AG, Showman AP et al. The temperature of Europa's subsurface water ocean. Icarus 2004; 168:498-502.
10. Hand KP, Chyba C. Empirical constraints on the salinity of the Europan ocean and implications for a thin ice shell. Icarus 2007;
11. Oro J, Squyres SW, Reynolds RT et al. Europa: prospects for an ocean and exobiological implications. In: Exobiology in solar system exploration, G. Carle, D Schwartz and J Huntington eds. NASA SP 1992; 512:102-125.
12. Reynolds R, Squyres S, Colburn D et al. On the habitability of europa. Icarus 1983; 56:246-254.
13. Chela Flores J. Possible degree of evolution of solar-system microorganisms. In: Exobiology: Matter, energy and information in the origin and evolution of life in the universe, J Chela-Flores and F Raulin Eds., Kluwer 1998; 229-234.
14. Souchez R, Jean Baptist P, Petit JR et al. What is the deepest part of the vostok ice core is telling us? Earth-Science Reviews 2002; 60:131-146.
15. Russell CT. Cassini Huygens Mission: Overview, Objectives and Huygens Instrumentarium. Dordrecht, The Netherlands Kluwer Academic Publishers: 2003.
16. Russell CT. The Cassini-Huygens Mission: Orbiter Remote Sensing Investigations. Dordrecht, The Netherlands, Kluwer Academic Publishers: 2005.
17. Sagan C, Khare BN. Tholins. Organic chemistry of interstellar grains and gas. Nature 1979; 277:102-107.
18. Lebreton JP et al. Huygens descent and landing on Titan: Mission overview and science highlights. Nature 2005; 438:758-764.
19. Niemann HB et al. The abundances of constituents of Titan's atmosphere from the GCMS instrument on the huygens probe. Nature 2005; 438:779-784.
20. Israël G et al. Evidence for the presence of complex organic matter in Titan's aerosols by in situ analysis. Nature 2005; 438:796-799. Complex organic matter in Titan's aerosols? (Reply), Nature 2006; 444:E6-E7.
21. Raulin F. Astrobiologiy and habitability of Titan. Space Science Reviews 2007.
22. Waite H et al. The process of tholin formation in Titan's upper atmosphere. Science 2007; 316:870-875.
23. Fortes AD. Exobiological implications of a possible ammonia-water ocean inside Titan, Icarus 2000; 146:444-452.

CHAPTER 4

Mars, the Astrobiological Target of the 21st Century?

Patrice Coll* and Fabien Stalport

4.1 Introduction

Mars is the fourth planet from the Sun in the Solar System and has approximately half the radius of Earth and only one-tenth the mass. It is also referred to as the "Red Planet" because of its reddish appearance as seen from Earth, due to the iron(III) oxide (more commonly known as hematite).

Mars is a terrestrial planet with a thin atmosphere mainly composed of CO_2, having surface features looking like both the impact craters of the Moon and the volcanoes, valleys, deserts and polar ice caps of the Earth. In addition to these geographical features, Mars' rotational period and seasonal cycles are likewise similar to those of Earth.

Dozens of spacecrafts, including orbiters, landers and rovers, have been sent to Mars by the United States, Soviet Union, Europe and Japan to study the planet's surface, climate and geology. Mars is currently the host of three functional orbiting spacecrafts: Mars Odyssey (NASA), Mars Express (ESA) and Mars Reconnaissance Orbiter (NASA). This is more than any planet except Earth. The surface is also the home of the two NASA Mars Exploration Rovers (Spirit and Opportunity).

All the current and former missions provide an amazing quantity of data, allowing us to better understand the physico-chemical processes taking place in this planet. We will present here a brief summary of scientific results relating to Mars, mainly obtained from spatial missions, showing if still necessary why Mars is an object of astrobiological interest.

4.2 Why Is Studying Mars Relevant to Astrobiology?

The question of a past and/or present Life on Mars remains under discussion. The first significant point to note is that in more than forty years of exploration of the Mars planet, only one mission had an astrobiological goal: the Viking mission in the middle of the 70s.[6,7]

The different measurements performed onboard Viking landers to gather evidence for biological activity gave positive answers at the two landing sites and thus seemed to prove the existence of a martian life. However, all these results were subsequently explained by chemical processes of oxidation due to the presence of oxidant materials at the atmosphere/surface interface of the planet.[8-15] More strikingly, the nondetection of any organic molecule (a type of molecule required for life as we know it) by the Gas Chromatograph—Mass Spectrometer experiment seems to be incompatible with the existence of life.[11] The conclusion of the mission at the end of the 1970s was that no life was—or has been—present on Mars and the search for biological activity was abandoned. Much later in the 1990s, scientists studied a meteorite supposed to have originated from Mars: ALH84001. Different molecules related to biological processes were detected and assigned to possible martian life,[16] even if their presence could also be explained by chemical processes or terrestrial contamination. These discoveries suddenly once again raised the interest in searching for life on Mars. The phase of intense exploration of Mars that started from these research studies, which still goes on to-day, has revealed that the history of Mars was compatible with the emergence of life, or at least with the existence of prebiotic chemical activity (see Section 4.3).

Then, coming back to the observations of astrobiological interest collected by Viking, which could not give any positive evidence for past or present biological activity on Mars, scientists determined that there was also no firm evidence for the absence of life on Mars. Why was organic material not observed by Viking in the pyrolysis-GC/MS experiment?[8,17,18] This could be due to experimental conditions not being adapted for amino acid (destroyed by heating) detection and having a relatively high detection limit for their decomposition products (amines). Furthermore, the Viking landers sampled only the first centimetres in depth of martian soil, where the organic material would have been oxidized/destroyed by atmospheric H_2O_2,[19,20] atomic oxygen[21] and superoxide ions.[22] Benner[23] showed that it might be degraded into salts of benzenecarboxylic acids that are highly refractory. Since Viking's ovens would be unable to pyrolyse such molecules, no signal could be expected. The "biological" experiments onboard the Viking landers that were intended to answer the question of the presence or absence of some form of current life on Mars thus provided partial results that are still under debate.[24-27] However, the most pessimistic verdict of these experiments is that there was no—at the end of 1970s—biological activity of the terrestrial type in the first ten centimeters of the soil within the few square meters of the explored surface on Mars.

Recently, methane was detected in the Mars atmosphere.[28,29] The presence of methane is at first surprising for a body the atmosphere of which was considered to be in thermodynamic equilibrium.[30] For example, methane in the Earth's atmosphere, which is considered to be off equilibrium, directly reflects biological activity. Although many inorganic sources could be the origin of methane on Mars, the presence of ecological niches can still be a logical possibility to consider. Anyway, it was pointed out recently that methane could also be produced by a nonbiological process called

*Corresponding Author: Patrice Coll—LISA, Universités Paris 12 et Paris 7, CNRS, 61 Av. du Général de Gaulle, 94010, Créteil, France. Email: pcoll@lisa.univ-paris12.fr

Figure 4.1. Mars (Credit: NASA).

surface is more than 2 billion years old on account of plate tectonics that have irremediably erased the prebiotic records of the origin of Life. It has even been suggested that Life could have appeared first on Mars for two principal reasons:

1. The small size of Mars (Mars' diameter is 6800 km, approximately 50% that of the Earth) would have been an advantage in the beginning, because its surface would have cooled down faster than the Earth's, thereby supporting an earlier appearance of liquid water.
2. Mars did not see a gigantic impact comparable with the one supposed to have caused the birth of the Moon and blown away the primitive atmosphere of the Earth.[40,41]

Consequently, while the question of potential biological activity on Mars is based on certain assumptions, it is also accompanied by solid arguments that justify our continued interest. Clearly Mars remains one of the main astrobiological targets in our Solar System: it is the only place apart from the Earth where it appears possible to discover traces of past or present Life.

4.3 A Past Environment Favorable to Life?

To-day no physicochemical process is known to definitively explain the appearance of Life on Earth. Without any meaningful point of comparison in the Solar System, we unfortunately miss out on information regarding the necessary conditions for the transition from inert chemical compounds to the simplest living organisms. Moreover, no direct clue on this transition is highlighted by any terrestrial geological records. We also do not know the evidence provided by the Earth represents a valid reference: for example, could Life have appeared under environmental conditions different from those on primitive Earth? However, we remain interested in the conditions that appear to be essential to the emergence of terrestrial life: the presence of liquid water preferably over a sustained period, the presence of organic matter and at least one available source of energy. We will thus discuss the geological records of Mars' surface, in particular the oldest geological surfaces of its southern hemisphere.

4.3.1 Liquid Water on Mars

4.3.1.1 Fluvial Flows

In 1972, the Mariner 9 probe provided photographs of the surface of Mars showing geomorphologic structures.[42,43] These structures pointed to ancient river beds, now drained, and outflow valleys. Four years later, new pictures of the surface of Mars taken by the Viking probes in orbit around Mars (Fig. 4.2) confirmed the presence of these geomorphologic structures whose formation seems to imply intense and transitory liquid water flows.[44,45]

More recently the THEMIS instrument (Thermal Emission Imaging System) of the Mars Odyssey mission made it possible

serpentinization, involving water, carbon dioxide and the mineral olivine that is known to be common on Mars.[31]

On the other hand, it is noteworthy that Mars might have possessed key ingredients for the emergence of Life according to our knowledge about the process. While the evolution pathways of Mars and of the Earth seem to have been separated very early in their histories, it is completely possible that similar conditions were present on the two planets during the first hundreds of millions of years following their formation. We know that these conditions had led to the emergence of the first terrestrial forms of life, which are suggested by the oldest sedimentary records to be more than 3.5 billion years old.[32-35] So, if the birth of Life on Earth was the result of reproducible physicochemical processes, there is no reason why these same processes could not have allowed the appearance of Life on Mars. Conversely, if it is demonstrated that Life did not appear on Mars even though all the conditions there were favorable, it would be of great interest to ask: why had Life emerged on the Earth but not on Mars?

Is our list of the essential parameters for Life incomplete? Is it erroneous? If so, what are the missing elements?

We have to keep in mind that Mars offers geological records that are much better preserved than the Earth's: half of its surface is more than 3,8 billion years old,[36-39] whereas only 1% of the Earth

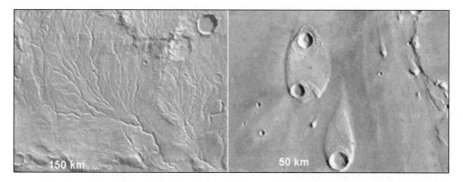

Figure 4.2. Images of the surface of Mars taken by the Viking probes in orbit around Mars. On the left, "ramified valley" in the area of Nirgal. On the right, "outflow valley". These geomorphologic structures seem to have been formed following catastrophic flows. (Source: NASA and ref. 45).

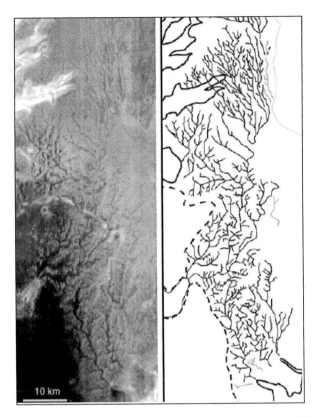

Figure 4.3. On the left, image in the thermal infrared obtained by the THEMIS instrument onboard the Mars Odyssey probe, in the Fallen Chasma area north of Valles Marineris. On the right, diagram of the fluvial networks. From Mangold, N., Quantin, C., Ansan, V., Delacourt, C. and Allemand, P. Evidence for Precipitation on Mars from Dendritic Valleys in the Valles Marineris Area. Science 2004; 305: 78-81. Reprinted with permission from AAAS.

to identify dendritic valleys in the plateau and the canyons of the Valles Marineris area. The geomorphologic characteristics of these valleys—especially those that present a strong density of ramifications comparable with the networks present on Earth—suggest a formation by streaming due to atmospheric precipitations (rains) (Fig. 4.3).[46] The channels and the maturity of the networks may indicate that the flows of liquid water were permanent over prolonged geological periods and suggest the existence of a hydrological cycle on Mars more than 3 billion years ago.

4.3.1.2 Sedimentary Deposits

The surface of Mars and in particular the southern hemisphere, is riddled with impacts of meteoritic objects. It was suggested that some of them could be temporarily occupied by lakes. The basin type morphology of the impact craters could have supported the formation of a water reservoir and the accumulation of sediments. Images of the surface of Mars show a fine stratification of clear and dark layers on the floors of some of the craters.[47-49]

Based on comparison with terrestrial morphologies, the arrangement of this type of layers is often characteristic of sedimentary deposits in a lake. These stratifications at the bottom of the craters were thus interpreted as evidence for the past existence of lakes on Mars (Fig. 4.4). Some of the deposits have a "delta" morphology that could be an indicator of the supply of liquid water to the lakes by rivers. However, other assumptions were also proposed to explain the formation of these structures not involving liquid water: deposits of volcanic lava and/or ashes, or dust brought by the wind.[50]

In 2004, the panoramic camera (Pancam) of the Opportunity rover, one of the two rovers of the Mars Exploration Rover mission, took the first photos of sedimentary deposits on a planet other than

the Earth. These deposits levelled within a few meters of the landing probe (Fig. 4.5) in a crater named Eagle[51] in the Terra Meridianni area. This discovery for the first time of sedimentary deposits on the martian surface represents a key discovery relating to the question of the presence of liquid water. The deposits appear to be fine intersected stratifications the geometry of which suggests formation in the presence of liquid water.

4.3.1.3 Ancient Ocean?

A highly discussed theory proposes that the low lands of the martian surface was occupied by a vast stretch of liquid water.[52-56] According to this theory, an ancient ocean of liquid water or mud could have been formed temporarily in the vast area of the plains of the northern hemisphere, named Oceanus Borealis. This ocean would have been several kilometers deep.[57] It was also proposed that the wide impact basins of the southern hemisphere, *Hellas Planitia* and Argyre Planitia, could have formed inland seas.[58,59]

One of the instruments of the Mars Global Surveyor probe, the altimetric laser MOLA, provided information on the topography of the surface of Mars. In the zone corresponding to the probable limits of this ocean (Fig. 4.6), this instrument revealed a practically constant altitude for a distance of several hundred kilometers, which could correspond, based on comparison with terrestrial geomorphological structures, to shore lines. However, the mineralogical composition of this area is dominated by the presence of volcanic rocks, which is surprising because the presence of such a quantity of water would have favored the formation of sedimentary deposits, particularly the formation of carbonates, above the volcanic rocks.

Many geomorphological structures thus seem to indicate the presence of water in the liquid state on the surface of Mars since the beginning of its history. They also suggest that water could have been a permanent presence over a time scale of several hundred million years.

4.3.1.4 Mineralogical Records

In year 2000, the TES instrument (Thermal Emission Spectrometer) of the Mars Global Surveyor probe made it possible for the first time to identify mineralogical clues favoring the presence of liquid water: grey or crystalline hematite in areas like Sine Meridianni, Valles Marineris and the bottom of the crater Aram Chaos (Fig. 4.7)[60,61] and carbonates in the dust particles of Mars.[62] However, the origin of these carbonates so far remains unexplained.

In 2004, two missions, the NASA Mars Exploration Rovers (MER) and the ESA Mars Express (MEx), revealed new records of minerals formed in the presence of liquid water. In addition to providing the first images of sedimentary deposits, the instruments onboard both MER rovers identified sulphates. On Earth, these minerals have the property of being formed only in the presence of liquid water.[51,63,64]

However, these sulphates can form very quickly, making it unnecessary that the water must remain stable over prolonged geological periods: these sulphates can indeed settle out when the water evaporates. The OMEGA instrument (Observatoire pour la Minéralogie, l'Eau, les Glaces et l'Activité) onboard the Mars Express probe made it possible to highlight not only outcrops of sulphates salts[65-67] but also outcrops of clays.[65,68] Both of them share the same property of being formed only in the presence of liquid water. However, the presence of clays carries another important piece of information: their formation most probably required an interaction between liquid water and the silicated rocks of the Mars crust over long geological periods (Fig. 4.8). The presence of clays at the bottom of the Holden crater was confirmed very recently by the CRISM instrument (Compact Reconnaissance Imaging Spectrometer for Mars) onboard the Mars Reconnaissance Orbiter.[69]

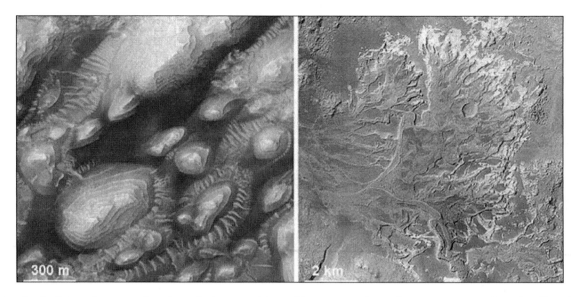

Figure 4.4. On the left, sedimentary deposits at the bottom of the Holden crater in the Arabia Terra area. On the right, alluvial deposits laid out in a delta. These structures seem to point to an environment of lakeside deposits, which would be an indicator of the presence of liquid water. (Source: ref. 49).

4.3.1.5 Water History on Mars

Considering all these geomorphologic and mineralogical data, it appears increasingly obvious that liquid water was indeed present at the beginning of Mars' history potentially in a durable way as long as the conditions necessary to its maintenance were established. Several geological periods based on the history of water on Mars were proposed to fit the description of the two principal types of Mars rocks that required the presence of liquid water for formation: clays and sulphates.[70] Clays were detected in the supposedly oldest surfaces, particularly where wind erosion or the impact of a meteoritic object revealed ancient deposits. The sulphates, although present in surfaces older than 3 billion years, seem more recent than the clays. Consequently, during the first 700 million years of the history of Mars, the planet probably knew several distinct episodes when liquid water was more or less abundant (Fig. 4.9):

- The first period, the "phyllosian," during which liquid water was probably abundant at the surface of the planet—probably deep and able to alter the surfaces over the long durations needed to form clays. It should be noted that this period probably coincided with the presence of a magnetic field able to attenuate the loss of the Mars atmosphere, which would correspond, at the maximum, to a period of 500 million years.[71,72]

- The second period, the "theiikian," during which the environment would have become drier, more acidic and then favourable to sulphate salts deposition with possibly great episodes of volcanic activity on the surface of Mars.[70] This period would coincide with the loss of the magnetic field.

- A third period, the "siderikian," between 3.8 billion years ago and the present: the conditions on the surface of Mars would have been degraded considerably and liquid water disappeared (except for some very short hypothetical episodes of catastrophic[45] or more classical[73] flows). Mars was then moving to the current environment we know to-day. This period would be dominated by the very slow weathering of the surface of Mars that led to the formation of anhydrous ferric oxides.

Figure 4.5. One of the first color pictures taken by the Pancam camera onboard the Opportunity rover. This image shows sedimentary deposits which level on the internal slopes of the Eagle crater at the bottom of which Opportunity landed. The instruments on the rover detected in particular the presence of sulphates which would be evidence for an aqueous environment. (Credit: NASA).

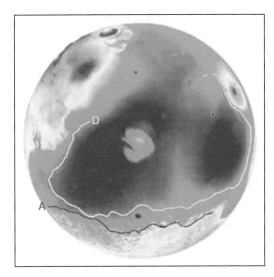

Figure 4.6. Possible ancient shore lines in the northern hemisphere in Arabia (A) and Deuteronilus (D). Reprinted by permission from Macmillan Publishers Ltd, Nature 447,840-843, copyright 2007.

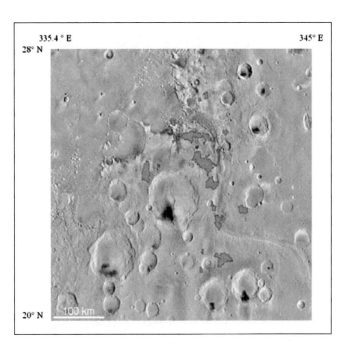

Figure 4.8. Infrared image of Mawrth Vallis and identification of clays by OMEGA instrument onboard Mars Express probe. In blue, clay deposit zones. Reprinted by permission from Macmillan Publishers Ltd: Nature 438, 623-627, copyright 2005.

4.3.2 Sources of Organic Molecules

The surface of Mars is dotted with impacts of meteoritic bodies. Moreover, the older are the surfaces, the more numerous are the impacts and more significant are their sizes. Consequently many bodies coming from the interplanetary medium have hit the surface of Mars, particularly at the beginning of its history.[36,37,74] Currently, several hundred tons of objects coming from the interplanetary medium fall on the Earth's surface[75-78] and also on the surface of Mars. Analyses of their molecular composition revealed the presence of organic molecules.[79-86] Consequently, it is extremely likely that the surface of Mars had received organic matter coming from the interplanetary medium since the beginning of its history (Fig. 4.10). It is also possible that endogenous organic syntheses, particularly atmospheric ones, could have taken place.[87-90] Many sources could have contributed to the organic matter delivered to the surface of Mars, especially over the first several hundred million years.

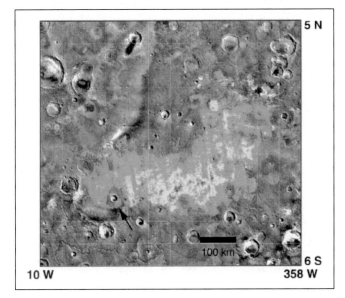

Figure 4.7. Infrared image of the Aram Chaos crater. In the center of the crater, a zone of 60 km of diameter containing between 10 to 15% of grey hematite was detected by the instrument TES of the Mars Global Surveyor mission. This crater could have contained a lake in the early beginning of Mars' history. (Source: ref. 61).

4.3.3 Energy Sources

Allowing that the history of water on Mars is still surrounded by some uncertainty, it is extremely likely that the presence of liquid water, particularly during the formation of clays, required sufficiently mild climatic conditions. The existence of such a climate would have been possible only if Mars, in its early years, had belonged to the habitable zone.[91-93] Its distance to the Sun and the weaker luminosity of the young Sun (between 25 and 30% less energetic than to-day) 4.5 billion years ago[94-96] would tend to lower the probability of presence of liquid water on its surface. Indeed, the average equilibrium temperature of the planet should have been approximately –75°C in the absence of an atmosphere.[91] Thus Mars could have benefited from many sources of energy to give rise to such favourable climatic conditions:

- A dense atmosphere mainly composed of greenhouse gases[97,98] and/or of clouds of carbonic ice.[99]
- A geothermic flow generated by a great quantity of energy (a) stored by accretion, (b) produced by the decay of radioactive elements and/or (c) produced by the release of latent heat from the progressive crystallization of the liquid core.
- Many bodies of the solar system which hit the surface of Mars could have temporarily and locally heated the surface of Mars. The impact of these bodies would then have provided energy for the fusion of rocks, etc.

4.3.4 Consequences for Potential Biological Activity

All the necessary conditions for the appearance and maintenance of biological activity were thus established on the surface of Mars during the first 700 million years of its history. Although the spontaneous chemical reactions which led to the birth of the first living organisms on Earth are not known and the establishment of Earth-like conditions does not necessarily lead to the appearance of Life, it appears justified to ask, faced with the presence of Earth-like environmental conditions on Mars over the same period, whether Life could have also appeared on Mars.

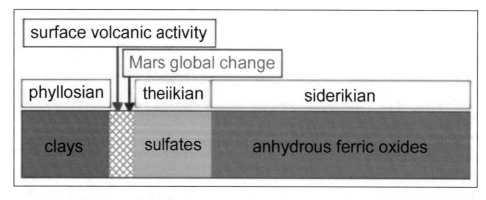

Figure 4.9. Plausible scenario of the history of water from the formation of Mars until to-day. Water would have been abundant at the beginning and would have led to clay formation during a period named the "phyllosian". Then, following environmental changes (i.e., volcanic activity ...), the environment would have become drier, more acidic and favourable to sulphate deposition during the period called the "theiikian". During the last 3.8 billion years, the Mars environment would have deteriorated considerably and water would not have been present any more in the liquid state in a durable way during the period called the "siderikian". From Bibring J-P, Langevin Y, Mustard JF, Poulet FA, Arvidson R, Gendrin A, Gondet B, Mangold N, Pinet P, Forget F et al. Global mineralogical and aqueous Mars history derived from OMEGA/Mars Express data. Science 2006; 312:100-404. Reprinted with permission from AAAS.

4.4 Where and How to Search for Life?

The distant past of Mars will shape the exploration of the red planet in the immediate future. The most recent discoveries of the European Mars Express and the American Spirit and Opportunity rovers will reorientate research into past Life towards Mars' oldest surfaces. Meteoritic bombardments and wind processes as liquid flows, probably violent and transitory, have contributed to unmasking hidden layers that can testify to events of ancient times, when liquid water could have remained on Mars. Currently, the scientific community is determining the location of the planned MSL rover landing site, taking into account the astrobiological vocation of the mission. Among the candidate sites, the clays identified by the spectro-imager OMEGA are the favourites. These surfaces could have preserved fossilized traces of a form of elementary life by protecting them over billions of years from a Mars environment that became very hostile.

In 2009, NASA will send this MSL (Mars Science Laboratory) rover to Mars; the aim of the scientific payload of this mission will be, in particular, to search for the presence of extinct or extant traces of life, or prebiotic chemistry that could have existed. It will use a drilling system able to attain depths of some decimeters and consequently able to sample zones where inorganic material might bear witness to epochs when Mars' atmosphere was capable of nurturing the development of life and where organic remnants might have been preserved from destruction.[100,101]

The SAM experiment onboard the MSL rover is intended for analysis of the martian atmosphere, ground and underground:

a. To determine the elementary composition of recovered samples of ground or underground, in order to compare with the results of other experiments on the Lander and to provide points of comparison for objectives (b), (f) and (g) below.

b. To seek, in the ground and underground, organic traces that could bear witness to a past prebiotic chemistry and, in the best cases, an organized life which could have developed at Mars' surface or underground. The chirality of organic molecules will give clues regarding their origin. In parallel with the search for organic molecules, the study of isotopic ratios will bring additional information.

c. To obtain information on atmospheric noble gases (Ar, Xe, Kr): isotopic abundances and ratios, as well as isotopic ratios of carbon (13/12) in CO_2, of nitrogen (15/14) in N_2 and of D/H in H_2O. This will make it possible to correlate

with other observations (on the lander and from the Earth) and to better understand crustal degassing and atmospheric escape in Mars' history.

d. To measure isotopic ratios of carbon, nitrogen, sulphur (34/32), oxygen (16/17/18) and D/H in samples of ground (or underground) after laser ablation or heating, for correlation with measurements performed on Earth and comparison with atmospheric measurements.

e. To search for atmospheric gases (CH_4, H_2S, etc.) whose origin could be organic and correlate with (b).

f. To obtain information on the presence of hydrates, carbonates, sulphates, clays in samples from ground (or underground); the presence of mineralogical indices of a dense and wet atmosphere that will support the search for organic traces in the ground (item b). One can also differentiate minerals from biominerals by simplified differential thermal analysis.

g. To obtain, if the drilling system allows sampling at different depths, a stratigraphy not only for items (a), (b), (d) and (f) above, but also of gases adsorbed on the solid phase (H_2O, CO_2, H_2O_2, etc).

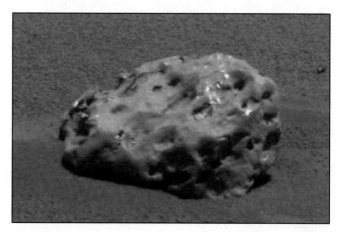

Figure 4.10. Image of the first meteorite identified at the surface of Mars, i.e., on the surface of a planet other than the Earth. This meteorite was located by the panoramic camera of the Opportunity rover close to its heat shield. It was named for this reason "Heat Shield Rock". Its chemical composition of iron and nickel is very close to the iron meteorites found on the Earth. (Credit: NASA/JPL/Cornell).

Figure 4.11. View of MSL 2009 (Credit: NASA MSL web site).

4.5 Ongoing Explorations

The current understanding of planetary habitability—the capacity to develop and sustain life—favors planets that have liquid water on their surface. Different lines of evidence suggest that Mars was significantly more habitable in the past than it is to-day, but whether living organisms ever existed there is still unclear. The Viking probes of the mid-1970s carried out experiments designed to detect microorganisms in martian soil at their respective landing sites and uncovered a temporary increase of CO_2 production upon exposure to water and nutrients. However, this sign of 'life' was later disputed by many scientists. A re-analysis of the now 30-year-old Viking data in the light of modern knowledge of extremophilic forms of life has led to the suggestion that the Viking tests were not sophisticated enough to detect such forms of life.

The NASA Phoenix Mars Lander, launched on August 4, 2007 arrived safely on 'Green Valley' in the north polar region of Mars on May 25, 2008 and succeeded in analyzing the first sample of martian soil in its Thermal and Evolved Gas Analyzer, or TEGA, weeks later on July 31. Upon heating the sample, the mass spectrometer in the instrument detected water vapor when sample temperature reached 0°C. "We have water," concluded William Boynton, lead scientist of TEGA, "We've seen evidence for this water ice before in observations by the Mars Odyssey orbiter and in disappearing chunks observed by Phoenix last month, but this is the first time martian water has been touched and tasted."

Following this historic discovery made by the Phoenix, explorations of Mars will be continued in 2009 with the Mars Science Laboratory (Fig. 4.11), a bigger, faster (90 m/hour) and smarter version of the Mars Exploration Rovers, and in 2012 with ESA's first Rover to Mars, the ExoMars Rover, which will be capable of drilling 2 meters into the soil to search for organic molecules. These and further missions will bring us closer to answering the big question: "Has there ever been Life on Mars?"

References

Further Readings

1. Kieffer HH, Snyder C, Jakosky BM et al. Mars. University of Arizona Press, 1977.
   - Still a reference ... This is THE book (the best and most technical): 1500 pages long and contributions from 114 authors.
2. Forget F, Costard F, Lognonné P. La planète Mars—Histoire d'un autre monde, Belin, 2006.
   - If you read French.
3. Rocard F. Dernière nouvelles de Mars. Planète Rouge. Dunod, 2006.
   - If you read French.
4. Zubrin R, Wagner R. The case for Mars. Simon and Schuster Trade, 1996.
   - If you have an interest in seeing humans make the trip to Mars....
5. Jakosky B. The search for life on other planets. Cambridge University Press, 1998.
   - If you have an interest in the Search for Life on other planets, with an important Mars section: the best book ever!

Specific References

6. Soffen GA, Snyder CW. The first Viking mission to Mars. Science 1976; 193:759-766.
7. Soffen GA. The Viking project. J Geophys Res 1977; 82:3959-3970.
8. Biemann K, Oro J, Toulmin P III et al. Search for organic and volatile inorganic compounds in two surface samples from the Chryse Planitia region of Mars. Science 1976; 194:72-76.
9. Horowitz NH, Hobby GL, Hubbard JS. The Viking carbon assimilation experiments—Interim report. Science 1976; 194:1321.
10. Klein HP, Horowitz NH, Levin GV et al. The Viking biological investigation: preliminary results. Science 1976; 194:99-105.
11. Biemann K, Oro J, Toulmin P et al. The search for organic substances and inorganic volatile compounds in the surface of Mars. J Geophys Res 1977; 82:4641-4658.
12. Klein HP. The Viking biological investigation: general aspects. J Geophys Res 1977; 82:4677-4680.
13. Levin GV, Straat PA. Recent results from the Viking Labeled Release experiment on Mars. J Geophys Res 1977; 82:4663-4667.
14. Oyama VI, Berdahl BJ. The Viking gas exchange experiment results from Chryse and Utopia surface samples. J Geophys Res 1977; 82:4669-4676.
15. Oyama VI, Berdahl BJ, Carle GC. Preliminary findings of the Viking gas exchange experiment and a model for martian surface chemistry. Nature 1977; 265:110-114.

16. McKay DS, Gibson EK, Thomas-Keprta KL et al. Search for past life on Mars: Possible relic biogenic activity in martian meteorite ALH84001. Science 1996; 273:924-930.

17. Klein HP, Horowitz NH, Biemann K. The search for extant life on Mars. In: Kieffer HH, Jakosky BM, Snyder C, Matthews MS, eds Mars. University of Arizona Press, Tucson 1992; 1221-1234.

18. Glavin DP, Schubert M, Botta O et al. Detecting pyrolysis products from bacteria on Mars. Earth Planet Sci Lett 2001; 185:1-5.

19. Bullock MA, Stoker CR, McKay CP et al. A coupled soil-atmosphere model of H_2O_2 on Mars. Icarus 1994; 107:142-154.

20. Zent AP. On the thickness of the oxidized layer of the martian regolith. J Geophys Res 1998; 103:31491-31498.

21. Kolb C, Lammer H, Abart R et al. The martian oxygen surface sink and its implications for the oxidant extinction depth. In: Proceedings of the Second European Workshop on Exo/Astrobiology 16-19 September 2002, Graz, Austria. Ed.: Huguette Lacoste. ESA SP-518, Noordwijk, Netherlands: ESA Publications Division 181-186.

22. Yen AS, Kim SS, Hecht MH et al. Evidence that the reactivity of the martian soil is due to superoxide ions. Science 2002; 289:1909-1912.

23. Benner SA, Devine KG, Matveeva LN et al. The missing organic molecules on Mars. PNAS 2000; 97:2425-2430.

24. Navarro-Gonzalez R, Navarro KF, Rosa Jdl et al. The limitations on organic detection in Mars-like soils by thermal volatilization-gas chromatography-MS and their implications for the Viking results. PNAS 2006; 103:16089-16094.

25. Biemann K. On the ability of the Viking gas chromatograph-mass spectrometer to detect organic matter. PNAS 2007; 104:10310-10313.

26. Mukhopadhyay R. The Viking GC/MS and the search for organics on Mars. Anal Chem 2007; 79:7249-7256.

27. Wu C. Secrets of the martian soil. Nature 2007; 448:742.

28. Formisano V, Atreya S, Encrenaz T et al. Detection of methane in the atmosphere of Mars. Science 2004; 306:1758-1761.

29. Krasnopolsky VA, Maillard JP, Owen TC. Detection of methane in the martian atmosphere: evidence for life? Icarus 2004; 172:537-547.

30. Lewis JS, Prinn RG. Planets and their atmospheres—origin and evolution. Publication by Orlando FL Academic Press Inc International Geophysics Series 1984; 33:480.

31. Oze C, Sharma M. Have olivine, will gas: Serpentinization and the abiogenic production of methane on Mars. Geophys Res Lett 2005; 32:10203.

32. Lowe DR. Stromatolites 3,400-Myr old from the Archean of Western Australia. Nature 1980; 284:441-443.

33. Walter MR, Buick R, Dunlop JSR. Stromatolites 3,400-3,500 Myr old from the North Pole area, Western Australia. Nature 1980; 284:443-445.

34. Byerly GR, Lower DR, Walsh MM. Stromatolites from the 3,300-3,500-Myr Swaziland Supergroup, Barberton Mountain Land, South Africa. Nature 1986; 319:489-491.

35. Allwood AC, Walter MR, Kamber BS et al. Stromatolite reef from the early archaean era of Australia. Nature 2006; 441:714-718.

36. Hartmann WK, Neukum G. Cratering chronology and the evolution of Mars. Space Sci Rev 2001; 96:165-194.

37. Neukum G, Ivanov BA, Hartmann WK. Cratering records in the inner solar system in relation to the lunar reference system. Space Sci Rev 2001; 96:55-86.

38. Frey HV, Roark JH, Shockey KM et al. Ancient lowlands on Mars. Geophys Res Lett 2002; 29:1384.

39. Frey HV. Impact constraints on the age and origin of the lowlands of Mars. Geophys Res Lett 2006; 33, doi:10.1029/2005GL024484.

40. Hartmann WK. Lunar origin: The role of giant impacts. Abstracts and program for the conference on the origin of the Moon Kona Hawaii. LPI Contribution 540, published by the Lunar and Planetary Institute 1984; 540:52.

41. Dickinson T, Newsom HE. A possible test of the impact theory for the origin of the moon. Lunar and planetary science XVI 1985; 183-184.

42. Masursky H. An overview of geological results from Mariner 9. J Geophys Res 1973; 78:4009-4030.

43 Baker VR, Milton DJ. Erosion by catastrophic floods on Mars and Earth. Icarus 1974; 23:27-41.

44. Carr MH. Formation of martian flood features by release of water from confined aquifers. J Geophys Res 1979; 84:2995-3007.

45. Baker VR, Carr MH, Gulick VC et al. In: Kieffer HH, Jakosky BM, Snyder C, Matthews MS Eds., Mars, University of Arizona Press, Tucson 1992:493-522.

46. Mangold N, Quantin C, Ansan V et al. Evidence for precipitation on Mars from dendritic valleys in the valles marineris area. Science 2004; 305:78-81.

47. Edgett KS, Parker TJ. Water on early Mars: Possible subaqueous sedimentary deposits covering ancient cratered terrain in western Arabia and Sinus Meridiani. Geophys Res Lett 1997; 24:2897.

48. Moore JM, Howard AD, Dietrich WE et al. Martian layered fluvial deposits: Implications for Noachian Climate Scenarios. Geophys Res Lett 2003; doi:10.1029/2003GL019002.

49. Moore JM, Howard AD. Large alluvial fans on Mars. J Geophys Res (Planets) 2005; 110:E04005 doi:10.1029/2004JE002352.

50. Leverington DW, Maxwell TA. An igneous origin for features of a candidate crater-lake system in western Memnonia, Mars. J Geophys Res (Planets) 2004; 109:E06006 doi:10.1029/2004JE002237.

51 Squyres SW, Arvidson RE, Bell JF et al. The Opportunity Rover's Athena science investigation at Meridiani Planum, Mars. Science 2004; 306:1698-1703.

52. Head JW, Kreslavsky M, Hiesinger H et al. Oceans in the past history of Mars: Tests for their presence using Mars Orbiter Laser Altimeter (MOLA) data. Geophys Res Lett 1998; 25: 4401-4404.

53. Parker TJ. Mapping of possible "Oceanus Borealis" shorelines on Mars: A status report. 29th annual lunar and planetary science conference, Houston, TX 1998, abstract no. 29; 1965.

54. Clifford SM, Parker TJ. The evolution of the martian hydrosphere: Implications for the fate of a primordial ocean and the current state of the northern plains. Icarus 2001; 154:40-79.

55. Sotin C, Couturier F, Bibring J. Analysis of gravity potential along paleo-shorelines on Mars: Implications for ocean on very early Mars. Eos Trans. AGU, 87(52) Fall Meeting 2006 Suppl, Abstract P31B-0139.

56. Perron JT, Mitrovica JX, Manga M et al. Evidence for an ancient martian ocean in the topography of deformed shorelines. Nature 2007; 447:840-843.

57. Öner AT, Ruiz J, Fairén AG et al. The volume of possible ancient oceanic basins in the northern plains of Mars. Lunar and Planetary Science XXXV 2004; Abstract no. 35: 1319.

58. Head JW. Exploration for standing bodies of water on Mars: When were they there, where did they go and what are the implications for astrobiology? Eos Trans. AGU, 82(47) Fall Meeting 2001 Suppl, Abstract P21C-03.

59. Hiesinger H, Head JW III. Topography and morphology of the Argyre Basin, Mars: Implications for its geologic and hydrologic history. Planet Space Sci 2002; 50:939-981.

60. Christensen PR, Bandfield JL, Clark RN et al. Detection of crystalline hematite mineralization on Mars by the Thermal Emission Spectrometer: Evidence for near-surface water. J Geophys Res 2000; 105:9623-9642.

61. Christensen PR, Morris RV, Lane MD et al. Global mapping of martian hematite mineral deposits: Remnants of water-driven processes on early Mars. J Geophys Res 2001; 106:23873-23886.

62. Bandfield JL, Glotch TD, Christensen PR. Spectroscopic identification of carbonate minerals in the martian dust. Science 2003; 301:1084-1087.

63. Christensen PR, Wyatt MB, Glotch TD et al. Mineralogy at Meridiani Planum from the Mini-TES Experiment on the opportunity rover. Science 2004; 306:1733-1739.

64. Klingelhöfer G, Morris RV, Bernhardt B et al. Jarosite and hematite at Meridiani Planum from Opportunity's Mössbauer Spectrometer. Science 2004; 306:1740-1745.

65. Bibring J-P, Langevin Y, Gendrin A et al. Mars surface diversity as revealed by the OMEGA/Mars Express observations. Science 2005; 307:1576-1581.

66. Gendrin A, Mangold N, Bibring J-P et al. Sulfates in martian layered terrains: The OMEGA/Mars Express view. Science 2005; 307:1587-1591.

67. Langevin Y, Poulet F, Bibring J-P et al. Sulfates in the north polar region of Mars detected by OMEGA/Mars Express. Science 2005; 307:1584-1586.

68. Poulet F, Bibring J-P, Mustard JF et al. Phyllosilicates on Mars and implications for early martian climate. Nature 2005; 438:623-627.

69. Milliken RE, Grotzinger J, Grant J, et al. Clay minerals in Holden crater as observed by MRO CRISM. Seventh International Conference on Mars, LPI Contributions 2007, Abstract no 1353:3282.

70 Bibring J-P, Langevin Y, Mustard JF et al. Global mineralogical and aqueous Mars history derived from OMEGA/Mars Express data. Science 2006; 312:100-404.

71. Acuna MH, Connerney JE, Lin RP et al. The magnetic field of Mars—A window into Mars' past. AGU Fall Meeting 2001; Abstracts no. P41A-06.

72. Acuña MH, Connerney JEP, Wasilewski P et al. Magnetic field of Mars: Summary of results from the aerobraking and mapping orbits. J Geophys Res 2001; 106:23403-23418.

73. Dickson JL, Head JW, Marchant DR et al. Recent gully activity on Mars: Clues from late-stage water flow in gully systems and channels in the antarctic dry valleys. Seventh International Conference on Mars 2007; 38:1678.

74. Chapman CR, Cohen BA, Grinspoon DH. What are the real constraints on the existence and magnitude of the late heavy bombardment? Icarus 2007; 189:233-245.

75. Halliday I, Blackwell AT, Griffin AA. The flux of meteorites on the Earth's surface. Meteoritics 1989; 24:173-178.

76. Love SG, Brownlee DE. A Direct measurement of the terrestrial mass accretion rate of cosmic dust. Science 1993; 262:550.

77. Bland PA, Smith TB, Jull AJT et al. The flux of meteorites to the Earth over the last 50,000 years. Mon Not R Astron Soc 1996; 283:551.

78. Taylor S, Lever JH, Harvey RP. Accretion rate of cosmic spherules measured at the South Pole. Nature 1998; 392:899.

79. Clemett SJ, Maechling CR, Zare RN et al. Identification of complex aromatic molecules in individual interplanetary dust particles. Science 1993; 262:721-725.

80. Maurette M, Brack A, Kurat G et al. Were micrometeorites a source of prebiotic molecules on the early Earth? Adv Space Res 1995; 15:113-126.

81. Brinton KLF, Engrand C, Glavin DP et al. A search for extraterrestrial amino acids in carbonaceous antarctic micrometeorites. Orig Life Evol Biosph 1998; 28:413-424.

82. Clemett SJ, Chillier XDF, Gillette S et al. Observation of indigenous polycyclic aromatic hydrocarbons in 'giant' carbonaceous antarctic micrometeorites. Orig Life Evol Biosph 1998; 28:425-448.

83. Maurette M. Carbonaceous micrometeorites and the origin of life. Orig Life Evol Biosph 1998; 28:385-412.

84. Botta O, Bada JL. Extraterrestrial organic compounds in meteorites. Surv Geophys 2002; 23:411-467.

85. Glavin DP, Matrajt G, Bada JL. Re-examination of amino acids in Antarctic micrometeorites. Adv Space Res 2004; 33:106-113.

86. Pizzarello S, Cooper GW, Flynn GJ. In meteorites and the early solar system II. Lauretta DS and McSween HY Eds. Published by The University of Arizona Press. Tucson 2006; 625-651.

87. Chyba C, Sagan C. Cometary and asteroidal delivery of prebiotic organics vs in situ production on the early Earth. Bulletin of the American Astronomical Society 1990; 22:1097.

88. Chyba CF, Thomas PJ, Brookshaw L et al. Cometary delivery of organic molecules to the early Earth. Science 1990; 249:366.

89. Chyba CF, Sagan C. Endogenous production, exogenous delivery and impact-shock synthesis of organic molecules: an inventory for the origins of life. Nature 1992; 355:125-132.

90. Bernstein M. Prebiotic materials from on and off the early Earth. Philos Trans R Soc Lond B Biol Sci 2006; 361:1689-1702.

91. Kasting JF, Whitmire DP, Reynolds RT. Habitable zones around main sequence stars. Icarus 1993; 101:108-128.

92. Kasting JF. Habitability of planets. Astrobiology workshop 1996 Abstract.

93. Kasting JF. Habitable zones around stars and the search for extraterrestrial life. Eos Trans. AGU, 82(47), Fall Meet. Suppl, Abstract P21C-01, 2001.

94. Newman MJ, Rood RT. Implications of solar evolution for the Earth's early atmosphere. Science 1977; 198:1035-1037.

95. Gough DO. Solar interior structure and luminosity variations. Solar Phys 1981; 74:21-34.

96. Gilliland RL. Solar evolution. Glob Planet Change 1989; 1:35-55.

97. Kasting JF, Brown LL, Acord JM et al. Was early Mars warmed by ammonia? In: Lunar and Planetary Inst., Workshop on the Martian Surface and Atmosphere Through Time 1992:84-85.

98. Kasting JF. Greenhouse models of early Mars climate. Eos Trans. AGU Fall Meeting 2002 Suppl, Abstracts no. P51B-0345.

99. Forget F, Pierrehumbert RT. Warming early Mars with carbon dioxide clouds that scatter infrared radiation. Science 1997; 278:1273.

100. McKay CP, Friedman EI, Wharton RA et al. History of water on Mars: a biological perspective. Adv Space Res 1992; 12:231-238.

101. Brack A, Clancy P, Fitton B et al. An integrated exobiology package for the search for life on Mars. Adv Space Res 1999; 23:301-308.

CHAPTER 5

Comets and Astrobiology

Hervé Cottin* and Didier Despois

5.1 Introduction

For a very long time, comets were regarded as bad omens, linked to superstitions, wars, deaths, diseases, etc... But nothing was really known about these wandering objects, suddenly appearing and disappearing in the sky, without prior notice. They were considered by Aristotle as an atmospheric phenomenon and not until the 17th century were they actually seen as astrophysical bodies by western scientists.

In a milestone paper, F. Whipple describes in 1950 comets as "dirty snowballs", a mixture of ices (dominated by water) and minerals.[7] This model has long since evolved: more is known about the nature of the nucleus from data collected with Earth-based observations, in situ investigations, sample returns and laboratory work. To date, our current knowledge gives comets a privileged location in astrobiology: they are at the crossroads of the origins. Origin of the Solar System, since they have sampled the matter of the Solar Nebula at the place of their accretion, but also, possibly, the origin of life on Earth since comets are rich in water and carbon, two essential constituents of terrestrial life. Therefore part of Earth's water and carbon might be of cometary origin.

Early last century, Chamberlin and Chamberlin proposed that infalling carbonaceous chondrite meteorites could have been an important source of terrestrial organic compounds.[8] J. Oró was the first in 1961 to suggest that comets may have played a similar role, from observations of carbon- and nitrogen-containing radicals in cometary comae:[9]

> "I suggest that one of the important consequences of the interactions of comets with the Earth would be the accumulation on our planet of relatively large amounts of carbon compounds which are known to be transformed spontaneously into amino acids, purines and other biochemical compounds".

An extended coverage of current knowledge about comets is given in the recent book Comets II,[10] to which readers that would like to find further information about comets are strongly encouraged to refer to. In the present chapter, Comets will mainly be considered as potential reservoirs of organic molecules for the early Earth. We first present their general characteristics, then the chemical composition of cometary matter as deduced from observations, in-situ exploration, sample returns and laboratory experiments. Anticipated results from the Rosetta mission are then presented and more specifically the instruments designed to probe the molecular composition of the cometary environment: COSAC, COSIMA, MIRO and VIRTIS. Finally, the various potential cometary contributions to the early Earth are addressed.

5.2 Comets in the Solar System

Comets are leftovers of the formation of planets in the Solar System. They formed in the first Myrs in the colder part of the protosolar nebula, where the temperature was low enough for water ice to condense, embedding "dust" particles made of organic and/or mineral material. The resulting "dirty snowball" is the nucleus of the comet, with typical sizes of 1 to 100 km: the nucleus of the famous comet 1P/Halley has an irregular shape, roughly $8 \times 8 \times 16$ km. It was measured from pictures taken during flyby of the ESA probe Giotto in 1986.

Most comets reside in two reservoirs: (i) a large (10^5 AU[a]) and spherical reservoir, the Oort cloud, with an estimated number of comets about 10^{11} to 10^{12}, for a total mass of 1 to a few tens Earth mass and (ii) a rather flat disk beyond Neptune orbit, the Kuiper Belt, smaller than the Oort Cloud (100 to 1000 AU at most) and containing also many 1000-km class objects (dwarf planets) like Pluto and the recently discovered Eris, Sedna and Quaoar trans-Neptunian objects (TNO). Part of the Kuiper belt objects were formed at the same distance from the Sun as they are located now, but another part of the Kuiper belt objects and the whole Oort cloud objects were formed closer to the Sun and later moved to their present location due to gravitational interaction with the giant planets Jupiter, Saturn, Uranus and Neptune.

Most of our information about the nucleus composition is indirect and comes from ground based or in situ observation of the gas and particles which are released by the nucleus when it comes close to the Sun. The typical comet loss of material for one return would represent a decrease by about 1 meter of its radius, if averaged over the whole surface.

Gas and dust particles expand in a more or less spherical shell, the coma, which may reach 1 million km for some species quite resistant to photodissociation and photoionisation by solar radiation. Comet tails result from the interaction of coma ions with the solar wind and of coma dust particles with the solar radiation pressure (a neutral sodium tail has also been observed). Tails can extend up to 100 million km, almost the Earth-Sun distance. Although much more visible than the nucleus, the coma and the tails have extremely low densities (below 10^4 atoms.cm^{-3} in most places (Figs. 5.1-5.2).

---

[a] 1 AU is the average Earth-Sun distance, nearly 150 million km (1 AU = $1.496 10^8$ km)

*Corresponding Author: Hervé Cottin—LISA, Universités Paris 12 et Paris 7, CNRS, 61 Av. du Général de Gaulle, 94010, Créteil, France. Email: cottin@lisa.univ-paris12.fr

Prebiotic Evolution and Astrobiology, edited by J. Tze-Fei Wong and Antonio Lazcano. ©2009 Landes Bioscience.

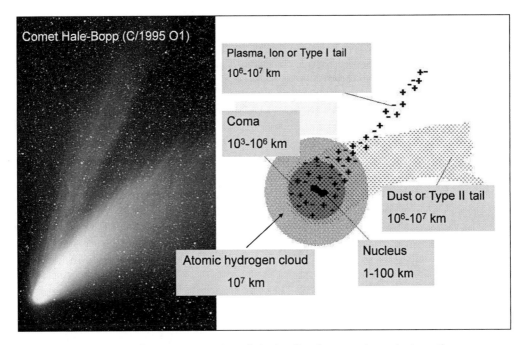

Figure 5.1. Structure of an active comet. The Sun is approximately in the direction opposite to the ion tail.

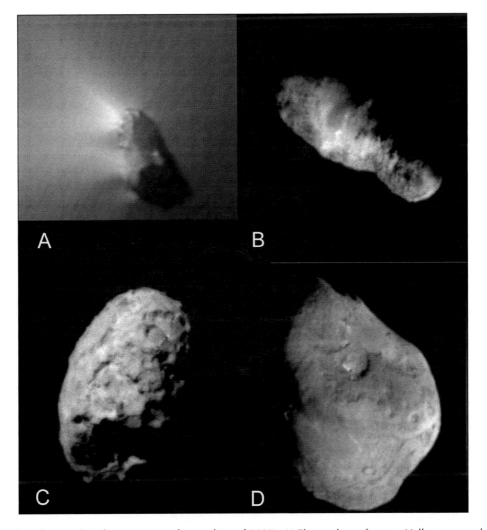

Figure 5.2. Picture of the four nuclei of comets ever observed (as of 2007). A) The nucleus of comet Halley, as seen by the ESA probe Giotto in 1986 from a few thousand km (© ESA and MPAe). B) The nucleus of comet P/Borrelly observed by NASA probe Deep Space 1 in 2001 from 3400 km (© NASA). C) The nucleus of comet P/Wild 2 observed from 236 km by Stardust, the NASA probe which has returned dust samples collected during the flyby (© NASA). D) The nucleus of comet P/Temple 1 observed by NASA probe Deep Impact from 3000 km (© NASA).

5.3 Chemical Composition of Comets

(i) Remote Sensing

Radio and Infrared spectroscopy of the coma with large telescopes has led to the confirmation of H_2O as the major cometary ice, but also to the detection of about thirty other less abundant molecular species: the very abundant CO (1-30% with respect to water) and CO_2 (5%) and also many species of interest for prebiotic chemistry—HCN, NH_3, H_2CO, H_2S (at the percent level), HC_3N, CH_3CN, NH_2CHO, CH_3CHO, H_2CS... (between 0.01 and 1%). Some of these species have important implications in aqueous solution: HCN is the key molecule for the synthesis of adenine and other nucleic bases, H_2CO for sugars (formose reaction), NH_3, HCN, H_2CO and other aldehydes allow Strecker synthesis (and with CO_2 Bucherer-Berg synthesis) of amino acids and related species. Table 5.1 lists the typical relative abundances of volatile molecules in comets as deduced from coma observations, together with the observed comet-to-comet variation.[11]

Concerning the mineral component of comets, infrared remote spectroscopy with the ISO satellite showed the presence of forsterite, a magnesium rich crystalline silicate.[12] Both amorphous and crystalline silicates are shown to be present. A recent reanalysis of these data[13] provided a tentative detection of carbonates, whereas the detection or not of polycyclic aromatic carbons (PAHs) from the same data remains controversial.

Recently, Deep Impact mission has provided a new original set of data about comets. On July 4th, 2005, a 370 kg impactor collided with comets 9P/Temple 1 with a relative rate of about 10 km/s. Both impactor and impactor-carrier spacecraft took pictures of the cometary nucleus, while most of the science measurements were performed from Earth telescopes. The impact was a success and resulted in a large amount of new information about the nucleus properties, but no new organic volatile compound was detected. The cometary activity of the comet during the impact looked in many ways like a natural outburst of the comet.[14]

(ii) In Situ Measurements

In 1986 mass spectrometers onboard the Giotto and Vega spacecrafts provided in situ information about dust particles composition during their Halley flyby. They analyzed organic refractory particles, later named CHON particles (from C,H,O, N atoms) and silicate grains. Mass spectrometry showed the presence of large molecular weight molecules, including possibly polymers of H_2CO.[15] A large amount of such heavy species was detected, but due to the rather low resolution of the mass spectra many different species are mixed in the same measurement channels. In most cases individual identification was not possible.[16,17]

Dust and ice abundances in a comet are of the same order, within a factor of 10—the comet to comet variation may reflect initial composition differences and/or comet evolution due to several orbits in the inner Solar System. Within the dust particle population, silicates and organic refractory material have comparable masses.

(iii) Sample Return Mission

The Stardust mission (NASA) was the very first spacecraft aimed to collect samples from a comet (81P/Wild 2) and return it to Earth.[18] This comet is relatively new in the inner Solar System, as it passed only a few times in the Sun vicinity. Dust grains have been trapped during the comet flyby in a very low density silicon-based aerogel (porosity >99%), designed to allow a progressive deceleration of the captured material, in order to minimize any alteration by heating and pyrolysis of organic molecules. Once brought back to Earth, physical properties and chemical composition of part of the grains have been analysed with the most recent and sensitive instru-

ments. Some of the material is kept untouched for future analyses by new methods to be developed in years to come.

More than 10,000 particles (from 1 to 300 μm) were captured in the coma of comet 81P/Wild 2 and returned to Earth, for a total mass of about $3\ 10^{-4}$g.[19] First analyses show an extreme complexity of the cometary material.

The Stardust samples contain amorphous and crystalline silicates such as olivine and pyroxene. The presence of crystalline minerals is an indication that some of the material has been processed at very high temperature (more than 1000 K), while part of the silicate is still in an amorphous form, an indication of a completely different thermal history. Those results suggest that the cometary material is a mixture of matter from different origins in a relatively well mixed Solar Nebula.

From an organic chemistry point of view, the first analyses of grains have enabled the detection of an organic component, which is rich in oxygen and nitrogen compared to the one found in carbonaceous meteorites. The samples are very heterogeneous and show N/C ratios ranging from $5\ 10^{-3}$ to 1. Aromatic molecules have been observed, such as naphthalene ($C_{10}H_8$), phenanthrene ($C_{14}H_{10}$) and pyrene ($C_{16}H_{10}$), but the samples tend to be poorer in aromatics than are meteorites and interplanetary dust particles. CH_2, CH_3, aromatic CH, OH and C = O groups have been identified by infrared spectroscopy. Detection of carboxyl, nitrile and amide functions are also reported after XANES (X-ray absorption near-edge spectroscopy) analyses.[20,21] More specific detection of methylamine (CH_3-NH_2), ethylamine ($CH_3CH_2NH_2$) and glycine ($NH_2CH_2CO_2H$) have also been reported, but are still very tentative until further specific isotopic measurements are conducted. Several years of extremely careful work are still ahead before it is possible to reach final conclusions from Stardust sample.

(iv) Laboratory Work

For a better insight into the most complex and less volatile material, one can also turn to experimental laboratory work. The principle of such experiments is the following: from observations of the most abundant species in comae and in the interstellar medium, one can infer the probable composition of the nucleus ices. A gaseous sample of the key species is deposited under vacuum on a cold substrate and irradiated during or after deposition by UV photons or charged particles. Condensed ices are sometimes simply warmed up slowly without irradiation. When the sample is warmed up for analysis a refractory organic residue remains on the substrate as the volatiles sublimate (Fig. 5.3). Mayo Greenberg, who conceived that kind of experiments, called this residue "Yellow Stuff".[22]

The diversity of organic compounds synthesized during such laboratory simulations is remarkable but their identification is seldom exhaustive.[23] The nature of the complex molecules depends on the ice composition and the nature of the energy source. The three kinds of energetic processing used during the experiment (thermal cycle, UV photolysis, energetic particles irradiations) can occur to ice mixtures either on icy coated dust grains in interstellar clouds (potentially precometary ices), or within the Solar Nebula during the accretion of icy planetesimals, or in the outer layers of comet ices in the Solar system. Constraining the degrees to which different processes affect cosmic ices is a highly convoluted problem. Differences between the products synthesized during processing, according to the energy sources, could give information on the history of cometary matter and comets. Investigations are still in progress to address this question.

From an astrobiological point of view, it must be noted that a great number of amino acids (such as glycine, alanine, sarcosine, valine, proline, serine etc...) are reported in residues obtained after UV irradiation of ice mixtures made of H_2O: NH_3: CH_3OH: HCN and

Table 5.1 Molecules detected in comets and some upper limits

Cometary Volatiles Category	Molecule	Hale-Bopp Abundance ($H_2O = 100$)	Intercomet Variation	Detected Comets + Upper Limits	
H	H_2O	100			Water
	H_2O_2	<0.03			Hydrogen peroxide
C,O	CO	23	<1.4-23	9 + 8	Carbon monoxide
	CO_2	6	2.5-12	4	Carbon dioxide
C,H	CH_4	1.5	0.14-1.4	8	Methane
	C_2H_6	0.6	0.1-0.7	8	Ethane
	C_2H_2	0.2	<0.1-0.5	5	Acetylene
	C_4H_2	0.05?			Butadiyne
	CH_3C_2H	<0.045			Propyne
C,O,H	CH_3OH	2.4	<0.9-6.2	25 + 2	Methanol
	H_2CO	1.1	0.13-1.3	18 + 3	Formaldehyde
	CH_2OHCH_2OH	0.25			Ethylene glycol
	HCOOH	0.09	<0.05-0.09	3 + 2	Formic acid
	$HCOOCH_3$	0.08			Methyl formate
	CH_3CHO	0.025			Acetaldehyde
C,O,H upper limits	H_2CCO	<0.032			Ketene
	c-C_2H_4O	<0.20			Oxirane
	C_2H_5OH	<0.1			Ethanol
	CH_2OHCHO	<0.04			Glycolaldehyde
	CH_3OCH_3	<0.45			Dimethyl ether
	CH_3COOH	<0.06			Acetic acid
N	NH_3	0.7	<0.2-1	4	Ammonia
	HCN	0.25	0.08-0.25	32 + 0	Hydrogen cyanide
	HNCO	0.1	0.02-0.1	4 + 2	Isocyanic acid
	HNC	0.04	<0.003-0.035	12 + 3	Hydrogen isocyanide
	CH_3CN	0.02	0.013-0.035	9 + 2	Methyl cyanide
	HC_3N	0.02	<0.003-0.03	3 + 7	Cyanoacetylene
	NH_2CHO	0.015			Formamide
N upper limits	NH_2OH	<0.25			Hydroxylamine
	HCNO	<0.0016			Fulminic acid
	CH_2NH	<0.032			Methanimine
	NH_2CN	<0.004			Cyanamide
	N_2O	<0.23			Nitrous oxide
	NH_2-CH_2-COOH	<0.15			Glycine
	C_2H_5CN	<0.01			Cyanoethane
	HC_5N	<0.003			Cyanobutadiyne
S	H_2S	1.5	0.13-1.5	15 + 5	Hydrogen sulfide
	OCS	0.4	<0.09-0.4	2 + 5	Carbonyl sulfid
	SO	0.3	<0.05-0.3	4 + 1	Sulfur monoxide
	SO_2	0.2			Sulfur dioxide
	CS_2	0.17	0.05-0.17	15 + 3	Carbon disulfide
	H_2CS	0.02			Thioformaldehyde
	S_2	0.005	0.001-0.005	5	Disulfur
	CH_3SH	<0.05			Methanethiol
	NS	0.02	<0.02-0.02	1 + 1	Nitrogen sulfide
P	PH_3	<0.16			Phosphine
Metals	NaOH	<0.0003			Sodium hydroxide
	NaCl	<0.0008			Sodium chloride

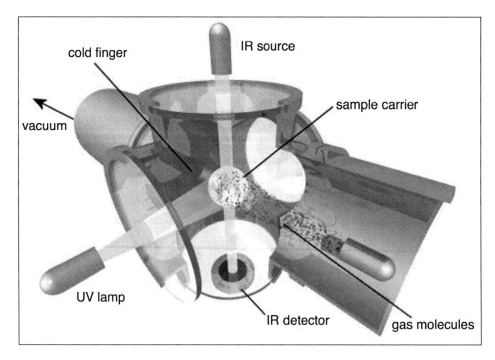

Figure 5.3. A typical experimental setup allowing the photolysis of cometary ice analogs made by deposition of a gas mixture on a cold sample carrier cooled down to 10 K in a cryostat (the UV lamp can be replaced in some setups by an ion or an electron gun). The ice evolution can be analysed in situ by infrared spectroscopy. The room temperature residue can be collected for further analysis such as GC-MS (Gas Chromatography coupled to Mass Spectrometry), HPLC (High Performance Liquid Chromatography) and many others. Picture courtesy of Jan Hendrik Bredehoft, University of Bremen, thoralf@uni-bremen.de.

H_2O : CH_3OH : NH_3 : CO : CO_2. Unhydrolyzed residues (without any liquid water introduced to the analysis protocol) produce only a trace of glycine. The detection of the other amino acids requires an acid hydrolysis of the residue under very strong conditions (HCl ≥ 6 M and T ≥ 100 °C).[24,25] Therefore it is not clear to date if (1) amino acids are present themselves in the laboratory residues and henceforth in cometary ices, or if "only" amino acids precursors are synthesized and (2) if the residues' processing (acid hydrolysis) is relevant to any chemistry which could have turned the amino acids' precursors imported by cometary impacts in the primitive oceans of the early Earth into actual amino acids.

The chirality issue has also been investigated through laboratory simulations. It has been reported that an asymmetric vacuum UV photolysis of a racemic mixture of leucine in the solid state results in the production of an enantiomeric excess of one form of the amino acid.[26] However, such an enantiomeric excess has not been detected yet when amino acids are directly synthesised within an ice mixture under circularly polarized light.[27] Further work on this topic is planned with the opening of the new French synchrotron SOLEIL in 2007 (beamline DESIRS).

Following remote sensing and in situ observations which can only probe the atmosphere of comets, sample return of a limited amount of cometary material and laboratory work on simulated cometary ices, an ambitious next step is the landing on a comet to study its composition. This will be achieved with the Rosetta mission from the European Space Agency (ESA).

5.4 Rosetta 2014—Rendezvous of a Laboratory with a Comet

The Rosetta mission is made of two parts: one orbiter revolving around the nucleus and a lander called Philae. It was launched on March 2nd 2004 and will reach its target, comet 67P/Churuymov-Gerasimenko in August 2014. The landing of Philae on the nucleus of the comet is expected in November 2014. The instruments onboard the orbiter will enable an unprecedented analysis of the composition of volatile and refractory compounds released from the nucleus. Two mass spectrometers on board Rosetta (ROSINA: Rosetta Orbiter Spectrometer for Ion and Neutral Analysis, COSIMA: COmetary Secondary Ion MAss spectrometer) will collect and analyse gas and dust as close as 1 km from the surface of the nucleus, hopefully close enough to study almost unaltered matter released from the nucleus. Four spectrometers will also probe the composition of the cometary environment on a broad spectral range of electromagnetic radiation (OSIRIS: Optical, Spectroscopic and Infrared Remote Imaging System, ALICE: UV spectrometer, VIRTIS: Visible and InfraRed Thermal Imaging Spectrometer, MIRO: Microwave Instrument for the Rosetta Orbiter). On board Philae, the most fruitful information from an astrobiological point of view will come from COSAC (Cometary Sampling and Composition experiment), a gas chromatograph coupled with a mass spectrometer. In this chapter, we emphasize some of these instruments (COSAC, COSIMA, VIRTIS and MIRO) because most of the new organic molecules should be detected thanks to these experiments.

(i) COSAC

The COSAC instrument is a gas chromatograph (GC) coupled with a mass spectrometer (MS, a linear time of flight spectrometer in this case). It consists of 8 chromatographic columns; each of them is connected to its own detector (TCD), but it is also possible to connect them to a mass spectrometer.[28] Previous results obtained thanks to direct mass spectrometry measurements with Puma, Giotto and Stardust spacecrafts gave "only" the mass spectrum of the mixture of all the molecules at the same time. COSAC will carry out a preliminary separation by chromatography, which will achieve a quasi definite identification of the compounds since they will be recognised both from their retention time and from their individual mass spectra. Samples will be collected after drilling the surface and heated at various temperatures before being injected into the analysis system. Pyrolysis (up to 600°C) is possible in order to degrade the most refractory components and enable gas phase analysis of the fragments. Out of the eight chromatographic columns, three are specifically devoted to the analysis of chiral molecules in order to distinguish enantiomers. The other five columns have been selected so that a maximum of molecules can be detected. Moreover, the simultaneous analysis of a single sample with several columns (up to four columns at the same time) will facilitate data analysis by comparison and thus increase the reliability of interpretation.

Many molecules were considered in the selection of the chromatographic columns (molecules already detected in the atmosphere of comets, in the interstellar medium, or from laboratory simulations of cometary ices). One must note that amino acids and other heavy compounds such as oxalic acid, urea, etc... are not detectable, since they are not volatile enough to be analysed in the gaseous phase. GC analysis of such compounds requires a preliminary stage of processing called derivatization (chemical reaction making the targeted compound more volatile). This procedure is not feasible with the COSAC instrument, but work is in progress to include derivatization in future Martian exploration experiments. Nucleus analyses by the COSAC instrument will be completed by CIVA (infrared analysis) and MODULUS (for isotopic measurements).

(ii) COSIMA

COSIMA is a time-of-flight secondary ion mass spectrometer (TOF-SIMS) instrument dedicated to in situ analysis of cometary dust grains.[29] Cometary grains will be collected on metallic targets (silver, gold, palladium and platinum) and an optical system (COSISCOPE) will locate cometary grains for further analysis with TOF-SIMS. COSIMA has a mass resolving power $(M/\Delta M)$ of about 2000 and will be able to analyse particles with a resolution around 50 µm. The TOF-SIMS technique is very sensitive to the composition of the surface and it only analyzes the very first mono-layers of the sample. TOF-SIMS spectra are difficult to interpret as they contain a very large amount of information, showing both elemental and molecular masses up to masses ~1000 amu.

Unlike analyses performed with COSAC, no separation of the different molecules is performed prior to analysis by mass spectroscopy. Therefore extremely complex spectra are expected. Quite a secure distinction between the organic and mineral components can easily be achieved (molecules rich in H such as organic compounds tend to have a mass a few fractions of decimal above the unit, $m = n + \delta$ with $n \in N$, while mineral masses tend to be below the unit, $m = n - \delta$). Any specific identification might be hazardous unless careful calibration is performed with a ground instrument, which is currently already in progress. The instrument has a very high resolution in mass and also a spatial resolution which can allow analysis of the composition of several spots on the same grain. Moreover, COSIMA can analyse non volatile compounds, which are not possible to detect with COSAC.

Analyses by the COSIMA instrument will be completed by ROSINA for the gaseous phase.[30] This instrument based on mass spectrometry measurements will determine the composition of the atmosphere and ionosphere of a comet thanks to two sensors. ROSINA has a wide mass range (1 to > 300 amu) and a very high mass resolution $M/\Delta M > 3000$ with the ability to resolve CO from N_2 and ^{13}C from ^{12}CH.

(iii) VIRTIS and MIRO

The two spectrometers VIRTIS and MIRO will observe the nucleus and the coma from the orbiter. They will provide information on the physical conditions and physical processes in the comet and on its chemical composition.

Virtis is a UV/Visible and infrared spectrometer.[31] Two channels cover respectively the 0.2-1 microns and 1-5 microns wavelength ranges. The infrared channel will provide information on the nature

of the comet surface and on molecules ("parent molecules") released in the coma before they are photodissociated, ionized or react with other species.

On the nucleus, the minerals and ices will be remotely mapped with high spatial resolution. Silicates and hydrates are expected; if present, phyllosilicates ("clay") will show up through their water and OH absorption bands. Water ice itself and other ices like NH_3, CO_2 or H_2S can be identified by their infrared spectral features. Hydrocarbon ices (CH_4, C_2H_2, C_2H_4, C_2H_6, C_3H_8) can also be identified in this spectral range. Some of these simple hydrocarbons could also have polymerized. Such carbonaceous material will redden the comet spectrum; a more precise identification requires however a high signal to noise ratio (>100).

In the coma, beside solid features coming from the grains, much narrower features coming from the gas phase are present. A major objective is the identification of the hydrocarbon emission in the 3-4 micron range, through the high ($\lambda/\Delta\lambda > 2000$) spectral resolution. The infrared spectrum is also rich in rovibrational emission of important parent molecules : H_2O, CO_2, H_2CO, CH_3OH, CO. Deuterated water HDO can also be detected with long integration times; the HDO/H_2O ratio is a key information for evaluating the importance of comets in the build up of Earth oceans. Current data are only available for 3 comets (Halley, Hyakutake and Hale Bopp) and point to a D/H ratio in water in comets (~3×10^{-4}) twice that in SMOW (standard mean ocean water ~1.5×10^{-4}). But the few observed comets are not thought to be representative of all comets and the homogeneity of this ratio at different locations in a given comet is also an open question.

The UV channel of Virtis is mainly sensitive to the products of coma processes ("daughter molecules"): radicals (OH, CN, C_2, C_3, NH, CH), ions (CO^+, CH^+, H_2O^+, N_2^+) and small molecules (CO). The nature of the parent can be constrained to some extent; this is especially interesting if one suspects these species to be the photo or thermal degradation products of complex organics expected in cometary grains. The presence of polymers like the H_2CO polymer polyoxymethylene (POM), or HCN polymers can thus be indirectly tested.

The MIRO instrument consists of two radio receivers operating at mm and submm wavelength (1.6 mm and 0.5 mm, or 190 and 562 GHz).[32] Due to the small antenna size (30 cm) and the absence of cryogenic cooling of the receiver, only major cometary species will be observed: H_2O (and its isotopes $H_2^{17}O$ and $H_2^{18}O$), CO, CH_3OH and NH_3. The precise physical data on the nucleus outgassing and the development of the coma deduced from these observations will be used for detailed modelling of the observations performed with the much larger ground-based instruments like the IRAM 30m in Spain, LMT 50 m in Mexico, GBT 110 m in Virginia or the ALMA interferometer under construction in Chile (64 antennas of 12 m diameter). NH_3 itself is quite difficult to observe from the ground and has only been detected in very few comets using cm radiowave or near IR. From its Earth orbit, the ODIN satellite provided tentative detections of NH_3 in two comets using the same 572GHz line as MIRO. The presence of NH_3 in Solar System icy objects lowers the water ice melting point and thus allows water to be liquid at temperatures lower than for pure water.

5.5 Comets and Life

(i) Delivery of Prebiotic Molecules

The exogenous source of prebiotic molecules involves impacts by large objects, comets and asteroids on the one hand and soft landing of micron to mm sized interplanetary dust grains of cometary or asteroidal origin on the other hand. The conditions for delivery by these two mechanisms are very different and the yields are very difficult to assess, but both may be quantitatively important to fueling the stock of prebiotic species.[33] We concentrate here on the cometary part.

A. Violent Delivery by Impacts

At first thought, impacts of comet nuclei on the Earth seem to be such highly energetic processes that all molecules should be destroyed. It has been shown however through laboratory experiments and numerical simulations that an impact, if in a grazing geometry, could have delivered to early Earth some amino-acids (like aspartic and glutamic acids) in comparable amounts to those produced by Miller-Urey synthesis in a CO_2 rich atmosphere.[34] If one considers smaller bodies like Mars, Europa or the Moon, despite a lower impact velocity favourable to molecule survival, the yield is lower as a large fraction of the impacting material can escape owing to the lower gravity; this results in impacts being still relatively efficient in the case of Mars and much less so for the smaller Europa or the Moon. However, even if Europa's formation history did not allow for the inclusion or in-situ formation of organics, Europa could have gained some organic molecules through such cometary impacts.

One should note that asteroids (some of which being organic-rich) do not have a similar organics delivery efficiency on account of their more rigid, less porous and higher density interior. Higher temperatures would be reached during the impact resulting in a higher rate of destruction of the complex molecules.

B. Decelerated Stratospheric Interplanetary Dust Particles (SIDPs) and Soft-Landing Micrometeorites

Small (10 microns) particles are efficiently decelerated by the atmosphere and "float around" long enough to allow their collection by airborne collectors. The fluffiest of these so-called Brownlee particles are thought to be of cometary origin. Somewhat larger 100-200 micron particles reach the ground and are called micrometeorites. They can be collected in polar ices, where they are more easily identified and less contaminated by Earth particles. SIDPs and micrometeorites are a regular source of extraterrestrial matter coming to the Earth; estimated at 10,000 to 40,000 tons/year, they are second in importance after the large impacts (when their contribution is averaged over geological times).

A fraction of these particles is carbon rich, even richer than carbonaceous chondrites. In a sample of C rich micrometeorites, two amino acids have been found, AIB and isovaline, the latter being absent in living organisms on Earth.

C. Atmospheric Shock Synthesis

Besides the organic molecules they carry and may deliver to Earth, infalling bodies input energy into the atmosphere. This leads to shock induced chemical syntheses, which are rather efficient in a reducing, Urey-Miller type atmosphere, but much less so in a weakly reductive atmosphere (where $H_2/CO_2 < 0.1$).

(ii) Comets as Life Frustrators?

Comets may represent, depending on the epoch (and on models), 1 to 50% of the km-size bodies impacting the Earth. Whereas such impacts may have been favourable to the appearance of life through the delivery or the shock-induced syntheses of organic molecules, they also may be detrimental and possibly lethal to incipient life. The possible cometary nature of two famous events is still being debated: the 1908 Tunguska event (a few ten meters body), which destroyed 2000 km² of Siberian forest and the K/T event 65 Myr ago, which has produced the Chixculub crater (from a >10 km body) and is a favourite explanation for the contemporary mass extinctions (including dinosaurs).[35]

(iii) Liquid Water in Comets?

Could advanced prebiotic chemistry take place in the comets themselves, leading to the possibility for the appearance of life in comets? Such a question makes sense especially if liquid water can exist in comets. Liquid water requires a high enough temperature and pressure. During comet formation, the accretion process heats mainly the surface of the comet. Two other mechanisms can generate heat inside the nucleus: radioactive heating and amorphous to crystalline ice (irreversible) transformation. Al^{26} is by far (by a factor of 500) the most important radiogenic heating element at early times in the Solar System. With a lifetime $t_{1/2}$ of 700,000 years, it is widespread in the interstellar medium and thought to have been present in the protosolar nebula. It can heat the interior of comets if they are large enough to avoid rapid cooling by heat conduction or gas diffusion; but this could happen only if these comets have formed very rapidly ($< a$ few $t_{1/2}$). According to numerical models, 10 km-size comets could typically harbour liquid water for 10^5 yrs.[36]

5.6 Conclusion

Comets can bring to the Earth elements like carbon, as well as water (in an amount which remains to be determined). Moreover, they contain simple organic molecules and it is probable that they contain more complex ones as well. Thus comets are likely contributors to early Earth enrichment in prebiotic molecules, through the dust particles they release (some of which end up encountering the Earth atmosphere), through atmospheric chemical syntheses and possibly also through direct delivery taking place when cometary nuclei impact the Earth. These same impacts may also strongly affect later developments of life, either locally or on a planetary scale.

Many questions are central to the current debates, which we hope will find a response in the coming years and for which space missions can bring important results:

- Where and when were the cometary matter and then the comets, formed?
- How close is cometary matter to interstellar matter?
- Do comets contain amino acids and other bricks of life?
- Do these molecules present an enantiomeric excess, or are they in racemic mixture?
- What is the precise budget of the cometary contributions to the Earth (fraction of the water of the oceans; fraction of carbon; fraction of prebiotic molecules)?
- What is the global effect of comets on life, its appearance and its development?
- Out of the Solar System, is the presence of an Oort cloud or a Kuiper belt an exceptional characteristic of our Solar System or a banal fact among main sequence stars? Given the crucial part potentially played by comets in the origin and the evolution of life on Earth, it appears fundamental to know the probability for a star with habitable planets to be also surrounded by comets. New research topics thus arise regarding these exocomets (comets orbiting around other stars). The properties of the spectra of a close star, β Pictoris—one of the rare main sequence stars around which a disc has been detected-currently are convincingly interpreted as being due to a constant infall of "evaporating bodies" on this star. Much is then to be expected from ambitious recent space experiments like the Corot mission, whose goal is to detect Earth-size planets around a wide range of other stars and which might make it possible also to detect comets directly.[37]

References

Further Readings

1. Crovisier J, Encrenaz T. Comet Science: The Study of Remnants From the Birth of the Solar System. Cambridge: Cambridge University Press, 2000.
   - A comprehensive and richly illustrated introduction to comets (for a Scientific American type readership).
2. Despois D, Cottin H. Comets. Potential sources of prebiotic molecules for the early Earth. In: Gargaud M, Barbier B, Martin H, Reisse J, eds. Lectures in Astrobiology. Berlin: Springer Verlag 2005; 1:289-352.
   - Where many topics of the present chapter are developed with greater details and with numerous references to original works in.
3. Irvine WM, Schloerb FP, Crovisier J et al. Comets: A link between interstellar and nebular chemistry. In: Manning V, Boss B, Russel S, eds. Protostar and Planets IV. Tuscon: University of Arizona Press 2000a:1159.
   - Presents comet chemistry and its relation either to protosolar nebula chemistry or interstellar chemistry.

Two recent multi-author books are highly recommended for detailed research work.

4. Thomas PJ, Hicks RD, Chyba CF et al. Comets and the Origin and Evolution of Life. 2nd edition. New York: Springer 2006.
   - Reviews in depth our present knowledge on cometary impact and delivery processes and the liquid water issue.
5. Festou MC, Keller HU, Weaver HA. Comets II. Tuscon: University of Arizona Press 2004.
   - Reviews much of cometary science beyond the astrobiological interest.
6. Newburn RL, Neugebauer M, Rahe JH. Comets the Post-Halley Era. New York: Springer 1991.
   - The two volumes of Comets in the Post-Halley Era, despite their age, are still a very good detailed introduction to cometary science.

Specific References

7. Whipple FL. A comet model I. The acceleration of Comet Encke. Ap J 1950; 111:375-394.
8. Chamberlin TC, Chamberlin RT. Early Terrestrial conditions that may have favored organic synthesis. Science 1908; 28:897.
9. Oro J. Comets and the formation of biochemical compounds on the primitive Earth. Nature 1961; 190:389-390.
10. Festou MC, Keller HU, Weaver HA. Comets II. Tuscon: University of Arizona Press, 2004.
11. Bockelée-Morvan D, Crovisier J, Mumma MJ et al. The composition of cometary volatiles. In: Festou M, Keller HU, Weaver HA, eds. Comets II. Tuscon: University of Arizona Press 2004:391-423.
12. Crovisier J, Leech K, Bockelée-Morvan D et al. The spectrum of Comet Hale-Bopp (C/1995 O1) observed with the Infrared Space Observatory at 2.9 Astronomical Units from the Sun. Science 1997; 275:1904-1907.
13. Lisse CM, Kraemer KE, Nuth III JA et al. Comparison of the composition of the Tempel 1 ejecta to the dust in Comet C/Hale-Bopp 1995 O1 and YSO HD 100546. Icarus 2007; 187:69-86.
14. Lara LM, Boehnhardt H, Gredel R et al. Behavior of Comet 9P/Tempel 1 around the Deep Impact event. A and A 2007; 465:1061-1067.
15. Huebner WF. First polymer in space identified in comet Halley. Science 1987; 237:628-630.
16. Kissel J, Krueger FR. The organic component in dust from comet Halley as mesured by the PUMA mass spectrometer on board Vega 1. Nature 1987; 326:755-760.
17. Mitchell DL, Lin RP, Carlson CW et al. The origin of complex organic ions in the coma of comet Halley. Icarus 1992; 98:125-133.
18. Brownlee D, Tsou P, Aleon J et al. Comet 81P/Wild 2 under a microscope. Science 2006; 314:1711-1716.
19. Horz F, Bastien R, Borg J et al. Impact features on Stardust: Implications for comet 81P/Wild 2 dust. Science 2006; 314:1716-1719.
20. Keller LP, Bajt S, Baratta GA et al. Infrared spectroscopy of comet 81P/Wild 2 samples returned by Stardust. Science 2006; 314:1728-1731.
21. Sandford SA, Aleon J, Alexander CMO et al. Organics captured from comet 81P/Wild 2 by the Stardust spacecraft. Science 2006; 314:1720-1724.
22. Greenberg JM. What are comets made of? A model based on interstellar dust. In: Wilkening LL, ed. Comets. Tucson: University of Arizona Press 1982:131-163.
23. Colangeli L, Brucato JR, Bar-Nun A et al. Laboratory experiments on cometary materials. In: Festou MC, Keller HU, Weaver HA, eds. Comets II. Tuscon: University of Arizona Press, 2004:695-717.
24. Bernstein MP, Dworkin JP, Sandford SA et al. Racemic amino acids from the ultraviolet photolysis of interstellar ice analogues. Nature 2002; 416:401-403.
25. Muñoz Caro GM, Meierhenrich UJ, Schutte WA et al. Amino acids from ultraviolet irradiation of interstellar ice analogues. Nature 2002; 416:403-406.
26. Meierhenrich UJ, Nahon L, Alcaraz C et al. Asymmetric vacuum UV photolysis of the amino acid leucine in the solid state. Angewandte Chemie-International Edition 2005; 44:5630-5634.
27. Nuevo M, Meierhenrich UJ, Muñoz Caro GM et al. The effects of circularly polarized light on amino acid enantiomers produced by the UV irradiation of interstellar ice analogs. A and A 2006; 457:741-751.
28. Goesmann F, Rosenbauer H, Roll R et al. Cosac, The cometary sampling and composition experiment on Philae. Space Science Reviews 2007; 128:257-280.
29. Kissel J, Altwegg K, Clark BC et al. COSIMA, a high resolution time of flight spectrometer for secondary ion mass spectroscopy of cometary dust particles. Space Science Reviews 2007; 128:823-867.
30. Balsiger H, Altwegg K, Bochsler P et al. Rosina—Rosetta Orbiter spectrometer for ion and neutral analysis. Space Science Reviews 2007; 128:745-801.
31. Coradini A, Capaccioni F, Drossart P et al. Virtis: an imaging spectrometer for the Rosetta Mission. Space Science Reviews 2007; 128:529-559.
32. Gulkis S, Frerking M, Crovisier J et al. MIRO: microwave instrument for Rosetta Orbiter. Space Science Reviews 2007; 128:561-597.
33. Chyba CF, Hand KP. Comets and prebiotic organic molecules on early Earth. In: Thomas PJ, Hicks RD, Chyba CF, McKay CP, eds. Comets and the Origin and Evolution of Life. 2nd edition. New York: Springer, 2006:169-206.
34. Pierazzo E, Chyba CF. Impact delivery of prebiotic organic matter to planetary surfaces. In: Thomas PJ, Hicks RD, Chyba CF, McKay CP, eds. Comets and the Origin and Evolution of Life. 2nd edition. New York: Springer, 2006:137-168.
35. Zahnle K, Sleep NH. Impacts and the early evolution of life. In: Thomas PJ, Hicks RD, Chyba CF, McKay CP, eds. Comets and the Origin and Evolution of Life. 2nd edition. New York: Springer, 2006:207-252.
36. Podolak M, Prialnik D. The conditions for liquid water in cometary nuclei. In: Thomas PJ, Hicks RD, Chyba CF, McKay CP, eds. Comets and the Origin and Evolution of Life. 2nd edition. New York: Springer 2006:303-314.
37. Lecavalier Des Etangs A, Vidal-Madjar A, Ferlet R. Photometric stellar variation due to extra-solar comets. A and A 1999; 343:916-922.

Meteorites and the Chemistry
That Preceded Life's Origin

Sandra Pizzarello*

6.1 Meteorites and Their Sources

Meteorites are pieces of extraterrestrial material that cross the Earth's orbit, enter the terrestrial atmosphere and reach the ground without being completely destroyed. Some are known to have been ejected from Mars, the Moon and even Mercury following local impacts in those bodies. For the most part however, meteorites are fragments of asteroids. These are small planets (<1000 km diameter) of odd size and shape that are found throughout the solar system but are concentrated in large number between the orbits of Mars and Jupiter (over 2000 have been characterized). According to the Titius—Bode law, by which planets are spaced at regular distances from the Sun, we would expect to find a planet occupying the asteroids' orbit and it was debated for a while whether these planetesimals were the remnants of an exploded planet, perhaps taken apart by the gravity of a nearby planet. Based on the current model of Solar System formation, it is agreed today that they represent a stillborn planet. When the giant planet Jupiter condensed nearby, it is believed that its gravity affected the surrounding smaller bodies on their way to form a planet, slinging some away to farther peripheral orbits as comets and leaving the rest circling in the asteroid belt unable to coalesce.[1]

The lucky result of the thwarted planet formation in the asteroid belt is that a good number of planetesimals there never suffered the geological upheaval that accompanies early planetary processes, such as loss of volatiles and rock melting among others and preserved a pristine record of early Solar System material.

However, this is a crowded place and collisions are common, as the poked surfaces of asteroids we have photographed show (Fig. 6.1) and it is the chunks produced by these collisions that can be put in Earth-crossing orbits and frequently reach the Earth as meteorites. Over 18,000 meteorites have been collected so far; a small part of these have been recorded as falls (taking the name of the location where they fell) and the rest, called finds, were found without anybody witnessing their fall. Most meteorites of the latter group have come in the last two decades from Antarctica, where both Japan and the USA own sectors of the continent and have sent scientific expeditions for their search.

Meteorites are classified in a variety of groups and subgroups of broadly similar types. The major distinction is between the groups of iron, stony and stony iron meteorites. The iron meteorites are chunks of iron/nickel alloy and come from asteroids that became large enough and derived sufficient heat from their gravitational collapse to separate matter on the basis of density and make a denser iron core. The portion of the asteroid belt closer to the Sun seems to be the prevalent region of origin for these meteorites. The stony meteorites have the appearance of rocks, composed of various silicates and other minerals and are grouped according to their mineralogical composition and level of geochemical alteration.

6.2 Carbonaceous Chondrites and Their Organic Content

Carbonaceous chondrite (CC) meteorites are a subgroup of the stony meteorites that take their name from round glassy inclusions, named chondrules and the distinction of containing organic carbon in various percent abundances (about 1-4%). CC have a primitive elemental composition that is similar to that of the Sun as well as that of the Cosmos overall, since the Sun is a rather average star. They have the appearance of aggregate rocks (Fig. 6.2), i.e., of material packed together without signs of having been transformed by extreme heat or pressure. Their major constituent is a fine-grained silicate matrix, which contains larger mineral inclusions and, interspersed within it, a complex and heterogeneous organic material.[2]

Because of the particularly pristine nature of CC and the general availability of meteorites, their organic material offers to direct analysis a unique natural sample of abiotic organic chemistry as it came to be in the Solar System shortly before the onset of life.

Table 6.1 shows the general organic composition of the Murchison meteorite, which fell in a large size (over 100 kg) in 1969[a] and has been extensively studied since. The larger portion of meteoritic organic carbon is sometimes described as kerogen-like because, like terrestrial kerogens, it is an insoluble macromolecular material made up of N, O and S besides C and H and has a complex composition that is not known in much molecular detail. The bulk of this material has been inferred mainly from spectroscopic analyses[b] and decomposition studies, where it is pyrolyzed by heat or oxidized into its fragments. It is believed to have the hypothetical structure shown in Figure 6.3 and to be composed of clusters of aromatic rings, bridged by aliphatic chains containing S, N, O, with peripheral branching and functional groups. The insoluble carbon is not homogeneous, however and ~ 10% is found as self-contained nanostructures, spheres and nanotubes of diverse elemental composition; also found within it are minute amounts of "exotic" carbon, so called because it was likely formed in the envelopes of stars (i.e., prior to the formation of Solar System). In addition, it was discovered recently that the material is not completely

*Sandra Pizzarello—Department of Chemistry and Biochemistry, Arizona State University, Tempe, Arizona 85287-1604 USA.
Email: pizzar@asu.edu

Prebiotic Evolution and Astrobiology, edited by J. Tze-Fei Wong and Antonio Lazcano. ©2009 Landes Bioscience.

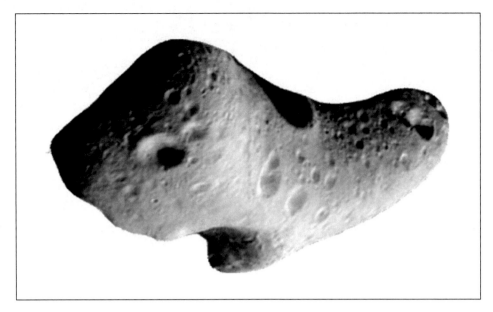

Figure 6.1. The Eros meteorite, rendezvoused by a NASA probe on Valentine's day 2001.

1 cm

250 μm

Figure 6.2. An interior fragment of the Murchison meteorite and a microscopic view of its matrix and inclusions. Courtesy of Dr. Michael E. Zolensky, NASA Johnson Space Center.

insoluble after all and can release some individual compounds under conditions of high T and P similar to those of terrestrial hydrothermal vents (300°C, 100 MPa).

For many carbonaceous meteorites of the same type as Murchison as well as others, the remaining portion of the organic carbon is made up of soluble compounds that are released from meteorite powders by extraction with water and solvents (Table 6.2).[3] As it can be seen by their names and structures, many of the types of individual compounds known to us throughout the Biosphere are found present in meteorites.

Table 6.1. The organic composition of the Murchison meteorite

A. Carbonate minerals (CO_3^{2-})	2-10%
B. Insoluble organic carbon	65-80%
C. Soluble organic compound	10-20%

Meteoritic organic compounds, although in amounts of only parts per million (ppm, micrograms/gram of meteorite), are found in a large variety of forms, which often contain all the possible structures for a given chemical group and up to the limits of their solubility.

For example, while the whole of terrestrial proteins is made up of just twenty amino acids, all of the α-amino type, over one hundred kinds of meteoritic amino acids are found, which vary over many, sometimes all, of the possible chemical structures involving an alkyl chain, one or two carboxyl- and one or two amino-groups (Scheme 6.1). Ten amino acids are found in both CC and terrestrial proteins: glycine, alanine, serine, threonine, valine, leucine, isoleucine, proline, aspartic acid and glutamic acid (serine and threonine have been detected only recently in the GRA 95229 Antartica meteorite from the Johnson Space Center).

Other examples of organic compounds distributed in both meteorites and the biosphere are nicotinic acid (which, in the form of

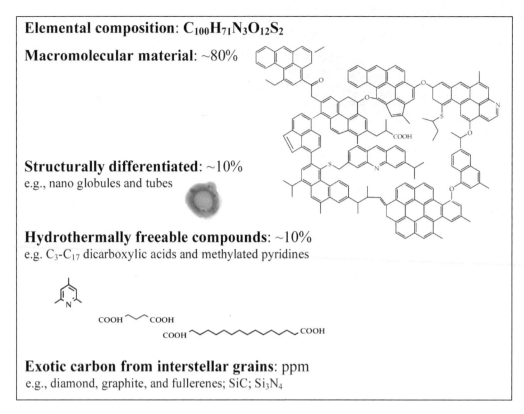

Figure 6.3. Make-up of the insoluble organic carbon in the Murchison meteorite.

Table 6.2. *The soluble organic compounds of the murchison meteorite*

Compound	Typical Molecule and Structure	Typical Molecule
Carboxylic acids	$H_3C-COOH$	Acetic acid
Amino acids	$H_3C-\overset{NH_2}{\underset{\mid}{CH}}-COOH$	Alanine
Hydroxy acids	$H_3C-\overset{OH}{\underset{\mid}{CH}}-COOH$	Lactic acid
Ketoacids	$H_3C-\overset{O}{\overset{\parallel}{C}}-COOH$	Pyruvic acid
Dicarboxylic acids	$OOC-(CH_2)_2-COOH$	Succinic acid
Sugar alcohols and acids	$H_2C-\overset{OH\ OH}{\underset{\mid\ \ \mid}{CH}}-COOH$	Glyceric acid
Alcohols, aldehydes and ketones	H_3C-CH_2OH	Ethanol
Amines and amides	$H_3C-CH_2NH_2$	Ethyl amine
Pyridine carboxylic acids		Nicotinic acid
Purines and pyrimidines		Adenine
Hydrocarbons:		
Alyphatic	$H_3C-CH_2-CH_3$	Propane
Aromatic		Naphthalene
Polar		Isoquinoline

nicotinamide, is the co-enzyme of a large group of oxido-reductases), several monocarboxylic acids (such as formic, acetic and butyric acids that take their names from being found in ants, vinegar and spoiled butter, respectively), sugar-like compounds (components of many terrestrial polysaccharides) and a complex amphiphilic material (having both polar and hydrophobic moieties) that is extracted from meteorite powders by chloroform and forms membrane-like vesicles in water.[4]

The most interesting similarity between meteoritic and biological compounds is that some of the chiral amino acids of meteorites display L-enantiomeric excesses (ee) that, if not as large, are similar to the ones found in the amino acids of terrestrial proteins.

So far, the ee have been determined unequivocally for the subgroup of the α-amino-α-alkyl amino acids.[5] They vary from 0-15% and represent the only example of molecular asymmetry outside the biosphere (see Chapter 7 for details on molecular chirality and its distribution in the biosphere). The finding of ee in meteoritic amino acids seems to suggest that the trait of chiral asymmetry was at least in part available to prebiotic chemistry. The origin of these ee is not known and could have been the result of asymmetric influences either during or after chiral amino acid formation. For example, circularly polarized light (CPL) is a form of radiation where the electric and magnetic vectors rotate along the axis of transmission. This rotation can take a clockwise or counter-clockwise turn; therefore there will be a right- and left-CPL that are mirror images of each other, chiral and will each interact differently with the two enantiomers of a chiral molecule (Chapter 7). If this radiation is "one-handed" and has high enough energy to be destructive of the molecule (as in the case of UV CPL), the two enantiomers will be destroyed at different rates and may leave one enantiomer in a larger amount than the other after exposure.[c] It has been proposed that such an effect could have taken place during the formation of meteoritic amino acids, possibly at the stage of precursor molecules and before reaching the shielding of the asteroidal parent body (see following).

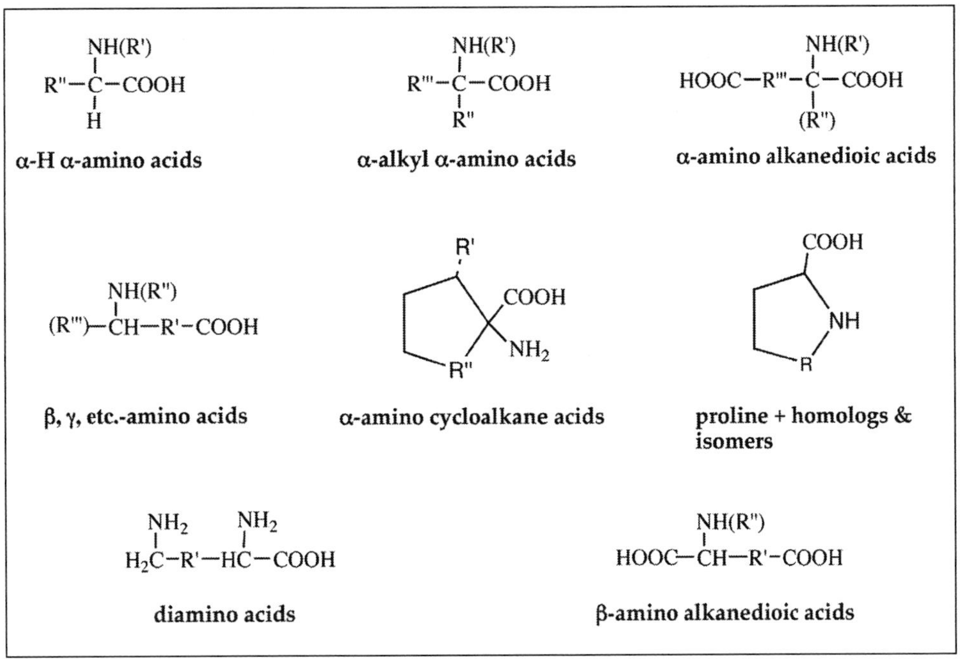

α-H α-amino acids

α-alkyl α-amino acids

α-amino alkanedioic acids

β, γ, etc.-amino acids

α-amino cycloalkane acids

proline + homologs & isomers

diamino acids

β-amino alkanedioic acids

Scheme 6.1. Diverse molecular composition of the organic compounds in meteorites.

6.3 Complex Cosmic History of Meteorite Organics

Overall, the analyses of meteorites' organic compositions tell us that abiotic syntheses are capable of producing a large variety of organic materials of remarkable complexity, including compounds having identical counterparts in terrestrial biomolecules. The question of how they were formed is of importance to Astrobiology because its answer would tell us which abiotic synthetic regimes outside the Earth could produce biomolecules and this knowledge would in turn allow us to elaborate on the cosmic distribution of organic molecules as well as their possible importance for the origin of life on Earth and elsewhere.

Given the complex and diverse molecular composition of the organic compounds in meteorites, the question of their origin is still open and currently being investigated for many of them. Meteoritic amino acids are probably the ones that have been studied in greatest detail, as they have terrestrial counterparts in terrestrial biopolymers (the proteins), display ee and could have been prebiotically significant upon their delivery to the early Earth. Even these compounds, however, are of diverse composition, as seen in Scheme 6.1 and the mechanisms for their formation might have been diverse as well. A possibility proposed for the α-amino acids is that they might have formed from the reaction of aldehydes and ketones with ammonia and HCN during a period when the asteroidal parent body was exposed to liquid water (Scheme 6.2).

This synthesis does not lend itself to the direct formation of asymmetric products or to the formation of non α amino acids, where pathways need to be postulated. For β-amino acids, it has been proposed that they could be derived from unsaturated nitriles by the addition of ammonia to the double bond(s), as from acrylonitrile in the case of β-alanine (Scheme 6.3).

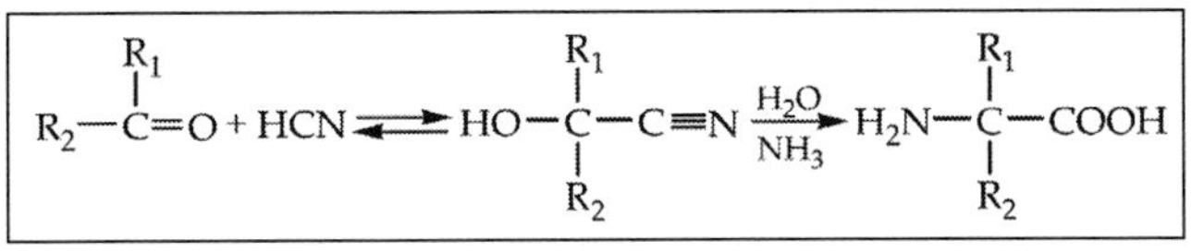

The study of the isotopic composition of meteorite organics has shed much needed light on the subject of meteoritic compound formation, for the reason that the isotope distribution in the molecules represents a good tracer of their synthetic history.

Isotopes are forms of an element that have different masses. The nuclei of all elements are composed of protons and neutrons; the protons, equal in number to the electrons, determine the identity of an element, but an element, may be present with different numbers of neutrons and therefore nuclear masses. All the biogenic elements, C, H, O and N, have isotopes, which are stable and are represented by the common symbol of the element with the mass at the upper left. They are ^{13}C, ^{2}H (which has a name and a symbol of its own, deuterium and D), ^{15}N, ^{17}O and ^{18}O; the isotopes ^{3}H (tritium) and ^{14}C are also known but both undergo radioactive decay.

The isotopes of an element, having different masses, form bonds with other elements or isotopes that are endowed with different energies. Therefore, during reactions where bonds are broken and formed (e.g., between H and C), the isotopes of these elements may reach in the products a different distribution than that in the reactants, depending on the type and conditions of those reactions. This change in distribution is referred to as mass dependent isotopic fractionation and often renders the isotope composition of molecules a good tracer of their synthetic history.[d] Isotopic fractionation is measured in terms of differences in the ratios between the heavier and lighter elements (D/H, $^{13}C/^{12}C$, etc.); it is directly proportional to the difference in masses between the isotopes and inversely proportional to the reaction T. It is expected to be largest for D/H at very low T because D/H is the largest of stable isotope mass ratios ($2/1 = 2$, compared to $13/12 \approx 1.08$ for C and $15/14 \approx 1.07$ for N).

A striking confirmation of these expectations is given by the spectroscopic measurements of large D enrichment for molecules in the dense clouds of the interstellar medium (ISM). These clouds are the vast expanses of gas and dust found, as the name says, in the space between the stars, where they are held loosely together by gravity and can span up to light years. The ISM was

Scheme 6.2.

$$CH_2=CH-C\equiv N \xrightarrow{NH_3} NH_2-CH_2-CH_2-C\equiv N \xrightarrow{H_2O} H_2N-CH_2CH_2-COOH$$

Scheme 6.3.

built up through millions of years by the material of generations of stars now dead and it is from these clouds that new generations of stars are constantly formed. Because chemical reactions also take place within these clouds, the ISM represents a focal point for chemical evolution in the universe, a sort of cosmic recycling center through which not only each generation of stars would start with a different pool of elements from the one that preceded them, but the medium would also continuously evolve in its elemental, isotopic and molecular composition (see also Chapter 7).

The temperatures of the ISM can be extremely low (5-10 K) and reactions between the elements (hydrogen, mainly, but also oxygen, nitrogen, carbon and sulfur) are restricted to those that do not need activation energies and are exothermic, such as reactions involving ions and molecules in the gas phase of molecular clouds. Such reactions are driven preeminently by cosmic rays, which produce the ions and directed by the fractionation rule mentioned above: the larger the energy difference between the isotopes and the lower the temperature, the greater is the enrichment of heavy isotope in the organic molecules formed under these conditions. In the ISM, in fact, D/H ratios have been observed to be far higher than those on the Earth (for example interstellar DHO/H_2O of 0.01 while the average Earth value is $1.5\ 10^{-4}$).

Measurements of the isotopic composition of meteorite organics have shown that meteoritic compounds are also enriched in D, ^{13}C and ^{15}N to various degrees. The D/H values of some individual compounds are particularly high and close to those determined spectroscopically for interstellar molecules (Table 6.3).[3] These isotopic data have been interpreted in terms of enrichment of

interstellar molecules and led to the following general theory for their formation.

The interstellar-parent body hypothesis proposes that icy planetesimals formed at asteroidal distances from the Sun and incorporated abundant volatile compounds, water and deuterium-rich interstellar organics. Warming of these bodies, possibly due to isotope decay or impacts and the subsequent aqueous phase chemistry yielded the large variety of organic compounds and matrix water alteration features of meteorites. Loss of volatiles and retention of nonvolatile material eventually followed.

This outline, although general, agrees with many of the molecular and isotopic features of the organic compounds of meteorites. The scenario has been refined on the basis of the isotopic composition of individual organic compounds (Table 6.3) with the understanding that planetary processes seem to have affected some groups of molecules more than others, i.e., the stage of interstellar formation was more advanced for some compounds, such as the branched-chained amino acids.

6.4 Astrobiological Questions

We do not know as yet how life came to be. On the other hand, we have learned a great deal of the chemical processes that followed and preceded the origin of life and we know that both were evolutionary processes. We have learned from paleontological records and comparative studies of macromolecules that life started much simpler, evolved through the ages and is still evolving today. We have also come to know that the biogenic elements have a long cosmic history of chemical evolution, throughout which they were often present in molecular forms sometimes quite complex. It is reasonable, therefore, to ask whether these evolutionary processes were ever linked. Had this been the case, the meteorites would have been the last recipients of cosmic chemical evolution prior to the onset of life and are most suitable to answer the many questions we may have about its "chemical" origin.

What does the chemistry of meteorites tell us? As it can be easily seen by comparing the data in Tables 6.1-6.3 with even elementary biochemistry, there seems to be a basic difference between the distributions of organic compounds in meteorite and the biosphere. Meteoritic compounds are characterized by large diversity (all isomeric compounds often being present) and appear to be products of random syntheses; biomolecules, instead, are the result of strict compositional selection (e.g., as mentioned, only twenty amino acids make up the whole of terrestrial proteins) and display functional specificity. It is hard to propose that such a varied "soup" of organics could have had any advantage in prebiotic chemical evolution without "some" aid or induction toward molecular selectivity.

The possibility of forming cell-like enclosures may lead to containment of useful molecules and is the basis for several hypotheses of molecular evolution.[6] Quite possibly, however, the finding of ee in meteoritic amino acids represents the most desirable prebiotic trait ever found in any sample of abiotic materials, be it extraterrestrial or the product of laboratory syntheses. This is because chiral homogeneity of biopolymers is an essential property of extant life (Chapter 7) and it is sensible to suggest that its development was important to the origin and/or evolution of life. The finding of

Table 6.3. Averaged deuterium enrichment of some of the soluble organic compounds in the Murchison meteorite

Compound	δD (‰)* Values
α-amino acids, linear	1800
α-amino acids, branched	3200
β-, γ-, etc. amino acids, linear	760
β-, γ-, etc. amino acids, branched	1800
Carboxylic acids, linear	650
Carboxylic acids, branched	1900
Dicarboxylic acids	1000

*While astrochemists measure D enrichments of interstellar molecules in terms of D/H abundances, biochemists and geologists prefer to measure the difference between the D/H ratio of a sample and that of a standard and express that difference as permil of the standard. The procedure is similar to the one used to measure percent values, i.e., $[(R_{sample}-R_{standard})/R_{standrd}]\times 1000$. This measurement is called the δ for the heavy isotopes of a given element and allows the estimation of smaller isotopic differences without the use of cumbersome ratios. The standard D/H is mean ocean water, which is 0 by definition. With this representation, the δD value of the above mentioned deuterated interstellar water becomes 65,667 and the ratios of all measured interstellar molecules vary from 5,400 to 38,440. The less studied meteorite CR2 found on the Antarctica ices has recently shown a value of 7,200 for one of its amino acids.

ee is far from definitive, however, and itself leads to several questions: what is the ee scope in meteorites? How were they formed? How could the delivery of asymmetric molecules have helped in molecular evolution?

All are the basis of current astrobiological investigations and the subject of suggested reading.[7] The first two questions have been addressed in part in the previous section. As for the third, several points have been made on the possible advantage of the nonracemic compounds in meteorites regarding the molecular evolution pathways on early Earth.

1. The delivery of extraterrestrial organic material by meteorites, micrometeorites and cometary fragments is abundant today (10^7 kg carbon year^{-1}) and was even larger during the period of early Earth bombardment; it is expected that a significant amount of these compounds could reach suitable Earth environments.

2. The ee-carrying amino acids of meteorites do not racemize in water (Chapter 7) and one of the difficulties in foreseeing the development of chiral homogeneity in prebiotic scenarios is the ease with which some essential molecules, such as protein amino acids and sugars, racemize in water. Thus these meteoritic compounds would have had an advantage in the early phases of molecular evolution by preserving their chiral asymmetry.

3. Amino acids are known catalysts and have been shown to transfer their asymmetry to sugars during syntheses from simple precursor molecules;[8] this effect is suggestive of possible prebiotic pathways toward the dissemination and amplification of chirality.

6.5 Synopsis

The study of carbonaceous chondrites has been important to Astrobiology because these meteorites offer a unique sample of extraterrestrial material for the direct analyses of the abiotic organic chemistry that preceded the onset of terrestrial life, its range and its complexity. They also allow an evaluation of the possible advantages this exogenous material might have brought to prebiotic molecular evolution upon its delivery to the early Earth.

Notes

a. This is the year when NASA laboratories in the USA were preparing for the return of lunar samples and, although the analysis of this meteorite was considered at first only a "practice", the instrumentation was ready and up to date to reveal quickly the importance of the organic composition of meteorites (lunar samples instead turned out to be devoid of any organics).

b. Spectroscopy is an analytical procedure that employs radiation (light as well as particle) absorbed or emitted by a molecule/molecular system to measure quantized energy transitions that characterize them.

c. The differential photolysis by UV CPL is expressed by the anisotropy factor **g**, given by the ratio between the difference in CPL absorption by the two enantiomers (indicated by their extinction coefficients) and the average total absorption, i.e., the larger this difference and the **g** for a molecule, the larger will be the difference in the extent of photolysis between the two enantiomers and the resulting ee. The **g** is a strict physical parameter that dictates the extent of enantiomeric excesses achieved by a given chiral molecule over the duration of the reaction and before both enantiomers are completely decomposed. For amino acids **g** is quite low, 0.02, restricting the possible enantiomeric excesses by UV CPL to a maximum of about 10%.

d. In searching for fossil records of ancient life forms, for example, biological material is recognized by the low $^{13}C/^{12}C$ ratios of the carbon formations under analysis because life is known to be rather lazy and to prefer the lighter ^{12}C in its metabolic reactions or build-up of biopolymers.

References

The following references are quoted in the text to provide further reading in the subjects indicated in their title as well as specific references to analytical papers in the literature.

Further Readings

1. Taylor SR. Destiny or chance, our solar system and its destiny in the cosmos. Cambridge University Press, Cambridge UK, 1998.
2. Meteorites and the Early Solar System II, DS. Eds. Lauretta and HY McSween Jr., University of Arizona Press, Tucson, 2006.

Specific References

3. Pizzarello S, Cooper GW, Flynn GJ. The nature and distribution of the organic material in carbonaceous chondrites and interplanetary dust particles. Ibidem 625-651.
4. Deamer D. Boundary structures are formed by organic components of the Murchison carbonaceous chondrite. Nature 1985; 317:792-794.
5. Cronin JR, Pizzarello S. Enantiomeric excesses in meteoritic amino acids. Science 1997; 275:951-935.
6. Bachmann PA, Luisi PL, Lang J. Autocatalytic self-replication micelles as model for prebiotic structures. Nature 1992; 357:57-59.
7. Pizzarello S. The chemistry of life's origin: A carbonaceous meteorite perspective. Acc Chem Res 2006; 39:231-237; and references therein.
8. Pizzarello S, Weber AL. Prebiotic amino acids as asymmetric catalysts. Science 1997; 303:1151.

Chirality, Homochirality and the Order of Biomolecular Interactions

Sandra Pizzarello*

7.1 Definition and Historical Background

Chirality is defined formally as the property of objects that cannot be brought into congruence with their mirror image by translation or rotation. The word derives from the Greek χειρ that means hand and, in fact, the familiar name of "handedness" may explain this property more readily. Like the hands, chiral objects are found in two forms that are mirror images of each other and are identical in their component parts but do not match when super-imposed because these parts have a different spatial distribution. Trying to put a right hand in a left hand glove or a left foot in a right foot shoe will give a convincing demonstration. Chirality is widespread in nature in a variety of two- and tri-dimensional forms (Fig. 7.1) and can also be associated with motion, for example, when an object describes a helix that can equally cross left to right or right to left such as in a screw.

A script looked in a mirror gives a good illustration of chirality, as well as a hint on how to detect it (Fig. 7.1A). Letters that are symmetrical about a line or a point (H, O, I, etc.) will be the same on the script or the mirror (i.e., not chiral), while the ones that are asymmetric (C, L, R, etc.) look different in the mirror and are chiral. Of the three dimensional forms shown in the figure, tartrate crystals (Fig. 7.1B) are chiral for the different distribution of their asymmetric surfaces and have historical importance, as they were the first organic molecules discovered to give chiral structures by Louis Pasteur (1822-1895).[1] Pasteur was an eclectic and pioneering experimental scientist in many fields and, early in his career, came to analyze the crystals that commonly deposit at the bottom of wine barrels as tartar, as it was called, which was known to be made up of tartaric acid salts. He found that these salts could crystallize in the two mirror-image forms shown in the figure when synthesized in the laboratory but gave only one of the two when formed naturally during wine processing. This appeared at first contradictory, because both types of crystals had the same chemical formulas, but Pasteur had the insight to frame his observations on the basis of the discoveries of polarized light and its interaction with matter that had taken place during the 17th and early 18th centuries.

If we consider light as a wave motion of magnetic and electric fields oscillating in planes perpendicular to each other and to their line of propagation (Fig. 7.2A, H and E respectively), these oscillations will occurs in innumerable planes when light is not filtered (Fig. 7.2B, viewed from a point along this line). However, light is found to emerge polarized on a single plane after passing through a transparent crystal of the mineral calcite (Fig. 7.2C). It has also been shown that the plane of this polarized light is rotated by passage through certain "optically active" substances. By experimenting with tartrate crystals, Pasteur came to demonstrate that solutions of the natural, one-handed tartrate crystals were the only ones to rotate polarized light while a mixture of the two forms did not. In doing so, he made two profound discoveries: that organic molecules, not just crystals, can have two spatial structures and that terrestrial biomolecules show only one handedness—that is, they are chirally homogeneous or homochiral as it was later determined.

7.2 Chiral Asymmetry of Organic Compounds and Biological Homochirality

As we know now, in fact, a common and most interesting case of chirality is that displayed by the compounds of carbon (C). This element has four binding valences and the spatial distribution of its bonds is such that a C in a molecule can be visualized as a regular tetrahedron with C at its center and the four bonded groups at the vertices (Fig. 7.1C,D). When all these groups differ, the C becomes asymmetric (C*) and the molecules that contain such a carbon atom are chiral, i.e., they can exist in two "handed" forms called enantiomers. The enantiomers of a chiral molecule have matching chemical properties and will interact equally with nonchiral molecules and structures. However, they do not have identical physicochemical properties, because of the different spatial distribution (configuration) of the chiral carbon substituents groups and the different polarizability of its electrons that this difference entails. For example, solutions of the two enantiomers of a chiral molecule will rotate polarized light in opposite directions and it is only when they are present in equal amounts and compensate each other's effect that solutions become optically inactive (this equal mixture of enantiomers is called racemic, as the synthetic tartaric acid Pasteur analyzed). Figure 7.1D, gives the Fisher projection formula of glyceraldehyde, which is the simplest chiral sugar whose enantiomers are designated by convention as D- and L-, from being dextro-rotatory and levo-rotatory for polarized light, respectively. Glyceraldehyde is also the reference compound in the designation of other sugar and amino acid enantiomers.

One difference between enantiomers, which derives from the above properties and is most important to biochemistry, is that they will interact differently with other chiral molecules. Even for the same compound, enantiomers will behave differently upon encountering enantiomers of the same or opposite configuration, for example, the interaction/reaction of an L-amino acid with its L- or D- enantiomer will differ and the interacting DD and DL complexes will have different energies.

*Sandra Pizzarello—Department of Chemistry and Biochemistry, Arizona State University, Tempe Arizona 85287-1604 USA.
Email: pizzar@asu.edu

Prebiotic Evolution and Astrobiology, edited by J. Tze-Fei Wong and Antonio Lazcano. ©2009 Landes Bioscience.

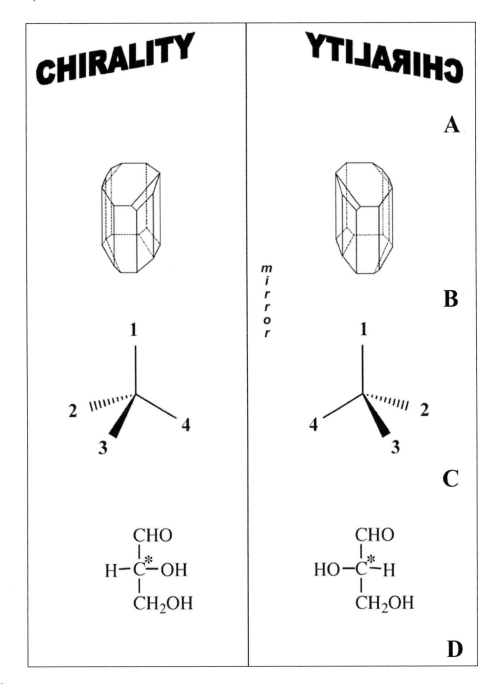

Figure 7.1. Chirality.

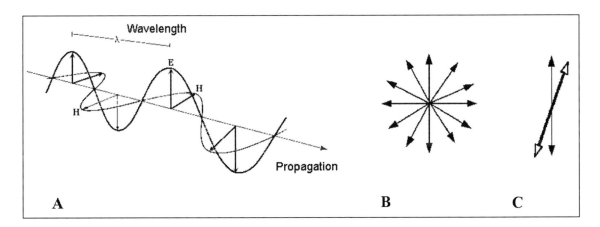

Figure 7.2. Light's electromagnetic vectors, their motion and directions.

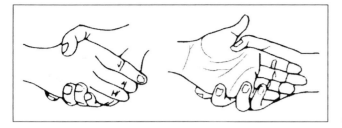

Figure 7.3. "Right" and "Wrong" handshakes.

Again, an analogy with the hands may be helpful. Disregarding their common difference in strength, the left and right hand will pick up a symmetrical object such as a ball equally. However, when confronted with a chiral object, another hand (Fig. 7.3) or a rotary valve for example, they will act with different ease towards the two forms of this object depending on their spatial match.

Essentially, therefore, chirality is an interactive property of molecules as well as of objects and it is as such that it becomes a prerequisite of life's molecules. Because extant life is based on carbon chemistry[a], many among the myriads of organic compounds that make up the biosphere and are a result of the unique ability of carbon to bond to itself and other atoms have at least one asymmetric carbon. It also happens, as discovered by Pasteur, that extant life makes a precise use of chiral compounds by building its polymers such as proteins, RNA, DNA, polysaccharides with monomers of only one handedness, L-amino acids and D-sugars. Many metabolic activities are also dependent on precise chiral interactions, with the result that chiral configuration determines many of the chemical communications within our body and also with the environment. The latter include the effect of medications, of which there are innumerable examples that vary from the tragic, when only one enantiomer of the drug thalidomide taken by pregnant women caused birth defects in the early '60s, to the amusing as in the case of the compound limonene that smells like lemon-orange in the R- and pine in the S-enantiomer configuration (only sugar and amino acid enantiomers take the D- and

L-notations, whereas all other organic compounds are indicated by R- and S- from the Latin rectus and sinister[b]).

This chiral homogeneity of biomolecules, called homochirality, appears essential to life as we know it. If we stipulate that life depends on the function of its polymers (see Chapter 6), it is easy to understand this essential role. Since the activity of a biopolymer is dependent on its spatial structure, the functional specificity of its superstructure is determined by the homochirality of the monomeric constituent molecules. The helices of proteins and DNA, the sheets into which proteins may organize and the precise lock-key combinations of substrate and enzyme of extant life are achieved thanks to one-handed monomers. Could life function with a different chiral organization of biopolymers—for example, could protein enzymes be built with D- and L-amino acids instead of being homochiral?[2] As long as protein sequences are dictated in a precise and reproducible way, as L-amino acids are in contemporary enzymes, we may assume that D- and L-monomers could also yield some exact and appropriate spatial structure, with a matching active site for the targeted molecule on which they perform their function. Such structures may turn out to be less economically built than extant homochiral ones, but there seems to be no reason to believe that they would be disruptive to a life system. However, this is not always the case, with DNA offering the primary example (Scheme 7.1). This biopolymer is formed by double stranded molecules twisted into a helix, with each strand comprised of a D-sugar-phosphate repeat-backbone to which the four bases of adenine, thymine, cytosine and guanine are attached in precise complimentarity. The bases face and exchange H-bonds with each other across the strands and maintain the integrity of the helix. Changing the chirality of any of the sugars along the strands would disrupt this regular ordering and complimentarity, making the double helical structure impossible. Because the fundamental biological processes involving DNA, such as transcription and replication, rely upon the maintenance of this specific structure, any alteration of its structure would also impede its function. For life as we know it, therefore, we have to conclude that homochirality is an essential trait of terrestrial biomolecules.

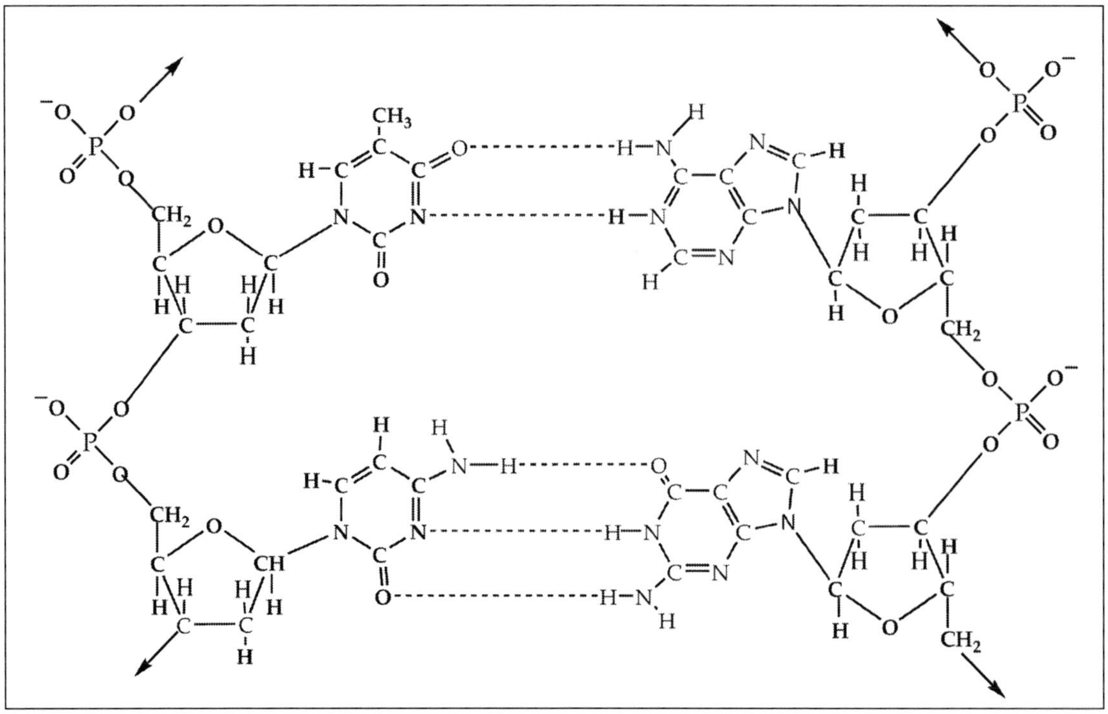

Scheme 7.1. The homochirality of the sugar components of DNA and RNA allow the bases to face and exchange H-bonds with each other across strands and maintain the integrity of the helix.

7.3 Hypotheses on the Origin of Terrestrial Homochirality

How did this homochirality originate? Was homochirality indispensable to the origin and/or development of life? Was it first widespread or limited? These are all Astrobiological questions that have been debated since the time of Pasteur and parallel in mystery those concerning the origin of life. Both biotic and abiotic theories have been put forward.[3] Biotic theories propose that life formed from achiral and/or racemic molecules and that chiral selection developed and evolved through time, while abiotic theories propose that a tendency toward homochirality was inherent in abiotic chemical evolution prior to terrestrial life (the finding of L-enantiomeric excesses in meteoritic amino acids appear to support, albeit not prove, the latter proposal—see Chapter 6).[2,3]

Biotic theories are by their very nature difficult to prove because of the uncertainties surrounding the origin of life and the early Earth environments that fostered it. The main rationale for proposing an evolutionary path toward homochirality is that both amino acids and sugars racemize easily in water[c] and small initial enantiomeric excesses would be easily lost unless life itself possessed mechanisms for producing chiral homogeneity.[c] It should be noted here that the meteoritic amino acids with enantiomeric excesses do not racemize because they lack a H next to the carboxyl group, whose removal and re-insertion in water would lead to racemization and that a laboratory model has been constructed with achiral nucleic acid analogues that can carry a linear array of bases and form helical structures.[6]

Abiotic theories face a different problem in that no molecular asymmetry has been measured outside the terrestrial biosphere, with the exception of the already mentioned enantiomeric excesses of meteoritic amino acids. The proposals for a prebiotic origin of homochirality therefore have either to rely on the abundant exogenous delivery of meteorites and comets to the early Earth (Chapter 6) or to postulate means for both symmetry breaking and the amplification of this initial imbalance. Causes intrinsic to matter and chance or determinate events have been invoked by the latter group of proposals, some of which are listed in the following paragraphs.

The most attractive and certainly universal, idea is that of an intrinsic energy difference between enantiomers of a chiral molecule, which has been proposed on the basis of theoretical and analytical studies. As mentioned above, chiral molecules have identical chemical properties, meaning that the motions and interactions of molecules, which are governed by the electromagnetic force, do not vary upon spatial reflection and could take place with equal probabilities as their mirror images. It was found,[4] however, that this is not true for all the known forces acting at the subatomic level of a molecule, which are gravity, the above-mentioned electromagnetic force, the strong nuclear force (the strongest yet known that holds nuclei together) and the weak nuclear force.

It is with the much weaker latter force that a handed bias is found: its best-known effect is the production of beta-rays during radioactive decay—these rays contain more electrons with a left-handed spin than with a right-handed one.

This finding has been interpreted to mean[d] that the enantiomers of a molecule cannot have exactly the same energies, which generates two types of proposals. I) The energy difference between enantiomers could be magnified through the numerous abiotic and prebiotic reactions of chemical evolution, eventually leading to homochirality. The drawback of this type of proposals is that the theoretically predicted energy difference is expected to be infinitesimally small and has not yet been demonstrated for any specific molecule either theoretically with a satisfactory consensus, or analytically. II) That the electrons produced during beta-decay could affect the syntheses of prebiotic molecules to yield a left-handed bias (see below).

Although less than universal, there are other frequent cases in which enantiomers are spontaneously separated and then undergo discrete differentiating processes with the enhancement of one enantiomer over the other. Crystallization offers good examples. Depending on the physicochemical conditions affecting their supersaturated solutions, chiral compounds may give crystals containing both enantiomers or just one of the two enantiomers (as Pasteur found out for tartaric acid). There are also crystals such as quartz that are chiral not because of the asymmetry of their constituent molecules (SiO_2) but because of the asymmetric distribution of their faces.

It has been proposed that chance predominance of one type of crystals in localized environments may lead to chirally biased processes, such as the adsorption of other chiral molecules (for example, amino acids). It is also known that during the initial stages of the crystallization process the initial type of chiral crystal formed (by chance again) may "nucleate" the successive crystallization of the solution and lead to an essentially homochiral batch of crystals.[7] Please remember that chance events could give an equal advantage to either enantiomer and lead globally to a stochastic distribution of localized environments of each handedness (where eventually one form will have to "win" over the other form to bring about global homochirality).

Other proposed effects for the enantiomeric enhancement of a racemic chiral compound depend on its interaction with chiral structures, molecules and forces. Examples are the asymmetric adsorption on clays and crystals such as quartz, as mentioned; kinetic resolution during polymerization, where molecules exploit the different energies of homochiral and heterochiral intermediates for homochiral enhancement (see Section 7.2); and encounters with circularly polarized light and beta-rays. While most of the hypotheses based on these effects are still indirectly related to chance (equal numbers of d- and l-quartz crystals, equal production of D- and L-proteins, etc.), the effects of the chiral forces may be more relevant to broader abiotic environments. Beta rays we know have a universal applicability and circularly polarized light has been observed in the vast expanses of the interstellar medium (see Chapter 6).

Intermediate between the biotic and abiotic theories stands the theory of Panspermia, meaning literally universal seeding, by which life was brought to the Earth from extrasolar bodies (with the help of solar winds, as first proposed by Arrhenius in 1908)[5]. If these "seeds" or spores had a homochiral make-up, they would solve the puzzle of the origin of homochirality on Earth. However, the more general puzzle still remains because panspermia simply pushes it back to another time and place along with other questions about the origin of life.

7.4 Synopsis

Chirality, or handedness, is a property of objects as well as molecules that lack a center of symmetry and differ only in the spatial distribution of their identical constituent parts. Great many of the molecules of carbon are chiral and their inherent asymmetry has provided extant life with special interactive advantages for the build-up, structure and function of its biopolymers, which are homochiral. This chiral homogeneity is essential to terrestrial life, whose onset it also might have helped to initiate. The search for its origin is important to the study of Astrobiology.

Notes

a. The term organic given to the chemistry of carbon, in fact, is based on the 18th century notion that only live organisms could produce its compounds.

b. The absolute configuration of organic compounds is obtained by the "sequence rule": (I) rank the four substituents of C* (see Fig. 7.1.d) by their atomic numbers (lowest number, lowest rank); (II) figuratively orient the chiral molecule in space so as to look down the C bond with the smallest rank (you will see the remaining three substituents radiating like spokes from the asymmetric C); (III) now if you can connect from the top through the middle to the lowest rank substituents with a counterclockwise (left, sinister) route the molecule is S; if the route is clockwise (right, rectus), it is R.

c. The carbon atom next to a carboxylic (COOH) acid function is slightly acidic, because of electron withdrawal by the nearby oxygen atoms and tends to loose a H^+ in water. The ion that forms can then re-acquire an H, a process that may result in a change of configuration. With time and appropriate conditions, this process will lead to racemic mixtures.

d. The interpretation of a difference in energy between enantiomers (referred to as parity-violating energy difference or PVE) derives from a theory of physics that unifies the electro magnetic and weak forces. The force resulting from this theory will be able to interact with both nucleus and electron and distinguish between "left" and "right".[4]

References

Further Readings

The following references are quoted in the text to provide further reading in the subjects indicated in their title as well as specific references to analytical papers to be found in the literature.

1. Geison GL. The private science of Louis Pasteur. Princeton University Press, Princeton, New Jersey, 1995.
2. Cronin JR, Reisse J. Chirality and the origin of homochirality. In: Gargaud, M. ed. Lectures in Astrobiology. Vol. 1; Berlin: Springer-Verlag, 2005:473-514.
3. Bonner WA. The origin and amplification of biomolecular chirality. Origin Life Evol Biosphere 1991; 21:59-111.
4. Hegstrom RA, Kondepudi DK. The handedness of the universe. Scientific American 1990:108-115.
5. The web-site: www.daviddarling.info/encyclopedia/P/panspermia.html from "The Encyclopedia of Astrobiology, Astronomy and Spaceflight" offers a descriptive entry on the subject of Panspermia.

Specific References

6. Nielsen PE. Peptide-Nucleic Acid (PNA)-A model structure for the primordial genetic code. Origins Life Evol Biosphere 1993; 23:323-327.
7. Kondepudi DL, Kaufman RJ, Singh N. Chiral symmetry breaking in sodium chlorate crystallization. Science 1990; 250:975-976.

CHAPTER 8

The Primitive Earth

James F. Kasting*

8.1 Introduction

Understanding how life may have originated necessarily involves understanding the environment of the early Earth. This statement is true even if life originated somewhere else and was transported to Earth—an idea referred to as "panspermia". Even in this, to some extreme, version of life's origin, it is of interest to know what the early environment of Earth might have been like, if only to predict how life might have adapted once it got here. For more conventional models, in which life originates somewhere on Earth's surface, it is even more important to understand what that surface was like. Here, we review current thinking about the origin of the Earth, along with its atmosphere and oceans. We then discuss what is known and what is not known, about early atmospheric composition and climate. Finally, we touch briefly on the effects of early life on the atmosphere. This last topic is not directly relevant to prebiotic evolution. However, it fits in with the overall picture, as one of the best ways of investigating the origin of life is to try to work backwards from the earliest organisms and to do so we need to understand how they interacted with the world in which they lived.

8.2 Formation of the Earth, Its Atmosphere and Oceans

(i) Historical Overview

The nature of the early atmosphere and oceans has long been of interest to researchers studying life's origin, and popular ideas of the early Earth are still dominated by some of the early thinking on this topic. A good place to begin is the book by the Russian biologist A. I. Oparin, first published in Russian in 1924 and later expanded in an English version in 1938.[6] Similar ideas were published in 1929 by J.B.S. Haldane in England, and so the resulting model for life's origin and for the nature of the early Earth is often termed the "Oparin-Haldane" hypothesis. In this model, the early Earth formed from the condensation of superheated gases torn from the Sun by a near-collision with another star—a mechanism that is by now considered highly unlikely. Without going into the details, which are by now irrelevant, this model predicted that Earth's earliest atmosphere was highly reduced, that is, it was rich in gases such as molecular hydrogen (H_2), methane (CH_4) and ammonia (NH_3), each of which can react with oxygen. The early atmosphere was hot in Oparin's model and the thermal energy from the atmosphere caused these reduced gases to react with each other to form a thick broth, or "soup", of organic compounds, from which life was thought to have emerged.

Although his theory for the Earth's formation was subsequently shown to be implausible, Oparin's model of a highly reduced primitive atmosphere received additional support several years later from work done by the University of Chicago geochemist, Harold Urey and his then-graduate student, Stanley Miller. (Stanley Miller passed away, unfortunately, as this chapter was being written. Those of us who knew him will miss him greatly.) Urey was a reknowned geoscientist whose research interests included spectroscopic observations of Jupiter and Saturn. Urey demonstrated from these observations that the atmospheres of these giant planets were rich in methane and ammonia. Urey was also aware that, because these planets are very massive, they have retained virtually all of the original hydrogen that they acquired during their formation. By contrast, smaller planets like Earth, Venus and Mars can lose hydrogen to space, and this causes them to become more oxidized with time. Urey reasoned that, before Earth had sufficient time to lose its hydrogen, its atmosphere should have resembled those of Jupiter and Saturn. So, when his graduate student Miller suggested that he study the effects of electrical discharges (simulating lightning) on such a gas mixture for his Ph.D. thesis, Urey gave him the go-ahead. The rest of the story is widely known: Miller's experiments yielded a whole host of plausible prebiotic organic compounds, including amino and nucleic acids,[7] and the "Miller-Urey" model of life's origin became firmly entrenched in both the scientific literature and the popular scientific consciousness.

At almost the same time, however, another geochemist named William Rubey was developing an alternative model for the origin of Earth's atmosphere.[8] Rubey studied the composition of gases emanating from modern volcanoes. He found that they were composed largely of CO_2 and H_2O, rather than CH_4 and NH_3. If Earth's atmosphere formed from volcanic outgassing, as was thought then to be the case and if volcanic gas composition has remained unchanged, then the earliest atmosphere should have been dominated by CO_2 and N_2, rather than by CH_4 and NH_3. (N_2 is not easy to measure in modern volcanic emanations because it is difficult to differentiate volcanic N_2 from air, but Rubey's theory predicted that both carbon and nitrogen would have been outgassed in their more oxidized states.)

Rubey's theory was further elaborated by other workers, especially Heinrich Holland and James Walker. Holland developed a two-stage model, in which volcanic gases were at first more reduced and then subsequently became more oxidized after the Earth's core

*James F. Kasting—Department of Geosciences, Penn State University, University Park, Pennsylvania 16802, USA. Email: kasting@geosc.psu.edu

formed. We now believe that core formation occurred during the accretion process itself (see next section) and so Holland's highly reduced 'Stage 1' atmosphere may not have existed (or, if it did, it was for different reasons). Walker, in his papers and books,[1] developed the model for what is now termed a weakly reduced primitive atmosphere. In his view, the early atmosphere was dominated by CO_2 and N_2, as Rubey had suggested, but it was still virtually free of O_2, at least at ground level. Molecular hydrogen, H_2, was present at concentrations of roughly 100-1000 ppmv (parts per million by volume). Walker estimated the H_2 abundance by balancing the outgassing of hydrogen from volcanoes, where it is a minor constituent, with escape of hydrogen to space. This idea of balancing the atmospheric hydrogen budget, or redox budget, is critical to understanding early atmospheric composition and we shall return to it below.

If Walker was right and the early atmosphere was only weakly reduced, then our ideas of how life formed had to be modified. Miller-Urey type synthesis of prebiotic compounds is much less efficient in such atmospheres.[9] Small molecules such as formaldehyde (H_2CO) and hydrogen cyanide (HCN) may be synthesized photochemically in weakly reduced atmospheres;[10,11] however, formation rates of complex organic compounds such as amino and nucleic acids are very low. This perceived problem with the Miller-Urey model for life's origin led various researchers to propose alternative models. Prominent amongst these are the "hydrothermal vent" model, in which life originates from reduced compounds synthesized within midocean ridge hydrothermal circulation systems and the "seeding-from-space" model, in which complex organics come in preformed as components of meteorites and interplanetary dust particles (IDPs).

(ii) Formation of Atmospheres by Impact Degassing of Planetesimals

How secure is our understanding of early atmospheric composition? Do we really know that the early atmosphere was only weakly reduced? And, if so, what exactly does this mean?

The truthful answer to these questions is that large uncertainties still remain. The reason is that the formation of the Earth was a complex process. Perhaps most importantly, the atmosphere may not have formed primarily by outgassing from volcanoes. In Rubey's day, the Earth was thought to have formed slowly by accretion of small planetesimals. (Accretion is the name given to the process by which small solid bodies orbiting a young star collide and stick together to form larger bodies.) We now are fairly certain that this was not the case. Rather, the Earth formed relatively quickly, on a time scale of tens of millions of years and the latter stages of accretion involved impacts of planetesimals that were Moon-sized, or even larger. (The Moon's mass is about 1/80th that of Earth.) Indeed, the Moon itself is thought to have been formed by the glancing impact of a Mars-sized body with at least 1/10th of Earth's mass, or perhaps twice that amount.

When large planetesimals impact the surface of a growing planet, tremendous amounts of energy are released. Some of this energy would have been used to melt the Earth's surface. Indeed, there is convincing geochemical evidence for the existence of one or more magma oceans during Earth's earliest history. Such extensive melting may have been instrumental in allowing metallic iron to separate from molten silicates, thereby allowing it to percolate downwards to form Earth's core. (That is one reason why we now think that core formation occurred simultaneously with planet formation. Various geochemical tracers of core formation, such as tungsten isotopes[12] yield similar results.) The energy dissipated in the impacting body is even greater and so larger planetesimals were probably vaporized on impact. Any volatiles that they contained would thus have been injected directly into the atmosphere. (Volatiles are compounds, such as H_2O, CH_4 and NH_3, that have low boiling points and hence are easily vaporized.) This process is termed impact degassing. Impact degassing almost certainly played an important role in formation of both Earth's atmosphere and its oceans. The oceans would not have formed immediately, however. Because of the extreme heating of Earth's surface, H_2O would initially have been in the vapor phase, forming a dense steam atmosphere with a surface pressure 10 or even 100 times higher than today.[13]

But the atmospheric formation story is still more complex. Many volatile compounds, including water, are soluble in silicate melts and impacting bodies that arrived when the proto-Earth was small would not have been fully vaporized. Hence, some water and other volatiles would have been incorporated into the growing Earth, to be released later by volcanic outgassing. This prediction is confirmed by the observation that primordial (nonradiogenic)[3] He is being outgassed at the midocean spreading ridges. So, Rubey was at least partly correct. Furthermore, large impacts can also remove atmosphere by directly blowing it off into space—a process termed impact erosion. Determining whether a given impactor would have contributed to volatile accumulation or volatile loss is a tricky problem. This is especially true for the largest impacts. For example, the Moon-forming impact may have blown off Earth's entire pre-existing atmosphere and ocean, requiring that the whole process begin anew. Surprisingly, blowoff is most effective if the ocean was present in condensed form when the impact occurred, because shock waves from the collision could then have efficiently transfered energy to it, vaporizing the water and allowing the superheated vapor to blow off the atmospheric gas above it.[14]

What would have been the composition of an impact-induced atmosphere? At high temperatures, the gases produced by impact vaporization would have consisted of small molecules, such as H_2, H_2O, CO, NO and N_2. But as impact plumes cool, larger molecules become favored thermodynamically. The impact plumes themselves would have been highly reduced environments and so much of the carbon may have been converted to CH_4.[15] This is also true for smaller impactors that were buried at shallow depths without being vaporized.[16] The amount of CH_4 that may have been produced initially is staggeringly large. The total amount of carbon in Earth's current surface inventory, including carbonate rocks, is about the same as on Venus—enough to produce a CO_2 partial pressure of 60-80 bars. (Venus' CO_2 partial pressure is 90 bars.) If most of this carbon was originally present as CH_4, then the CH_4 partial pressure could in theory have been of the order of 20-30 bars. (One has to divide by ~3 to compensate for the lower molecular weight of CH_4 compared to CO_2.) This estimate is almost certainly too high, as some of the carbon from the impactors would have gone initially into CO and CO_2 and because CH_4 would immediately have started to be consumed by photochemistry. Some of it would have been oxidized to CO and CO_2, while some of it may have polymerized to form higher hydrocarbons. Even so, it seems clear that large amounts of CH_4 could have persisted for some time following accretion and again following subsequent large impacts.

This leads us to a difficult question: How frequent were large impacts during the early part of the Earth's history? As described in Box 8.1, we do not know the answer to this question, because it depends on unresolved issues concerning the rate of bombardment of the terrestrial planets during the first several hundred million years of Solar System history. Consequently, it is difficult to come to firm conclusions about what the atmosphere may have looked like at that time. In the next section, we attempt to summarize what we know and what we don't know.

Box 8.1. Was the Heavy Bombardment Continuous, or Was It a Pulse?

Scientists have known since the days of the NASA Apollo manned lunar missions (1969-1973) that many of the Moon rocks have age dates around 3.8-3.9 Ga. This was initially interpreted as a pulse of impacting bodies that all arrived around that time and it was termed the *late heavy bombardment*. Graham Ryder of the Lunar and Planetary Institute in Houston was a leading proponent of this idea and he remained so until his death.[17] Subsequent theorists, though, had difficulty understanding why such a pulse of impacts should have occurred at this relatively late date in Earth's evolution, when the bulk of the Earth was known to have formed by 4.4 Ga, or earlier. (U-Pb age dating of meteorites gives a uniform age of 4.55 Ga for the early solar nebula. The formation of the Earth by accretion is thought to have been complete within 100 million years after that time.) So, the revisionists suggested instead that the entire period between 4.5 Ga and 3.8 Ga represented a *heavy bombardment period*, during which the impact flux tailed off exponentially following the main accretion period. If so, then the continued, sporadic influx of incoming bodies may have affected atmospheric composition throughout its early history.

Within the last two years, however, a new dynamical model of Solar System formation has provided support for the original "pulse" theory of the late heavy bombardment.[18,19] The model is sometimes termed the *Nice model*, because several of the coauthors work in laboratories situated near the city of Nice, in southern France. (The name of the city is pronounced "neese", not "nice".) These authors used a sophisticated, celestial mechanics code to simulate the latter stages of planetary accretion. They started their simulation with a more-or-less evenly distributed swarm of Moon-sized planetesimals and then calculated their mutual gravitational interactions and collisions as they grew into planets. The 4 giant planets—Jupiter, Saturn, Uranus and Neptune –were assumed to be fully formed at the beginning of the simulation; however, they were placed at locations that were different from their present orbital radii. In the model, Jupiter was assumed to have started slightly farther away from the Sun than it is now and Saturn started slightly closer in. Uranus and Neptune were assumed to have formed just beyond Saturn's orbit, with Neptune being closer to the Sun than Uranus (the opposite of the situation today). All of these assumptions are plausible, although they are by no means a unique starting point for generating the present Solar System.

The results of the simulation are interesting and they just may be close to what actually happened. Jupiter migrated inwards in this scenario, as a consequence of its gravitational interaction with smaller planetesimals, while Saturn migrated outwards. At some time around 3.9 Ga, Saturn passed through the 2:1 resonance with Jupiter, where Saturn's orbital period was exactly twice that of Jupiter. (It is just slightly greater than that now.) The resonance excited both planets into highly eccentric (elliptical) orbits and this in turn affected the orbits of the two less massive giant planets, Uranus and Neptune. Both of these planets were pushed farther away from the Sun and Neptune moved from inside Uranus' orbit to beyond it. Although this complicated scenario may sound somewhat *ad hoc*, it is a direct consequence of applying Newton's laws of motion to the initial dynamical system and it is consistent with what we have learned about giant planet migration by studying extrasolar planets. The net result of the simulation was that Uranus and Neptune were suddenly thrown into the outer Solar System, which at this time was still filled with icy planetesimals that had not yet had sufficient time to accrete into larger bodies. Most of these planetesimals were subsequently scattered out of their original orbits, with some of them passing through the inner Solar System and a few of these impacting the Moon and the terrestrial planets. If this scenario, or something akin to it, is correct, then the original interpretation of the Moon rocks as representing a pulse of bombardment may well have been correct.

Fortunately, this is one debate that is likely to be settled some time in the relatively near future. NASA has plans to send astronauts back to the Moon within the next decade. With a larger collection of Moon rocks from a wider variety of locations, it should be possible to definitively answer the question of whether the heavy bombardment was continuous, or whether it was a relatively sudden, catastrophic event.

8.3 Prebiotic Atmosphere and Environment

The discussion above sets the stage for describing the composition of the prebiotic atmosphere. Obviously, given the uncertainties that have been mentioned, especially concerning Earth's bombardment history, it is not possible to provide definitive answers. However, we can still draw some tentative conclusions, which we list below in numbered form:

1. A conventional, highly reduced atmosphere composed principally of methane, ammonia and other reduced gases is unlikely. Ammonia, in particular, is unlikely to have been abundant because it is rapidly photolyzed and converted to N_2 and H_2.[20] The hydrogen escapes to space, leaving stable, triply-bonded N_2 as the major nitrogen-bearing gas. UV shielding by hydrocarbon haze appears unable to prevent this from happening.[21] To be sure, this result depends on the distribution of particle sizes and so it could change as more sophisticated haze models are developed. (More small particles would cause better UV shielding.) But, for the time being, the models suggest that Miller-Urey type synthesis would *not* have been an efficient method of making of prebiotic organic compounds.

2. Despite point 1, atmospheres rich in CH_4, CO and CO_2 remain plausible. CH_4 has a potentially large abiotic source from impacts, although this thought should be tempered by the discussion of Earth's impact history in Box 8.1. CH_4

also has a plausible abiotic source from the interaction of CO_2-rich seawater with ultramafic (Fe- and Mg-rich) rocks. This process, termed serpentinization, occurs within mid-ocean ridge hydrothermal circulation systems. Currently, this source of CH_4 is thought to be modest—enough to produce an atmospheric CH_4 concentration of only a few ppm on an anoxic Earth.[22] It could, however, have been larger in the past if more ultramafic rock was exposed as a consequence of higher geothermal heat flow and associated changes in plate tectonics. It is difficult to be quantitative about this because we do not really understand how plate tectonics operated in the distant past. An atmosphere containing CH_4 and H_2O would, of necessity, have also contained substantial amounts of CO and CO_2, because the methane would have been oxidized by the by-products of water vapor photolysis. This process is quasi-irreversible because the hydrogen escapes to space and because CH_4 is not reformed photochemically at any appreciable rate. Surface interactions, like the serpentinization reactions mentioned above, are required to regenerate CH_4. Some CH_4 could also have been photochemically converted into higher hydrocarbons, if the atmospheric CH_4/CO_2 ratio exceeded unity.[21] It is not clear whether this limit was ever achieved on the prebiotic Earth. On the postbiotic Earth, it is likely that this limit *was* reached and that hydrocarbon haze was indeed formed.[21]

3. As pointed out earlier, an atmosphere containing even a modest amount of CH_4 (10-100 ppmv), which is entirely plausible on the prebiotic Earth, should have generated HCN and H_2CO through photochemistry. These compounds are suitable starting points for the synthesis of amino acids, nucleic acids and sugars, although the details of how these complex compounds may have formed are in general not well understood. CO, which should also have been reasonably abundant (100 ppmv or more), is also a useful starting compound for prebiotic synthesis because its thermodynamic free energy is very high.[23] So, the prebiotic atmosphere should have provided a generous source of small precursor molecules that may have contributed to organic synthesis and the origin of life.

4. Exactly when all of this happened makes a difference, as atmospheric composition almost certainly evolved during the first several hundred million years of Earth history. Both CH_4 and CO_2 and perhaps H_2 as well, should have been more abundant at first, as a consequence of impact degassing during planetary formation. The concentrations of both gases should have subsequently declined, as CH_4 was converted into CO_2 (and perhaps into higher hydrocarbons) and as CO_2 was converted into carbonate rocks. Surface temperatures may have declined as well during this period, as the atmosphere became thinner. We return to this topic in the next section.

5. In all of these scenarios, large impacts should have occurred as late as 3.8 Ga. In the Nice model (Box 8.1), they were clustered around this time; in other models, they were spread out between 4.4 Ga and 3.8 Ga. By "large", we mean impactors that were comparable in size to those that formed the lunar craters Imbrium and Orientale, both of which are about 1000 km across. Forming a crater of this size requires an impactor that is roughly 100 km in diameter.[2] This is approximately 10 times the diameter, or 1000 times the mass, of the impactor that is thought to have killed off the dinosaurs at the end of the Cretaceous Period, 65 million years ago. A 100-km diameter impactor would evaporate the uppermost 100 m of the oceans, including the entire photic zone.[2] Hence, if life did evolve prior to 3.8 Ga, which seems entirely possible, it would likely have experienced multiple global catastrophes that would have preferentially killed off organisms living in the near-surface environment. Organisms living in subsurface environments, such as the midocean ridge hydrothermal vent systems, would have been largely unaffected by such events. As has been discussed frequently at origin of life meetings, this provides a possible explanation for the predominance of hyperthermophiles (organisms with preferred growth temperatures >80°C) near the base of the ribosomal RNA (rRNA) evolutionary tree.[24]

8.4 Archean Atmosphere and Climate

The Earth's rock record really begins at ~3.8 Ga. The period prior to that, which we have been discussing so far, is called the *Hadean Era*. The period between 3.8 Ga and 2.5 Ga is called the *Archean Era*. It is still poorly understood compared to more recent time periods; however, there are rocks of Archean age on all 7 continents and there is a vast literature describing what those rocks tell us about the Archean atmosphere and climate. We shall not, in this brief chapter, attempt to summarize all of this evidence. Instead, we focus on a few topics that may help the reader to understand what we do know

about the Archean Earth, as this may guide our thinking about the earlier period for which the rock record is absent.

(i) Archean Climate

The basic climate issue regarding the Archean and the Hadean, eras is often called the faint young Sun problem. The Sun is thought to have been about 30 percent less luminous at the time when the Solar System formed and to have brightened more or less linearly since that time (Fig. 8.1). This is a theoretical prediction from stellar evolution models, but it is nonetheless considered to be a robust result. (This claim is supported by the fact that the prediction has not changed much in the last 25 years.) The consequences for Earth's climate are clear: If atmospheric composition had not changed over the Earth's history, the mean global surface temperature would have been below freezing prior to about 2 Ga. This can be seen from the upper dashed curve in Figure 8.1. The lower dashed curve in Figure 8.1 represents the effective radiating temperature of the Earth, which can be thought of as the temperature that Earth's surface would have in the absence of an atmosphere. The shaded area between the two dashed curves represents the greenhouse effect of the model atmosphere, which in this case has been assumed to consist of present-day levels of N_2 and CO_2, along with variable amounts of H_2O. The greenhouse effect increases with time in this calculation because the atmosphere holds more water vapor as the surface temperature rises.

Possible solutions to the faint young Sun problem have been reviewed elsewhere.[3] The answer almost certainly involves higher concentrations of greenhouse gases in the past. Greenhouse gases are those that absorb and emit thermal infrared radiation between about 5 µm and 100 µm. The most likely candidates are CO_2, CH_4 and possibly C_2H_6 (ethane). The latter gas has only recently been suggested, but it is a powerful infrared absorber that may have provided 10 degrees or more of greenhouse warming. H_2O is also a powerful greenhouse gas, but its concentration is limited by its saturation vapor pressure and so it is a feedback on climate, as opposed to being a climate forcer.

The argument for higher CO_2 levels in the past is as follows: on long time scales, atmospheric CO_2 concentrations are controlled primarily by the inorganic carbon cycle, or carbonate-silicate cycle. CO_2 is injected into the atmosphere by volcanoes and it is removed by the weathering of silicate rocks on land, followed by deposition of carbonate sediments in the ocean (Fig. 8.2). The term weathering refers to the physical and chemical alteration of rocks through erosion and interaction with dissolved species in rainwater. The process of silicate weathering requires liquid water in order to proceed at an appreciable rate. Hence, if Earth's mean surface temperature were to fall below the freezing point of water, as suggested in Figure 8.1, silicate weathering should slow, or stop entirely and volcanic CO_2 should accumulate in the atmosphere. The accumulation would continue until the greenhouse effect of the added CO_2 became high enough to melt the ice and allow silicate weathering to proceed once again. Calculations[4] show that a CO_2 partial pressure of ~0.3 bars would have been sufficient to compensate for a solar flux reduction of 30 percent. The present atmospheric CO_2 concentration, by comparison, is approximately 300 ppmv, or 3×10^{-4} bars. Hence, the amount of CO_2 required to compensate for the faint young Sun is roughly 1000 times the present atmospheric level (PAL). Although this may sound high, it is only a small fraction of the available CO_2. Earth has the equivalent of 60-80 bars of CO_2 tied up in carbonate rocks on the continents. Only a tiny fraction of this inventory would have been needed to compensate for the faint young Sun.

On the other hand, CO_2 may not have been the only greenhouse gas whose concentration was high on the early Earth. We saw in the previous section that CH_4 could also have been abundant. This is

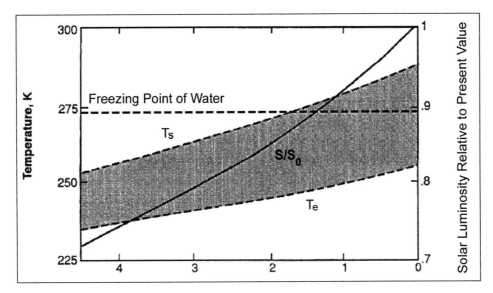

Figure 8.1. Diagram illustrating the faint young Sun problem. The solid curve is solar luminosity relative to the present day value. The lower dashed curve is the effective radiating temperature, T_e. The upper dashed curve is the mean global surface temperature, T_s. The shaded area in between the two dashed curves shows the magnitude of the greenhouse effect. (adapted from Scientific American, Vol 256, Kasting.[30] Copyright (1988) with permission from Scientific American)

particularly true after the origin of life, when organisms may have produced copious amounts of methane. The organisms that do so today are called methanogens. (Higher plants actually produce some methane, as well, but we shall ignore that because higher plants were not around during the Archean.) Methanogens are not true Bacteria; rather, they are all found on one branch of the the Archaeal domain on the rRNA tree (see Chapter 2). They are considered by most biologists to have evolved very early and so they were probably extant during the Archean era. Atmospheric O_2 levels were low at that time (see below), causing the photochemical lifetime of CH_4 to have been at least 1000 times longer than today (10,000 years instead of 10 years). The likely source of CH_4 in the Archean was from methanogens living in the oceans and in marine sediments.[22] Calculations suggest that the CH_4 source was comparable to today's source (ibid.). Hence, the CH_4 concentration, instead of being 1.6 ppmv, as today, may well have been over 1000 ppmv. At these concentrations, CH_4 is an effective greenhouse gas, generating approximately 10 degrees of greenhouse warming.[25] Furthermore, photolysis of methane in a low-O_2 environment forms ethane, C_2H_6, and ethane is an even better greenhouse gas. Hence, a combination of CO_2, CH_4 and C_2H_6 appears to be a plausible mechanism for keeping the Archean Earth warm.

Other proposed constraints on Archean climate come from geochemistry and geology. Oxygen isotopes in cherts (SiO_2) have been

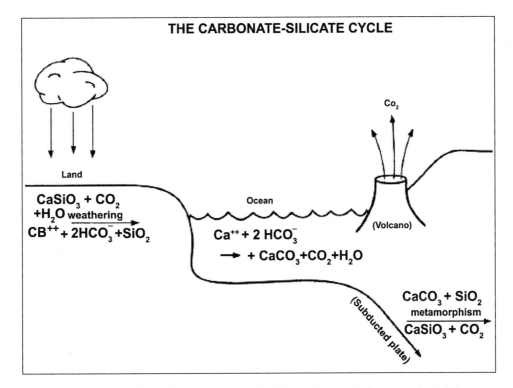

Figure 8.2. Schematic diagram illustrating the carbonate-silicate cycle. This cycle controls the atmospheric CO_2 concentration over long time scales.

used to argue that the Archean climate was hot (~70°C).[26] Indeed, the chert data and also O isotope data from carbonates, suggest that the Earth remained quite warm until as recently as 400 million years ago. On the other hand, geomorphic evidence indicates that the Earth experienced glaciations at 2.9 Ga, 2.4 Ga and 0.6-0.7 Ga.[27] These data are not at all consistent with the O isotope data for a hot early Earth. The likely resolution to this problem, in our opinion, is that the oxygen isotopic composition of the oceans has changed with time (ibid.). This topic remains highly controversial and it may be that this story will change over time. However, we would argue that, while the Archean climate was warm, it was nowhere near as hot as the O isotope data suggest.

(ii) Archean O_2 Concentrations

For many years, both geologists and biologists have been intrigued by the question of how much O_2 was present in the Archean atmosphere and when atmospheric O_2 first rose to appreciable concentrations. In the late 1960s, the geologist Preston Cloud used a variety of geologic evidence, including redbeds, detrital minerals and banded iron-formations (BIFs), to show that the atmosphere first became O_2-rich sometime around 2.0 Ga. Cloud's hypothesis was further developed and elaborated by Heinrich Holland,[5] whose name was mentioned earlier. The date of the Great Oxidation Event (GOE), as Holland termed it, was revised back to ~2.4 Ga, but Cloud's basic idea remained secure. Various geochemists have questioned this hypothesis, notably Hiroshi Ohmoto of Penn State University, but their ideas have failed to gain traction.

About 7 years ago, a significant new development occurred that all but clinched the debate over the rise of atmospheric O_2. James Farquhar and colleagues published a paper[28] in which they used multiple sulfur isotopes in ancient rocks to test Cloud's hypothesis. Sulfur isotopes (^{32}S, ^{33}S, ^{34}S and ^{36}S) normally fractionate, or separate, along a predictable line that depends on their relative masses. (^{33}S differs from ^{32}S by 1 mass unit, whereas ^{34}S differs from ^{32}S by 2 mass units; hence, they normally fractionate along a line with a slope of ~½.) In all rocks younger than about 2.0 Ga, the S isotopes plot along this mass fractionation line; however, in rocks older than 2.4 Ga, the S isotopes fall off it. (Rocks between 2.0 Ga and 2.4 Ga are close to the mass fractionation line, but not quite on it.) As part of their study, Farquhar et al photolyzed sulfur dioxide,

SO_2, in the laboratory in the absence of O_2 and they were able to show that the products of the reaction had mass independent S isotope signatures, i.e., they fell off the mass fractionation line. The fact that the S isotope data show essentially the same pattern as the conventional geologic data of Cloud and Holland provides strong support for their O_2 evolution hypothesis.

This does not mean that there was zero O_2 in the Archean atmosphere, nor in the Hadean atmosphere. O_2 should continually be formed in such atmospheres from reactions such as

$$CO_2 + h\nu \rightarrow CO + O$$
$$H_2O + h\nu \rightarrow H + OH$$
$$O + OH \rightarrow O_2 + H$$

Here, '$h\nu$' represents a UV photon capable of splitting CO_2 or H_2O. However, O_2 should not have accumulated in the early atmosphere, as it would also have had photochemical sinks. In addition to CO_2 and H_2O, which are their major components, volcanic gases also contain smaller amounts of reduced gases, such as CO and H_2. As pointed out in Section 8.1, James Walker showed that the H_2 content of the early atmosphere would have been determined by the balance between outgassing from volcanoes and escape of hydrogen to space. This concept of atmospheric redox balance has since been refined to include rainout of oxidized and reduced species that are soluble in water.[3] A typical photochemical model calculation that includes these processes is shown in Figure 8.3. O_2 is formed in the stratosphere by photolysis and it is consumed in the troposphere (the lower atmosphere) by reaction with H_2, catalyzed by the by-products of H_2O photolysis. Such calculations predict that the O_2 concentration near Earth's surface should have been of the order of 10^{-13} atm. Though finite, this is far too small to have any influence on biology or on prebiotic synthesis of organic compounds. Thus, both models and data suggest that atmospheric O_2 concentrations were low throughout most of the Earth's early history.

At some time at or before 2.7 Ga, oxygen-producing cyanobacteria evolved.[3] The evidence for their appearance comes from complex organic compounds preserved in rocks: 2-alpha methyl hopanes, thought to come from the cell walls of cyanobacteria and steranes (derived from sterols) that are presumed to come from eukaryotes. Eukaryotes, which occupy the domain Eucharya on the rRNA tree, are organisms that have cell nuclei. The biosynthesis of sterols in eukaryotes requires free O_2; hence, the presence of steranes in 2.7-Ga

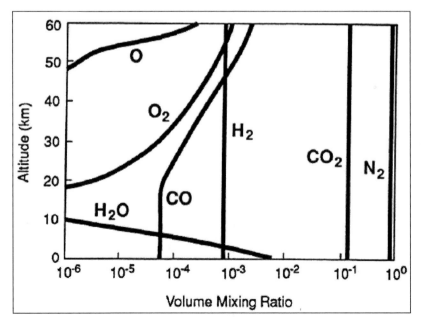

Figure 8.3. Vertical profiles of major atmospheric gases in a weakly reduced primitive atmosphere. (adapted from Science, Vol. 259, Kasting.[4] Copyright (1993) with permission from Science).

rocks suggests that O_2 was being produced at that time. Some early cyanobacteria probably lived together in microbial mats, later being preserved as laminated stromatolites in rocks. Other cyanobacteria may have lived in the surface ocean, especially in upwelling regions that were rich in nutrients like fixed nitrogen and phosphorus. Plumes of O_2 from these local oxygen oases may have wafted into the otherwise anoxic Archean air, gradually being consumed, just as plumes of reduced pollutants are gradually consumed in today's oxidizing atmosphere. This "backwards" atmosphere persisted until about 2.4 Ga, at which time O_2 became dominant and H_2 and CH_4 were relegated to less important roles.

(iii) Connection with the Glacial Record

One of the strongest arguments in support of the scenario described above is its success in explaining the glacial record. In the previous section, we mentioned that there were glaciations at 2.9 Ga and at ~2.4 Ga. The 2.4-Ga, or Paleoproterozoic, glaciation is the better documented of the two. Glacial diamictites (or tillites) from this period are found on at least 3 continents. In North America, they appear as part of the Huronian Formation in southern Canada, just north of Lake Huron. There are 3 diamictites in the Huronian (Fig. 8.4): the Ramsey Lake (lowermost), the Bruce (middle) and the Gowganda (uppermost). Below the Ramsey Lake diamictite in the underlying Matinenda Formation, one finds detrital uraninite (UO_2) and pyrite (FeS_2)—two reduced minerals that are evidence of low oxygen. (The term "detrital" means that they were weathered out of their parent rock without dissolving. As these are reduced minerals, this could only have happened in the near-absence of O_2.) Above the Gowganda diamictite, the Lorraine formation is a well-developed redbed. (A redbed contains oxidized iron in the form of hematite, Fe_2O_3 and is therefore evidence of high O_2.) So, it appears that this series of glaciation occurred at the same time when

atmospheric oxygen levels first rose. To the Canadian geologist, Stuart Roscow, who first mapped this sequence, this appeared to be a coincidence. However, if methane and ethane were important contributors to the greenhouse effect during the Archean, but much less so during the Proterozoic, then it is not at all surprising that the climate became cold when atmospheric O_2 increased.

There are some new wrinkles to this story. Sulfur isotope data between 2.8 Ga and 3.2 Ga have smaller mass-independent fractionation (MIF) signals than do data from before and after that time.[29] This may signal a transient increase in atmospheric O_2 (ibid.) or, alternatively, it may reflect other changes in atmospheric composition. If an organic haze layer developed at this time, it might explain both the low-MIF values and the 2.9-Ga glaciation.[27] This story is still clearly speculative. Here, we simply point out that present models of the Archean atmosphere and climate are more or less consistent with the existing geologic evidence. This does not necessarily imply that they are correct, but it means that they may provide adequate explanations until new data, or new models, come along.

8.5 Conclusions

We have presented here a brief overview of atmospheric and environmental conditions on the primitive Earth. The perceptive reader will note the contrast between Sections 8.2 and 8.3, where we described the Hadean era and Section 8.4, which described the Archean. The latter discussion is reasonably well developed, as it is based on comparisons between models and data. The earlier sections are much more speculative, as they are based almost entirely on models. Our models for the Earth's formation and for the Hadean Earth do have observational support, but they rely on data (e.g., the lunar cratering record) that bear only indirectly on conditions on the primitive Earth. And, yet, the Hadean is in all likelihood the time period when life originated. This suggests that the search for life's origin will continue to be difficult, as it is only poorly guided by knowledge about conditions on the primitive Earth. This is just one more hurdle to be overcome in solving this preeminently interesting problem.

Acknowledgements

The author would like to thank Y. Watanabe for help with the figures and J. Siefert for encouragement to write this chapter. Funding for this research was provided by NASA's Astrobiology and Exobiology programs.

References

Further Readings

1. Walker JCG. Evolution of the atmosphere. New York: Macmillan 1977.
   - Although it is now 30 years old, this book is still the best place to learn about hydrogen escape processes and weakly reduced atmospheres.
2. Sleep NH, Zahnle KJ, Kasting JF et al. Annihilation of ecosystems by large asteroid impacts on the early Earth. Nature 1989; 342:139-142.
   - This paper provides a quantitative assessment of the effects of large impacts on the early Earth. Bear in mind that it was written long before the nice model was developed (Box 1) and so the authors presume that the heavy bombardment was more or less continuous.
3. Kasting JF, Catling D. Evolution of a habitable planet. Ann Rev Astron Astrophys 2003; 41:429-463.
   - This is a review paper that discusses various aspects of planetary habitability. It includes discussions of the faint young sun problem, the rise of atmospheric oxygen and the atmosphere redox budget.
4. Kasting JF. Earth's early atmosphere. Science 1993; 259:920-926.
   - This is an earlier review paper that covers much of the same territory as reference 3, along with quantitative estimates of how much CO_2 would have been needed to keep the early Earth warm, if CH_4 was absent.
5. Holland HD. Early proterozoic atmospheric change. Early life on Earth. Vol Bengtson, S. New York: Columbia Univ Press 1994:237-244.

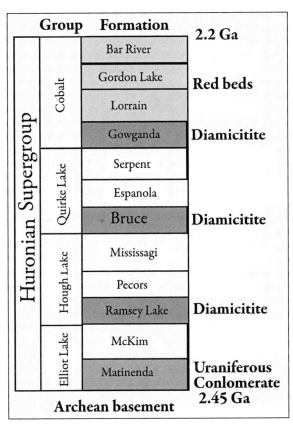

Group	Formation	
	Bar River	**2.2 Ga**
Cobalt	Gordon Lake	**Red beds**
	Lorrain	
	Gowganda	**Diamicitite**
Quirke Lake	Serpent	
	Espanola	
	Bruce	**Diamicitite**
Hough Lake	Mississagi	
	Pecors	
	Ramsey Lake	**Diamicitite**
Elliot Lake	McKim	
	Matinenda	**Uraniferous Conlomerate 2.45 Ga**

Huronian Supergroup

Archean basement

Figure 8.4. Stratigraphic sequence of the Huronian Supergroup in southern Canada. (Modified from ref. 31).

- This book chapter summarizes the "conventional" geologic evidence for the rise of atmospheric oxygen. It appeared well before the sulfur MIF evidence (ref. 28) was discovered.

Specific References

6. Oparin AI. The Origin of Life. New York: MacMillan 1938.
7. Miller SL, Urey HC. Organic compound synthesis on the primitive Earth. Science 1959; 130:245-251.
8. Rubey WW. Geological history of seawater. An attempt to state the problem. Geol Soc Am Bull 1951; 62:1111-1148.
9. Stribling R, Miller SL. Energy yields for hydrogen cyanide and formaldehyde syntheses: the HCN and amino acid concentrations in the primitive ocean. Origins of Life 1987; 17:261-273.
10. Pinto JP, Gladstone CR, Yung YL. Photochemical production of formaldehyde in the Earth's primitive atmosphere. Science 1980; 210:183-185.
11. Zahnle KJ. Photochemistry of methane and the formation of hydrocyanic acid (HCN) in the Earth's early atmosphere. J Geophys Res 1986; 91:2819-2834.
12. Halliday AN, Lee DC, Jacobsen SB et al. Tungsten isotopes, the timing of metal-silicate fractionation and the origin of the Earth and moon. Origin of the Earth and Moon. Tucson, AZ: Univ of Arizona Press 2000:45-62.
13. Matsui T, Abe Y. Evolution of an impact-induced atmosphere and magma ocean on the accreting Earth. Nature 1986; 319:303-305.
14. Genda H, Abe Y. Enhanced atmospheric loss on protoplanets at the giant impact phase in the presence of oceans. Nature 2005; 433:842-844.
15. Kress ME, McKay CP. Formation of methane in comet impacts: Implications for Earth, Mars and Titan. Icarus 2004; 168:475-483.
16. Schaefer L, Fegley JB. Outgassing of ordinary chondritic material and some of its implications for the chemistry of asteroids, planets and satellites. Icarus 2007; 186:462-483.
17. Ryder G. Bombardment of the hadean Earth: Wholesome or deleterious? Astrobiol 2003; 3:3-6.
18. Tsiganis K, Gomes R, Morbidelli A et al. Origin of the orbital architecture of the giant planets of the solar system. Nature 2005; 435:459-461.
19. Gomes R, Levison HF, Tsiganis K et al. Origin of the cataclysmic late heavy bombardment period of the terrestrial planets. Nature 2005; 435:466-469.
20. Kuhn WR, Atreya SK. Ammonia photolysis and the greenhouse effect in the primordial atmosphere of the Earth. Icarus 1979; 37:207-213.
21. Pavlov AA, Kasting JF, Brown LL. UV-shielding of NH_3 and O_2 by organic hazes in the archean atmosphere. J Geophys Res 2001; 106:23,267-223,287.
22. Kharecha P, Kasting JF, Siefert JL. A coupled atmosphere-ecosystem model of the early archean Earth. Geobiology 2005; 3:53-76.
23. Huber C, Wächtershauser G. Peptides by activation of amino acids with CO on (Ni,Fe) surfaces: implications for the origin of life. Science 1998; 281:670-672.
24. Gogarten Boekels M, Hilario E, Gogarten JP. The effects of heavy meteorite bombardment on the early evolution—the emergence of the three domains of life. Origin Life Evol Biosph 1995; 25:251-264.
25. Haqq Misra JD, Goldman SD, Kasting PJ et al. A methane/ethane greenhouse for the early Earth. Astrobiol; in press.
26. Knauth P, Lowe DR. High archean climatic temperature inferred from oxygen isotope geochemistry of cherts in the 3.5 Ga swaziland supergroup, South Africa. GSA Bull 2003; 115:566-580.
27. Kasting JF, Howard MT. Atmospheric composition and climate on the early Earth. Phil Trans Royal Soc Lond B 2006; 361:1733-1742.
28. Farquhar J, Savarino J, Jackson TL et al. Evidence of atmospheric sulfur in the martian regolith from sulphur isotopes in meteorites. Nature 2000; 404:50-52.
29. Ohmoto H, Watanabe Y, Ikemi H et al. Sulphur isotope evidence for an oxic archaean atmosphere. Nature 2006; 442:908-911.
30. Kasting JF, Toon OB, Pollack JB. How climate evolved on the terrestrial planets. Scientific Am 1988; 256:90-97.
31. Young GM. Stratigraphy, sedimentology and tectonic setting of the huronian supergroup. Toronto: Geological Assoc. Canada; Annual Meeting, Toronto, Field Trip B5, Guidebook 1991.

CHAPTER 9

Biomolecules

J. Tze-Fei Wong*

9.1 Introduction

The triple convergence of genetic code structure, atmospheric amino acid synthesis and meteoritic amino acids establishes the heterotrophic nature of the first living cell (Section 1.4). To develop a heterotrophic living cell, a range of biomolecules would have to be sourced from the environment, including building blocks of RNA, proteins, cell membrane and the energy resources to piece them together. Therefore prebiotic chemistry has to identify feasible routes for sourcing organic compounds from either prebiotic synthesis on Earth, or extraterrestrial delivery via meteorites or interstellar dust. The criteria for a meaningful prebiotic synthesis are as follows:[6]

a. The proposed prebiotic synthesis must be plausible with the starting materials being present in adequate amounts at the site of synthesis;
b. The conditions employed must be consistent with the prebiotic atmosphere and environment (Section 8.3);
c. The reaction must occur in water or in the absence of a solvent;
d. The yield of the reaction must be significant.

These criteria help to decide which model abiotic reactions can be prebiotic and which cannot based on current knowledge. In practice, sometimes optimism may be justified retroactively. For example, it is well known that the formose reaction can give rise to sugars including ribose for the production of RNA, but the yield of ribose is usually so low that doubts have arisen whether ribose synthesis by the formose reaction could be regarded as 'prebiotic'. However, recent discoveries of high yields of ribose by the formose reaction have increased the 'prebiotic-ness' of this plausible source of ribose.

9.2 Amino Acids

(i) Electric Spark Synthesis

In the historical experiment of Stanley Miller, water vapor was produced by heating to simulate evaporation from the oceans and, upon mixing with methane, ammonia and hydrogen, mimic a water vapor-saturated primitive atmosphere. When the mixture was subject to electrical sparking, glycine was detected in the water reservoir. After one week, the inside of the sparking flask was coated with an oily material and the water turned yellow-brown, which was shown by analysis to contain a variety of amino acids and carboxylic acids (Table 9.1).[7,8]

The appearance of HCN and aldehydes in the electric spark system suggests that a plausible mechanism for the production of α-amino acids might be Strecker synthesis (Fig. 9.1). Although the atmosphere on early Earth likely contained less reduced gases than those employed in the initial electric spark synthesis (Section 8.3),

the occurrence of atmospheric synthesis of amino acids remains supported by a number of findings:[8-11]

a. Amino acids comparable to those synthesized by electric discharge are also produced by other physical means including ultraviolet, high energy radiation and heat, confirming the basic feasibility of abiotic synthesis of amino acids on primitive Earth;
b. The profiles of amino acid products obtained from the carbonaceous Murchison meteorite were comparable to those from electric spark synthesis with respect to both the protein amino acids and nonprotein amino acids such as α-aminobutyric acid and β-alanine;
c. A mildly reducing atmosphere also can give good yields of amino acids.

(ii) Meteorites

The abiotic synthesis of amino acids under simulated prebiotic conditions on Earth is supported by interstellar and meteoritic chemistries. Many interstellar organic compounds are being discovered at an average rate of about four new molecules per year (Table 9.2).[12] So organic synthesis is very much an ongoing process in space where a variety of organic compounds can be synthesized by e.g., ultraviolet photons. Most matter from space comes to the surface of the Earth as interplanetary particles in the form of micrometer sized comet and asteroid dust. Since there was much more floating debris in the early Solar System, the quantities of extraterrestrial infall would be one million fold that arriving to-day.[13] It is estimated that the total organic carbon brought to Earth by extraterrestrial infall over 100 million years of late accretion in the solar system amounted to 10^{16} to 10^{18} kgs, which is several orders of magnitude greater than the total organic carbon now circulating in the biosphere.[14]

The infall of comet and asteroid dust is spread over the surface of the Earth, adding to the organic supplies on primitive Earth but relatively difficult to collect by primitive life forms as nutrients. Collection would be easier in the case of meteorites. Molecular analysis of meteorite samples[15] has yielded all ten canonical amino acids suggested by genetic code structure to be Phase 1 amino acids available from the early Earth environment.[16,17] Complete agreement between these two lines of findings with the results of atmospheric amino acid synthesis[10,11] has led to a triple convergence that establishes a heterotrophic instead of autotrophic origin of life (Section 1.4).

The triple convergence indicates not only the prebiotic availability of the ten Phase 1 amino acids, viz. Gly, Ala, Ser, Asp, Glu, Val, Leu, Ile, Pro and Thr from the prebiotic environment, but also the prebiotic unavailability of the ten Phase 2 amino acids, viz. Phe, Tyr,

*J. Tze-Fei Wong—Applied Genomics Center, Fok Ying Tung Graduate School and Department of Biochemistry, Hong Kong University of Science and Technology, Clear Water Bay, Hong Kong, China. Email: bcjtw@ust.hk

Prebiotic Evolution and Astrobiology, edited by J. Tze-Fei Wong and Antonio Lazcano. ©2009 Landes Bioscience.

Table 9.1. Electric spark products obtained with a methane-ammonia-hydrogen-water mixture[8]

	Yield (Mols × 10⁵)
Glycine	63
Alanine	34
Sarcosine	5
β-Alanine	15
α-Aminobutyric acid	5
N-Methylalanine	1
Aspartic acid	0.4
Glutamic acid	0.6
Iminodiacetic acid	5.5
Iminoacetic-propionic acid	1.5
Formic acid	233
Acetic acid	15.2
Propionic acid	12.6
Glycolic acid	56
Lactic acid	31
α-Hydroxybutyric acid	5
Succinic acid	3.8
Urea	2
Methylurea	1.5
Sum of yields of compounds listed	15%

Arg, His, Trp, Asn, Gln, Lys, Cys and Met. Since all Phase 1 amino acids have in fact been obtained by irradiation of a mildly reducing CO-N_2-H_2O atmosphere,[10,11] the quest for atmospheric amino acid synthesis has been successfully accomplished. The Phase 2 amino acids were not adopted as earliest building blocks of proteins because of inadequate synthesis or excessive instability. Instead they had to be produced later by biosynthesis (Section 14.3). Those amino acids that were available from the environment yet did not make it into the universal genetic code, such as α-aminobutyric acid or β-alanine (Table 9.1), were either not utilized by the life forms, or tried out and rejected by the life forms. For β-alanine, even if it was utilized at first, eventual rejection is hardly surprising, since the insertion of a β-amino acid into an otherwise all α-amino acid polypeptide is bound to create disruption.

(iii) Hydrothermal Synthesis

Compared to the mildly reducing atmosphere on primitive Earth, the highly reducing CO_2 and H_2-enriched gases issued from hydrothermal vents are even more favorable for organic synthesis. Abiotic syntheses of amino acids were observed under simulated hydrothermal vent conditions when a 3:1 gas mixture of CO_2:H_2, in equilibration with a neutral aqueous solution containing potassium cyanide, ammonium chloride and formaldehyde, was exposed to 150°C at ten atmospheres of pressure. The amino acids obtained included Ser, Gly, Ala, Asp, Ile and Glu, again uniformly Phase 1 amino acids.[18]

(iv) Chirality

The chiral L-amino acid constituents are essential to protein structure and function and the origins of this chirality have long posed a challenge to prebiotic evolution. The discovery of an enantiomeric excess, or *ee*, among meteoritic amino acids points to an extraterrestrial origin of chirality (Sections 6.2, 7.3). Selective adsorption on calcite ($CaCO_3$), an abundant marine mineral, was also found to produce a nonracemic distribution in amino acids. When aspartic acid was adsorbed to calcite, the D/L ratio of the adsorbed aspartic acid indicated that the *ee* could exceed 10%. Interestingly, while *ee* values were observed favoring either the D or L form depending on the face type of the calcite crystal, the larger *ee* values favored the L-form.[19] Thus terrestrial processes could also contribute to the chirality of amino acids.

(v) Stability

When the abiotic synthesis of an organic compound under simulated prebiotic conditions proves to be elusive, further experimentation may yet bring about a positive outcome. In comparison, when a compound is found to be too unstable to survive in the prebiotic environment, its prebiotic unavailability is more difficult to overcome. Such is the case for some of the Phase 2 amino acids. On primitive Earth, on account of the absence of ozone, ultraviolet radiation was intense. Different amino acids differ greatly in their susceptibilities to UV. The Phase 2 amino acids Trp and His in pH 7 solution underwent over 90% degradation and Phe and Tyr over 60% degradation, after UV irradiation at 1.8 mW/cm² for just 48 hours.[20] On land or in the oceans, these amino acids might find protection from UV through shielding by rocks or deep water columns. However, in electric spark synthesis in the atmosphere, it would be difficult for them to escape UV damage prior to being washed down by rain. The result was low final yield even given significant abiotic synthesis.

Thermal degradation is even more difficult to hide from than UV. The most heat-sensitive amino acids are the two Phase 2 amides Gln and Asn. The concentration of any compound A in the prebiotic environment may be described as a balance between abiotic synthesis and degradation:

Rate of change in $[A] = v_f - k_d[A]$

At steady state, the rate of change is zero and therefore,

$[A] = v_f / k_d$

Where v_f represents the rate of formation and k_d the first order degradation rate constant. An extremely optimistic scenario that all UV photons of less than 260 nm wavelength reaching primitive Earth were utilized for abiotic synthesis could lead to an accumulation of 20M total amino acids in the hydrosphere. Even allowing that 2.0% total amino acids was Asp and 0.49% Glu (i.e., comparable to electric spark synthesis) and that all Asp and Glu were derived from Asn and Gln respectively, the maximum levels of Asn and Gln in the oceans assessed from v_f and k_d at 20°C, pH 7 are only:

Maximum $[Asn] = 2.4 \times 10^{-8}$ M

Maximum $[Gln] = 3.7 \times 10^{-12}$ M

Consequently, on account of their thermal instabilities, Asn and Gln could not possibly accumulate to any useful concentrations.[21] There was no Gln and Asn to be found in the prebiotic environment.

$$RCHO + HCN + NH_3 \rightleftharpoons RCH(NH_2)CN \xrightarrow{H_2O} RCH(NH_2)\overset{\overset{\textstyle O}{\|}}{C}-NH_2 \xrightarrow{H_2O} RCH(NH_2)COOH$$

aldehyde ⟶ amino acid

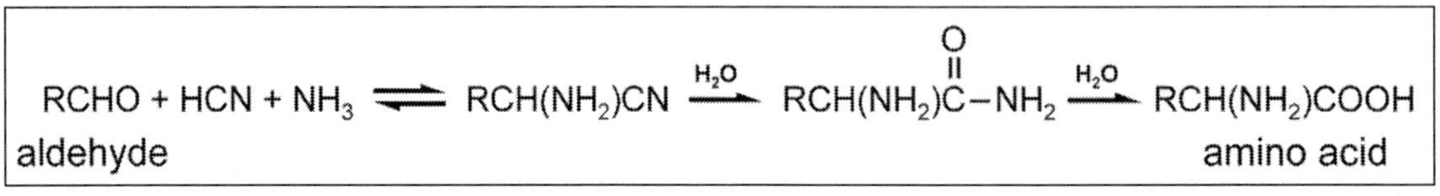

Figure 9.1. Strecker synthesis for the production of α-amino acids.[8]

Table 9.2. Interstellar molecules. After Thaddeus.[12]

Number of Atoms								
2	3	4	5	6	7	8	9	10 to 13
H_2	H_2O	NH_3	SiH_4	CH_3OH	CH_3CHO	$HCOOCH_3$	CH_3CH_2OH	CH_3COCH_3
OH	H_2S	H_3O^+	CH_4	NH_2CHO	CH_3NH_2	CH_2OHCHO	$(CH_3)_2O$	$CH_3(C\equiv C)_2CN$
SO	SO_2	H_2CO	CHOOH	CH_3CN	CH_3CCH	CH_3C_2CN	CH_3CH_2CN	$HOCH_2CH_2OH$
SO^+	HN_2^+	H_2CS	$HC\equiv CCN$	CH_3NC	CH_2CHCN	C_7H	$H=(C\equiv C)_3CN$	CH_2CH_2CHO
SiO	HNO	HNCO	CH_2NH	CH_3SH	HC_4CN	H_2C_6	$CH_3(C\equiv C)_2H$	$H(C\equiv C)_4CN$
SiS	SiH_2	HNCS	NH_2CN	C_5H	C_6H	HC_6H	C_8H	CH_3C_6H
NO	NH_2	CCCN	H_2CCO	HC_2CHO	$c\text{-}CH_2OCH_2$	CH_3CO_2H		$c\text{-}C_6H_6$
NS	H_3^+	c-CCCH	CH_2	$CH_2=CH_2$	CH_2CHOH_2	H_2C_3HCN		$H(C\equiv C)_5CN$
HCl	NNO	CCCS	$c\text{-}C_3H_2$	H_2CCCC		CH_2CHCOH		
NaCl	HCO	HCCH	CH_2CN	$HC_3NH^\bullet$				
AlCl	OCS	HCCN	SiC_4	HC_4H				
AlF	CCH	H_2CN	H_2CCC	C_5S				
PN	HCS^+	$c\text{-}SiC_3$	HCCNC	C_4H_2				
SiN	c-SiCC	CH_3	HNCCC	HC_4N				
NH	CCO	CH_2D^+	$H_3CO^\bullet$	$c\text{-}H_2C_3O$				
SH	CCS	AlNC						
HF	C_3							
CN	MgNC							
CO	NaCN							
CS	CH_2							
C_2	MgCN							
SiC	HOC^+							
CP	HCN							
CO^+	HNC							
CH^+	CO_2							
CH	SiCN							
N_2	AlCN							
	SiNC							
	KCN							

9.3 RNA Constituents

Prebiotic evolution, as concluded by Orgel, in all likelihood went through an RNA World:

> "The demonstration that ribosomal peptide synthesis is a ribozyme-catalyzed reaction makes it almost certain that there was once an RNA World."[6]

Accordingly extensive research has been directed to the plausible abiotic syntheses of the constituents of RNA.

(i) Nucleobases

The synthesis of adenine from hydrogen cyanide by Oro initiated the prebiotic chemistry of nucleic acids.[22] Refluxing of ammonium cyanide forms the HCN tetramer, which is further polymerized to a dark intractable solid from which adenine and guanine can be recovered.[6] Tetramer formation requires HCN concentrations higher than 10 mM. Since it would not be possible to reach such a concentration in the bulk oceans and concentration via evaporation also could not work with as volatile a molecule as HCN, a more plausible method for concentrating HCN is eutectic freezing: if a dilute aqueous solution of HCN is cooled below 0°C, pure ice crystallizes out and the solution becomes more concentrated until a eutectic point is reached at −23.4°C at 74.5 (moles)% which gives significant yields of adenine.[23,24] There are also HCN-independent routes of purine synthesis.[25] The feasibility of prebiotic synthesis of purines is confirmed by their occurrence on the Murchison meteorite (Table 6.2).

Cyanoacetylene is a major product formed when electric discharge is passed through a mixture of nitrogen and methane. It reacts with two molecules of cyanic acid to give cytosine. It also can be hydrolysed to cyanoacetaldehyde, which condenses with urea to give cytosine (Fig. 9.2). Hydrolysis of cytosine yields uracil.[6,26,27] Since cytosine is thermally unstable with a $t_{1/2}$ of only 17,000 years even at 0°C,[28,29] it would have to be replenished continually, or else utilized for RNA or RNA-like biopolymers only at a later stage.

(ii) Nucleosides

The polymerization of formaldehyde in the presence of simple mineral catalysts to form a mixture of sugars is known as the formose reaction. Discovered by Butlerow in the 19th Century, it is a unique cyclic autocatalytic process that converts the simple four-atom formaldedye, found even in interstellar gas (Table 9.1), into a complex mixture of sugars. Since the complexity of the formose reaction products results in a low fraction of ribose among the products, doubts have been voiced regarding the plausibility of prebiotic ribose synthesis by means of this reaction. However, in the presence of Pb^{+2} ion and catalytic amounts of an intermediate in the pentose pathway, pentoses including ribose can exceed 30% of reaction products, thus furnishing prebiotic ribose.[30]

Alkaline environments, favorable to the formose reaction but once thought to be rare, could be found in alkaline hydrothermal systems. This together with borate stabilization of ribose further enhance the probability of ribose production on primitive Earth.[31] As well, ribose permeates fatty acid and phospholipid membranes more rapidly than other 5- or 6-carbon aldo-sugars, which might be a significant factor favoring its selection as an RNA constituent.[32] The carbonaceous Murchison and Murray meteorites contain a variety of polyols including glycolic acid, glyceric acid, dihydroxyacetone, glycerol, erythritol, threitol and ribitol which can be converted into sugars. These meteoritic components, besides testifying to the feasibility of abiotic synthesis of polyols under prebiotic conditions, could furnish polyols to the primitive life forms for nucleic acid synthesis and intermediary metabolism.[33]

When D-ribose is heated with hypoxanthine in the presence of magnesium chloride or sea salts, up to 8% β-D-inosine is formed. With adenine, reaction with D-ribose under similar conditions followed by hydrolysis likewise yields up to 3% β-D-adenosine. No direct synthesis of pyrimidine nucleosides from ribose and uracil or cytosine has been reported, but α-cytidine may be obtained from reaction of ribose or ribose phosphate with cyanamide and cyanoacetylene in aqueous solution and photo-converted to β-cytidine.[6,34,35] Overall, the formation of nucleosides represents at present one of the weakest links in the chain of prebiotic reactions leading to oliogonucleotides.[6]

(iii) Nucleotides

In modern organisms, nucleosides are phosphorylated by kinases to form nucleotides using ATP as phosphorylating agent. The attractive properties of phosphate in energy transfers stem from the ability of phosphorus to form multiple bonds of moderate strength and the tribasic nature of phosphoric acid. Inorganic polyphosphates (polyP) such as trimetaphosphate and linear polyphosphates and to some extent pyrophosphate, could function as prebiotic phosphorylating agents for nucleotide synthesis. Although the concentration of inorganic phosphate in the hydrosphere is expected to be as low as 0.1 nM owing to the low solubility of calcium phosphate at neutral pH, an adequate supply of inorganic phosphate might be acquired via a number of routes including:[36]

a. Although many common phosphorus-containing minerals aside from schreibersite from meteorites are only slightly soluble in water, it is estimated that up to 5% of total crustal P on early Earth might be extraterrestrial schreibersite

added to the Earth by meteoritic impacts. Fe_3P, a model for schreibersite, reacts with water to yield orthophosphate and condensed phosphates.

b. Decreased alkalinity leads to the precipitation of the acid calcium salt brushite ($CaHPO_4 \cdot 2H_2O$) and high magnesium and ammonia concentrations favor the precipitation of struvite ($MgNH_4PO_4 \cdot 6H_2O$). Both brushite and struvite can form condensed phosphate upon heating.

c. Under strongly reducing conditions, phosphate is reduced to phosphite. Ammonium phosphite reacts with nucleosides to yield nucleoside-phosphites, which in the dimeric form is easily oxidized to a phosphodiester bond.

d. Pyrophosphate and polyP have been obtained from volcanic fumaroles.[37]

Nucleosides can be converted, not very efficiently, to nucleotides by heating in the solid state with acidic phosphates.[38] Heating ammonium phosphate with urea yields a mixture of linear phosphates in the absence of an organic component and nucleotides in excellent yields in the presence of nucleosides. For example, when uridine is heated with excess urea and ammonium phosphate at 100°C, 70% of the uridine is converted to a complex mixture of uridine nucleotides, with some possibility of structuring the reaction conditions to favor a particular form of uridine phosphate.[6] Reaction of trimetaphosphate with adensosine catalyzed by magnesium gives about 9% nucleotides, mainly 2′,3′-cyclic AMP, at neutral pH.[39] Introduction of wet-dry cycles and catalysis by Ni(II) enhanced the yield of nucleotides to 30%, producing up to 10% 2′3′-cyclic AMP as well as 13% ATP.[40] These studies point to the prebiotic availability of polyP, ATP as high-energy compounds and NTPs as substrates for RNA polymerization.

(iv) Non-Canonical Backbones

D-ribose is utilized in the construction of both cellular and viral RNA, suggesting that the use of this particular pentose has been a long standing one. Nonetheless, it does not necessarily follow that other forms of nucleobase-containing biopolymers might not have preceded RNA. In fact, a range of alternative nucleobase-containing biopolymers have been explored with respect to their potential as predecessor replicators/genes prior to the advent of the RNA World based on a canonical backbone where phosphate is joined to the 3′-OH and 5′-OH of adjacent ribose-furanose residues:

a. Owing to the relative difficulty of abiotic synthesis of nucleosides and the occurrence of enantiomeric cross-inhibition, it was suggested that ribose in the RNA backbone might be advantageously replaced by acyclic analogues derived from glycerol, acrolein or erythritol in order to overcome such adverse effects.[41]

b. Pyranosyl-RNA (p-RNA), with a backbone of phosphate joined to the 2′-OH and 4′-OH of adjacent ribose-pyranose residues, has a stronger and more selective base-pairing system than RNA.[42,43]

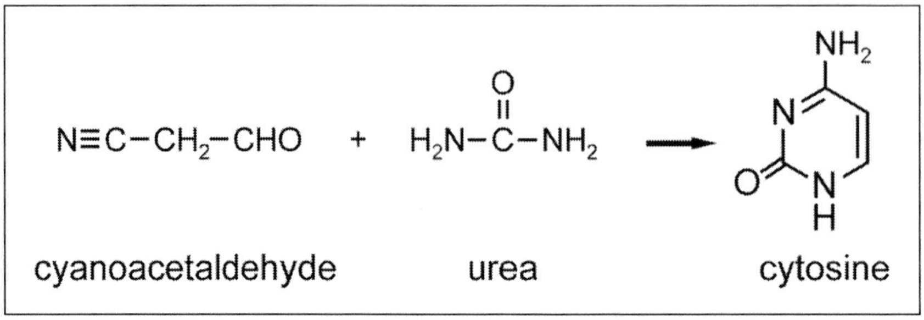

Figure 9.2. Cytosine synthesis from cyanoacetaldehyde.

c. Threofuranosyl nucleic acid (TNA) has a backbone where the five-carbon ribose is replaced by the four-carbon threose (Fig. 9.3). It forms stable heteroduplexes with both RNA and DNA.[44] One disadvantage with TNA is the absence of free OH-group from its backbone, which would reduce its capability to function as threozymes and explain the thorough displacement of what prebiotic TNA there was by RNA.

d. Peptide nucleic acids (PNA) contain the usual nucleobases and pair with nucleic acids, but are free of phosphate.[45] PNA has been considered as a prebiotic biopolymer,[46] but so far no straightforward prebiotic synthesis of PNA monomers has been reported.[6] Moreover, the lower global flexibility of PNA relative to RNA[47] suggests that PNA might experience greater difficulty than RNA in developing small-sized aptamers or PNA-zymes.

These alternatives serve the valuable purpose of focusing attention on the possibility of RNA replicators being preceded by other informational biopolymers in prebiotic evolution. TNA offers the advantage that its ability to form a TNA-RNA duplex might allow transition of an early usage of TNA, or co-usage of TNA and RNA, to a subsequent RNA World with limited disruption of the replicator system.

Nowadays, organisms employ DNA genes, whereas viruses employ both RNA and DNA genes. Evidence suggesting that DNA appeared after RNA[48] and even after proteins,[49] includes:

a. Greater stability of the phosphodiester backbone of DNA compared to RNA;

b. Absence of proofreading by RNA polymerases leading to higher mutation rates in RNA genomes;

c. Information in RNA degrades because deamination of cytosine to form uracil can be detected and repaired in DNA but not in RNA, thereby leading to higher mutation and error rates in RNA genomes compared to DNA genomes;

d. UV irradiation produces more photochemical changes in RNA than in DNA;

e. Deoxyribonucleotides are formed from reduction of ribonucleotides by ribonucleotide reductase;

f. Use of a free radical in the catalytic mechanism of ribonucleotide reductase that likely could be fashioned only by a sophisticated protein.

Accordingly, the evolution of genes likely began with RNA, TNA or some other predecessor replicators and evolved to an RNA World. Subsequently the RNA genes were replaced by DNA in cells, while viruses can choose between DNA (e.g., smallpox virus) or RNA (e.g., tobacco mosaic virus) genes. The primitive replicators had to either act as biocatalysts themselves or encode the formation of biocatalysts, which imposes important constraints on the nature of the first replicators. RNA meets the biocatalytic requirement through its ribozymic activities (Sections 11.2). Proteins are virtuoso biocatalysts but cannot readily self-replicate. One expects some biocatalytic capability from pRNA as 'p-ribozyme', TNA as 'threozyme' and PNA as 'PNzyme', but it remains to be determined just how well they perform in these capacities.

9.4 Lipids

Amphiphiles are compounds containing a hydrophilic head group that is attracted to water and a hydrophobic tail that is attracted to organic solvents immiscible with water. Lipoidal substances are bioamphiphiles consisting of long chain fatty acids or fatty alcohols where a carboxylate or alcohol moiety furnishes the hydrophilic head group of the elongated molecule and a long alkane chain the hydrophobic tail; or phospholipids where a phosphate-moiety furnishes the hydrophilic head group and alkane chains the hydrophobic tail. When placed in water, amphiphilic molecules tend to arrange themselves into micelles with their tails clustered together inside the micelle away from the water molecules they dislike and their heads orientated toward the surrounding water molecules to which they are attracted. The amphiphiles may also self-assemble into vesicles with a bilayer membrane, where two layers of the amphiphile are aligned with their tails toward one another and their heads facing water on either side of the membrane.

When samples of the Murchison meteorite were extracted with organic solvent and allowed to interact with an aqueous phase, self-assembled membrane vesicles were observed. This discovery strikingly points to the existence of extraterrestrial amphiphiles and the ease with which they can generate membrane vesicles. The prebiotic availability of amphiphiles, their self-assembly into vesicles and the capacity of such vesicles to grow and divide (Section 12.4) facilitated the development of vesicular life forms.[50]

Amphiphiles might also be produced at submarine hydrothermal vents and deep subterranean hot aquifers, where thermal energy, mineral surfaces and a reducing hydrogen-carbon dioxide gaseous composition promote the synthesis of a variety of organic compounds (Section 1.4). In a model prebiotic reaction using oxalic acid as starting material, aqueous Fischer-Tropsch type reaction at hydrothermal temperatures gave rise to more than 5% lipids. Both n-alkanols, which are

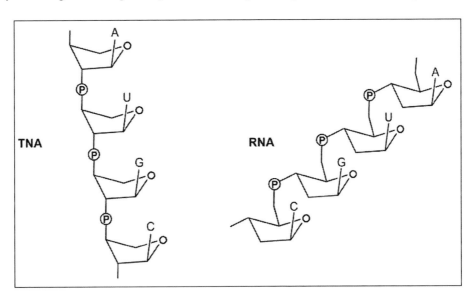

Figure 9.3. Comparison of threonucleic acid (TNA) and ribonucleic acid (RNA).

employed in archaeal lipids, and n-alkanoic acids, which are employed in bacterial and eukaryotic lipids, were obtained.[51]

9.5 Energy Resources

Energy inputs are important to prebiotic evolution at two different levels. First, the abiotic syntheses ongoing in the prebiotic environment, making biomolecules available to the heterotrophic life forms, were endergonic in nature and required energy supplied by electric spark from lightning, ultraviolet photons, shockwaves from thunder, cosmic rays, heat from volcanoes and hydrothermal vents, etc. Secondly, within the life forms, or their precursors in coacervates, vesicles or organic layers adsorbed to mineral surfaces, endergonic reactions including the polymerization of building blocks also had to be supported by high-energy compounds.

In modern organisms, the high-energy compounds employed to drive endergonic biosynthesis, transport against concentration gradient, electrical conduction, mechanical contraction, etc, consist of ATP, other high-energy phosphates and thioesters. These energy carriers are often built up through coupling to redox reactions or scission of covalent bonds. Extant organisms utilize a wide range of reductants, oxidants and energy substrates. However, the thermodynamic ground rules and the need for high-energy bonds are fundamental to all organisms and may be expected to apply equally to prebiotic systems. An important lesson from extant organisms is that high-energy compounds readily energize one another: e.g., the higher-energy creatine phosphate (−10.3 kcal/mole) stored in the muscle can rapidly transfer its phosphoryl group to the lower energy ATP (−7.3 kcal/mole) and ATP can cross-energize with the thioester acetyl-CoA (−7.5 kcal/mole). A number of potential mechanisms have been proposed for the prebiotic generation of high-energy compounds:

(i) High-Energy Phosphates

Thermally produced polyP could energize the synthesis of nucleoside triphosphates (NTPs) including ATP to drive RNA polymerization and other biosynthetic reactions in open reaction systems during the prevesicular phase of prebiotic evolution. Nowadays, polyP is found in all organisms examined and known to substitute for ATP in function.[52]

(ii) Thioester World

Redox (reduction-oxidation) reactions are basic to the generation of high-energy bonds through substrate-level phosphorylation or membrane-linked electron-transport. The chemical energy they release is usually captured and stored in ATP in organisms. However, the existence of a prebiotic Thioester World before the appearance of ATP, where thioesters prevailed as the high-energy compounds driving the development of metabolism, has been proposed.[53] According to this proposal, although the atmosphere lost its hydrogen early on, prebiotic ferrous iron could bring about the appearance of a variety of reduced compounds including the foul smelling hydrogen sulphide (H_2S), which gave rise to organic thiols (R-SH). The thiols can react with carboxylic acids (R'COOH) to form high-energy acyl-thioesters (R'COSR) via three routes:

    a. reaction at high temperature and acidic pH;
    b. coupling to oxidation of aldehydes to carboxylic acids using ferric iron as oxidant;
    c. coupling to oxidative decarboxylation of keto acids using ferric iron as oxidant.

The most important thiol compound in organisms is Coenzyme A (CoA), which contains a thiol group on its β-mercaptoethylamine moiety. Biologically, because ATP and acetyl-CoA each contain a high-energy bond of comparable energy content, they can cross-energize one another e.g., through acetyl-CoA synthetase.

Prebiotically, condensed phosphates including polyP and thioesters could likewise cross energize one another.[53-57]

(iii) Wood-Ljungdahl Pathway

Acetogens employ this pathway to produce acetic acid anaerobically employing hydrogen as reductant to form acetyl-CoA:

$$4H_2 + 2CO_2 + HSCoA \rightarrow CH_3COSCoA + 3H_2O$$

In prokaryotes, the core enzymes for this pathway include carbon monoxide dehydrogenase. The mechanism of the dehydrogenase is thought to entail the reduction of carbon dioxide to carbon monoxide with electrons supplied by hydrogen through an iron-nickle-sulphur cluster, which suggests a possible derivation of the enzymic pathway from an earlier geothermal pathway. On this basis it is proposed that an inorganically catalyzed analogue of this mode of acetyl-CoA synthesis making use of hydrothermal H_2 represented the first biochemical pathway and that, prior to the appearance of membrane vesicles, the pathway might be housed in mineral micro-compartments.[57]

(iv) Pyrite Formation

The surface metabolism theory envisages a chemoautotrophic origin of life at sites of reducing volcanic exhalations. The pioneer organism with its primordial metabolism underwent development on a positively charged mineral surface such as that of pyrite (FeS_2). The negatively charged RNAs could adsorb to and evolve on the surface, thereby overcoming the problem of insufficient local concentrations. To achieve carbon fixation, the central challenge in an autotrophic origin of life, it is proposed that an autocatalytic cycle related to the reductive tricarboxylic acid cycle could function with the required reducing power being obtained from the oxidative formation of pyrite.[58,59]

(v) Glycolytic Fermentation

The glycolytic pathway leading from glucose to pyruvate and further on to lactate, ethanol, etc. is a key energy generating pathway in organisms from all three biological domains. Based on the glycolytic sequence, the conversion of one glucose molecule to two lactic acid molecules generates two ATP:

$$glucose + 2ADP + 2\ Pi \rightarrow 2\ lactic\ acid + 2ATP$$

Because energy production by substrate-level phosphorylation in glycolysis occurs downstream to glyceraldehyde phosphate and dihydroxyacetone phosphate, the triose model proposes that these two trioses served as prebiotic energy substrates.[60] More generally, not only glycolysis but also other fermentation pathways utilizing abundant prebiotic compounds such as glycine might fulfil an energy-provider role.[61]

9.6 Energy Strategy

(i) Strengths and Weaknesses

In order to identify a plausible program of energy utilization over the course of prebiotic evolution, it is necessary to weigh the individual strengths and weaknesses of the array of potential energy resources.

Polyphosphates

The advantage of polyP resides in the availability of phosphate despite the vanishing solubility of its calcium salts,[36] the thermal formation of polyP by volcanism[37] and the generation of ATP from adenosine and trimetaphosphate.[40] The NTPs obtained from phosphorylation supported by polyP could be polymerized to produce replicators and both polyP and NTPs drove metabolism and biosyntheses. A disadvantage with polyP relates to its moderate instability,[62] which dictates that the site of its generation should not be too far removed from the site of its utilization. Another disadvantage is that, although extant organisms possess a battery of membrane transporters to import solutes from the surrounding medium, polyP and nucleotides are not taken up readily by cells.[63] There is no reason to suppose that prebiotic life forms could do better in this regard.

Thioesters

For a heterotrophic origin of life, the key attribute determining the significance of any energy substrate had to be its availability. In this regard, the soluble compounds from the Murchison meteorite (Table 6.2) suggest that meteorites could supply ketoacids, hydroxyacids and amino acids to the prebiotic environment. Table 9.1 also shows the production of these compounds from electric spark synthesis with hydrogen-methane-ammonia-water. Likewise, the largest yields from electric spark synthesis with methane-nitrogen-water-trace ammonia were: Ala, Gly, α-aminobutyric acid, α-hydroxy-γ-aminobutyric acid, norvaline, sarcosine and Asp.[64] Present day dehydrogenases can convert hydroxyacids to ketoacids, e.g., lactate to pyruvate, malate to oxaloacetate, isocitrate to α-ketoglutarate. Amino acid oxidases and transaminases can convert amino acids to ketoacids, e.g., Ala to pyruvate, Asp to oxaloacetate, Glu to α-ketoglutarate. Primitive inorganic catalysts and biocatalysts might catalyze similar conversions. Consequently ketoacids were among the most plentiful energy substrates on early Earth.[53] These ketoacids could produce acyl-thioesters, catalyzed by primitive biocatalysts analogous to pyruvate dehydrogenase, α-ketoglutarate dehydrogenase and thiolase, which catalyze respectively:

$$\text{pyruvate} + \text{CoA} + \text{NAD}^+ \rightarrow \text{acetyl-CoA} + CO_2 + \text{NADH}$$

$$\alpha\text{-ketoglutarate} + \text{CoA} + \text{NAD}^+ \rightarrow \text{succinyl-CoA} + CO_2 + \text{NADH}$$

$$\text{3-keto fatty acid-CoA} + \text{CoA} \rightarrow \text{fatty acid-CoA}$$
$$\text{(shortened by 2 carbons)} + \text{acetyl-CoA}$$

Therefore, based on the availability of ketoacids, hydroxyacids and amino acids in the prebiotic environment, thioesters could be eminent prebiotic high-energy compounds.

However, because metabolism could not evolve far by themselves without replicators and vice versa, a Thioester World devoid of ATP and therefore of RNA or RNA-like replicators, would not operate and evolve efficiently. Accordingly the case for an exclusive utilization of thioesters prior to polyP and NTPs is not strong.

Geothermal Energy

Owing to its use of hydrogen as reductant, the Wood-Ljungdahl type pathway is suited to an autotrophic origin of life at hydrothermal sites. Likewise, oxidative formation of pyrite might generate reducing power to drive autotrophic carbon fixation via thiocarboxylates. Although an autotrophic origin of life is rendered unlikely by the triple convergence of evidence for the heterotrophic utilization of environmental amino acids (Section 1.4) and the thermal instability of important biomolecules,[65,66] both the Wood-Ljungdahl and pyrite-formation pathways could play significant roles producing biomolecules as nutrients for the evolving heterotrophic life forms. Moreover, the overcoming of low reactant concentrations through adsorption to mineral surfaces or inclusion inside mineral micro-compartments could be equally attractive for a heterotrophic origin of life.

Sugars

Nowadays, photosynthesis is responsible for the bulk of carbon fixation in the living world and sugars, key participants in photosynthesis, are the foremost energy substrates. Prebiotically, sugars could be produced from formaldehyde through the formose reaction and the provision of ribose or threose was required for the synthesis of the RNA or TNA replicators. However, a pivotal role of sugars in primordial energy metabolism is not supported by the following observations:

a. Sugars are not the most stable of compounds. The formose reaction producing sugars also requires alkaline pH for high yields.
b. Glycolysis is a two-segment enzymic pathway. The upper segment from glucose to triose phosphates is a sugar pathway. The lower segment from 1,3-bisphosphoglycerate to pyruvate is a carboxylic acid pathway. The tricarboxylic acid cycle that metabolizes pyruvate is likewise a carboxylic acid pathway. There are two substrate-level phosphorylation steps in the lower carboxylic acid-segment of glycolysis, but none in the upper sugar-segment.
c. Enzymes in the lower segment are highly conserved compared to those in the upper segment, suggesting that the glycolytic pathway originally operated in the gluconeogenic direction from bottom to top, instead of the glycolytic direction from top to bottom.[67] Starting from the RNA World, gluconeogenesis could be essential for transforming carboxylic acids and amino acids to glucose-6-phoshate in order to produce ribose-5-phosphate via the pentose phosphate cycle and onward to 5-phosphoribosyl-1-pyrophosphate (PRPP), the precursor of both pyrimidine and purine ribonucleotides.
d. ATP generation by oxidative phosphorylation through membrane chemiosmosis utilizes carboxylic acids from the tricarboxylic acid cycle and the glyoxylate cycle as substrates, not sugars.

Therefore sugars might be less important than carboxylic acids and amino acids as energy substrates prior to the rise of photosynthesis.

(ii) PolypP-Thioester-ATP Strategy

In view of the individual strengths and limitations of various energy resources, it becomes necessary to identify an energy program that could ensure uninterrupted energy supply to the prebiotic life forms through the different phases of their development. A three-stage polyP-thioester-ATP energy utilization strategy might provide an adequate solution:

Stage 1: Pre-Vesicular

At the start, polyP energized the production of NTPs for RNA, TNA and other RNA-like polymerizations and both polyP and NTPs energized biosynthesis and metabolism. Among the NTPs, because abiotic synthesis gave higher yields of adenine than guanine and formation of pyrimidine nucleosides represented more of a supply bottleneck than purine nucleosides, ATP was more abundant than the other three NTPs. This led to the prebiotic preeminence of ATP, which has led to the present day cellular usage of ATP instead of the other NTPs as the preferred high-energy carrier. The NTPs polymerized to form TNA or RNA replicators which evolved threozymes and ribozymes under the direction of metabolites including the carboxylic acids and amino acids through REIM (replicator induction by metabolite) and REAS (replicator amplification by stabilization) (Section 1.3) to catalyze a growing network of metabolic reactions, including the production of thioesters from the ketoacids, hydroxyacids and amino acids supplied by the environment. Because of the abundance of these organic acids, the NTPs increasingly depended on thioesters instead of polyP as the energy source for their synthesis.

Stage 2: Early Vesicular

When the evolving replication-metabolism system was encapsulated into membrane vesicles, intra-vesicular biotic evolution began. Because exogenous polyP and NTPs were cut off on account of their poor entry into the vesicles, only prevesicular systems equipped with the generation of thioesters could have survived vesicularization. The reliance on thioesters as energy currency approached the description of a Thioester World[53] in terms of the bioenergetic preeminence of thioesters over ATP, but not in terms of a total absence of ATP. Thioesters would supply the energy to produce NTP for RNA synthesis and continued evolution of replication-metabolism. At this early vesicular stage, the membranes were still too immature to support efficient membrane-linked phosphorylation—the gradualness of membrane development is suggested by the absence of cytochrome genes from the genome of the Last Universal Common Ancestor, or LUCA (Appendix 15.1). Without membrane-linked ATP production, the bioenergetic leadership of thioesters prevailed during this stage, which is in accord with present day biochemistry:

a. Nowadays thioesters activate only carboxyl groups, whereas ATP can phosphorylate carboxyl, hydroxy, enol, phenolic and guanido groups. Since the high-energy bonds in acetyl-CoA and ATP are comparable in energy content, the persistence of the narrow-scope acetyl-CoA in biochemical pathways in parallel with the much more versatile ATP may be explained by a historical period of thioester dominance predating the modern period of ATP dominance. Such a functional layering of versatile ATP bioenergetics on top of more restricted thioester bioenergetics is analogous to the anatomical and functional layering of a neocortex over older neural structures in the human brain.

b. The puzzling redundancy of parallel thioesters and ATP usages is nowhere more striking than in peptide bond formation. Thioesters participate in peptide bond formation by nonribosomal peptide synthetases (NRPS) in the production of peptide antibiotics,[68] but not in ribosomal protein synthesis. Since ribosomal protein synthesis is a late invention that initiated the displacement of ribozymes by enzymes, NRPS was likely an older mechanism for peptide synthesis employed in the production of peptidyl-ribozymes that served as transitional biocatalysts between the ribozymes and enzymes (Section 14.2).[69] The use of thioesters in NRPS but not in ribosomal protein synthesis thus lends evidence to the existence of an older thioester era that witnessed the development of NRPS and a subsequent ATP era that witnessed the development of ribosomal protein synthesis.

c. Glyceraldehyde 3-phosphate dehydrogenase enables the use of sugars to form 1,3-bisphosphoglycerate, which phosphorylates ADP to ATP through an enzyme-bound thioester intermediate. Acetyl-CoA from glycolysis-derived pyruvate and from fatty acid oxidation condenses with oxaloacetate to form citrate to kick off the ATP-generating tricarboxylic acid cycle. In these instances, thioesters are opening up portals into important pathways for ATP production. In primordial times, as environmental ketoacids, hydroxyacids and amino acids were nearing depletion by the teeming heterotrophic life forms, these pathways would usefully capture sugars and fatty acids as additional energy substrates to fuel ATP production. In effect, the predecessor thioester energetics were paving the way for the newer ATP energetics.

Stage 3: Mature Membranes

Membrane-linked energy generation produces large quantities of ATP but no acetyl-CoA in present day organisms and this was expectedly also the case in the primitive life forms. Consequently, as the cellular membranes matured in their structure and function and membrane-linked energy production using proton-gradients became efficient, the fraction of high energy bonds generated by membranes steadily increased. In the human body, for example, the complete oxidation of one glucose molecule yields 2 ATPs in the cytosol and 28 ATPs through mitochondrial membranes.[70] Thus the arrival of mature membranes and membrane-linked energy generation established ATP as the unchallenged high-energy compound eclipsing thioesters. This final stage of ATP-dominant bioenergetics has endured to this day.

In conclusion, during the long years of prebiotic/early biotic evolution, the evolving systems confronted such potential discontinuities as exhaustion of environmental carboxylic acids and amino acids, encapsulation inside membrane vesicles and rise of membrane-based energy generation. The energy program employed by the life forms must respond effectively to these changes. The PolyP-Thioester-ATP strategy illustrates how the life forms could navigate these potential discontinuities by mobilizing different energy resources at different stages of development. Long distance human ground transport has switched from leg-power to horse-power to petroleum-power within two thousand years, with electric-power, biofuel-power and hydrogen-power still waiting for their turns. So optimal energy strategy is constantly dictated by the relative availability, efficiency and cost of competing energy resources, no less for prebiotic and early life forms than for humans.

9.7 Where on Earth?

Research into the abiotic synthesis of biomolecules of the living cell has yielded a range of building blocks that furnish a foundation for prebiotic evolution to begin its journey. Some of the syntheses, however, might require potentially contradictory conditions for optimization. For example, polyP and nucleoside productions are favored by high temperatures, whereas purine synthesis and RNA polymerization are favored by icy conditions. Since a heterotrophic origin of life depended on access to multiple environmental nutrients, the question of where an optimal primordial reactor best equipped for life's emergence might be found arises. There are a number of candidate sites.

(i) Surface Water

In atmospheric amino acid synthesis powered by lightning, the amino acids produced would be washed down by rain on to surface bodies of water, which would represent useful candidate sites as observed by Darwin:[71]

> *"If (and oh, what a big if) we could conceive in some warm little pond, with all sorts of ammonia and phosphoric salts, light, heat, electricity, etc., present that a protein compound was chemically formed, ready to undergo still more complex changes, at the present day such matter would be instantly devoured or adsorbed, which would not have been the case before living creatures were formed."*

In looking to the warm little pond, Darwin was in fact anticipating the lead focus of astrobiological explorations—always following the water.

(ii) Clays and Minerals

There are many varieties of clay, a class of negatively charged minerals, comprising the major groups of kaolinite, montmorillonite-smectite, illite and chlorite. Among them, montmorillonite has been most extensively investigated.[72-75] There are likewise numerous positively charged minerals such as the double-layer hydroxide (DLH), e.g., green rust, or $Fe(II)_2Fe(III)(OH)_6$. Both types of minerals could be widespread in occurrence and play important roles in prebiotic chemistry through catalysis of chemical transformations and concentration of biomolecules from dilute solutions.

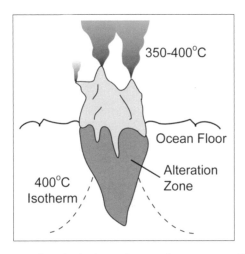

Figure 9.4. Seafloor hydrothermal vent. After Van Dover.[79]

Thus DLH could bring about the formation and concentration of tetrose phosphate and pentose phosphate for the production of TNA and RNA.[5,75] Montmorillonite catalyses the formation of RNA oligomers up to 50 bases long and introduces sequence-, regio- and homochiral selectivities into the oligomer products, yielding up to 13% of an isomer where only 0.2% might be expected on a random basis, thereby accelerating the appearance of specific oligomers.[76] Pre-vesicular prebiotic systems might also be adsorbed on mineral surfaces[58] or enclosed in mineral micro-chambers[57] to prevent excessive dilution of reactants and impart individuality to the system to make possible natural selection.

(iii) Hydrothermal and Impact Sites

The first hydrothermal vents were discovered along the Galapagos Rift in explorations induced by detection of ocean temperature anomalies.[77,78] These vents are fissures on the Earth's crust often in areas of volcanic activities where tectonic plate movements cause an upwelling of the magma, as in the case of the Mid Atlantic Ridge and East Pacific Rise. At these sites sea water is drawn into the hydrothermal system, and heated by the volcano edifice up to as high as 400°C (Fig. 9.4). When the super-heated water runs into chilly surrounding water, minerals precipitate out to build up chimney structures at a rate as high as 30 cm per day, towering as tall as 60 m. Chimneys that emit a cloud of black material rich in sulphur are called black smokers. White smokers in comparison are lower in temperature and emit lighter color materials rich in barium, calcium and silicon. The vents and comparable deep subterranean hot aquifers support the abiotic syntheses of a range of biomolecules (Section 1.4).

Besides the submarine hydrothermal vents, sites close to impacts of meteorites would benefit from organic compounds brought by the meteorites (Table 6.2) as well as phosphate-yielding schreibersite. Crater formation by asteroids and comets also could result in hydrothermal systems. The shock waves created by the impact could generate amino acids and other organic compounds and the impacting bodies could enrich their surroundings with compounds from space.[80]

(iv) Icy Pools

Water-ice mixtures facilitate template-directed RNA polymerization, which depends on base pairing and base stacking that are favored by low temperatures. In the presence of Mg(II)/Pb(II) mixtures at slightly below the freezing point, up to 90% quasi-equimolar incorporation of all RNA monomers into 5 to 17-mers was observed with traces of longer products (Section 10.4). Given the pivotal importance of RNA or RNA-like replicators, water-ice mixtures could be attractive sites.[81] Such mixtures would also accelerate purine synthesis.[23,24]

(v) Composite Reactors

Surface water, clays, minerals, hydrothermal systems and icy pools all have important advantages to offer. So where might the fastest reactor be found? Speed was important, for as Darwin pointed out, life forms that arrived late would be lunch for those who arrived early. Given a heterotrophic origin and the need sooner or later for RNA replication, access to multiple nutrient sources together with optimal conditions for RNA polymerization rank high among the factors enabling a speedy reactor.

These considerations favor the proposal of a composite reactor in a landscape where multiple nutrient sources converged to optimize prebiotic development.[5] Figure 9.5 illustrates a reactor of this kind at

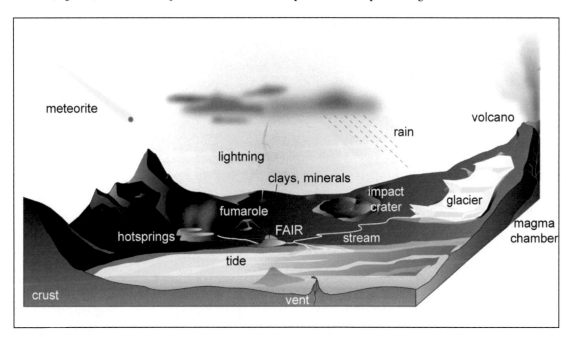

Figure 9.5. Fire And Ice Reactor (FAIR), one of the composite reactors of Mojzsis et al collecting biomolecules from diverse environmental sources, including in this instance near shore hydrothermal vents.[5]

a tidal pool in a volcanic sub-polar region, not too cold to be totally frozen but cold enough to have continual or intermittent presence of water-ice mixtures, with rain water mixing with sea water to result in a gradient of ionic compositions suitable for formose reaction and adequate solubility of phosphates, adorned with facilitator clays and minerals and next to hot springs, fumaroles, asteroid and meteorite craters and possibly even near-shore submarine hydrothermal vents. Organic compounds and condensed phosphates from the different sources would be fed into such a Fire And Ice Reactor (FAIR) via streams, seepage and tide to engage in prebiotic evolution. To-day, volcanic icy landscapes of this description may be found in the northern arc of the Pacific Ring of Fire such as Kamchatka Peninsula, Cook Inlet in Alaska, or in the Northern Atlantic such as Reykjavik (or 'Bay of Smokes' on account of its many hot springs). On primitive Earth, when volcanism was prevalent, FAIRs would be commonplace. In countless FAIRs as well as a multitude of other varieties of composite and noncomposite reactors, prelife assemblages would make continual attempts to surmount the barely surmountable hurdles they faced. It was a life or no-life primordial bingo that eventually accomplished at one of these reactors a success story that is still reverberating to-day across the galaxies.

References

Further Readings

1. Brack A ed. The molecular origins of life. Cambridge University Press, 1998.
2. Deamer DW, Fleischaker GR. Origins of life. The central concepts. Jones and Bartlett, 1994.
3. Hazen RM. Genesis. The scientific quest for life's origin. Joseph Henry Press, Washington DC, 2005.
4. Jortner J. Conditions for the emergence of life on the Earth: summary and reflections. Phil Trans R Soc B 2006; 361:1877-1891.
5. Mojzsis SJ, Krishnamurthy R, Arrhenius G. Before RNA and after: geophysical and geochemical constraints on molecular evolution. In: Gesteland RF, Cech TR, Atkins JF, eds. The RNA World, 2nd ed. Cold Sprong Harbor Laboratory Press 1999:1-47.

Specific References

6. Orgel LE. Prebiotic chemistry and the origin of the RNA world. Critical Review Biochem Mol Biol 2006; 39:99-123.
7. Bada JL, Lazcano A. Prebiotic soup—revisiting the Miller experiment. Science 2003; 300:745-746.
8. Miller SL. The formation of organic compounds on the primitive Earth. Annals NY Acad Sci 1957; 69:260-275.
9. Miyakawa S, Yamanashi H, Kobayashi K et al. Preiotic synthesis from CO atmospheres: implications for the origins of life. Proc Natl Acad Sci USA 2002; 99:14628-14631.
10. Kobayashi K, Tsuchiya M, Oshima T et al. Abiotic synthesis of amino acids and imidazole by proton irradiation of simulated primitive Earth atmosphere. Orig Life Evol Biosph 1990; 20:99-109.
11. Kobayashi K, Kaneko T, Saito T et al. Amino acid formation in gas mixtures by high energy particle irradiation. Orig Life Evol Biosph 1998; 28:155-165.
12. Thaddeus P. The prebiotic molecules observed in the interstellar gas. Phil Trans R Soc B 2006; 361:1681-1687.
13. Bernstein M. Prebiotic materials from on and off the early Earth. Phil Trans R Soc B 2006; 361:1689-1702.
14. Deamer DW. Prebiotic amphiphilic compounds. In: Seckbach J, ed. Origins, Genesis, Evolution and Diversity of Life. Kluwer Academic Publishers 2004:77-89.
15. Pizzarello S. Meteorites and the chemistry that preceded life's origin. In: Wong JT, Lazcano A, eds. Prebiotic Evolution and Astrobiology. Austin: Landes Bioscience, 2008.
16. Wong JT. Coevolution of the genetic code and amino acid biosynthesis. Trends in Biochem Sci 1981; 6:33-35.
17. Wong JT. Coevolution theory of the genetic code at age thirty. BioEssays 2005; 27:416-425.
18. Hennet RJC, Holm NG, Engel MH. Abiotic synthesis of amino acids under hydrothermal conditions and the origin of life: a perpetual phenomenon? Naturwissen 1992; 79:361-365.
19. Hazen RM, Filley TR, Goodfriend GA. Selective adsorption of L- and D-amino acids on calcite: implications for biochemical homochirality. Proc Natl Acad Sci USA 2001; 98:5487-5490.
20. Wong JT. Evolution and mutation of the aminoacid code. In: Ricard J, Cornish-Bowden A, eds. Dynamics of Biochemical Systems. Plenum Press 1984:246-257.
21. Wong JT, Bronskill PM. Inadequacy of prebiotic synthesis as origin of proteinous amino acids. J Mol Evol 1979; 13:115-125.
22. Oro J. Mechanism of synthesis of adenine from hydrogen cyanide under possible primitive Earth conditions. Nature 1961; 191:1193-1194.
23. Sanchez R, Ferris JP, Orgel LE. Conditions for purine synthesis: did prebiotic synthesis occur at low temperatures? Science 1966; 153:72-73.
24. Schwartz AW, Joosten H, Voet AB. Prebiotic adenine synthesis via HCN oligomerization in ice. Biosystems 1982; 15:191-193.
25. Miyakawa S, Murasawa K, Kobayashi K et al. Abiotic synthesis of guanine with high temperature plasma. Orig Life Evol Biosph 2000; 30:557-566.
26. Ferris JP, Sanchez RA, Orgel LE. Studies in prebiotic synthesis. 3. Synthesis of pyrimidines from cyanoacetylene and cyanate. J Mol Biol 1968; 33:693-704.
27. Sanchez R, Ferris JP, Orgel LE. Cyanoacetylene in prebiotic synthesis. Science 1966; 154:784-785.
28. Levy M, Miller SL. The stability of the RNA bases: implications for the origin of life. Proc Natl Acad Sci USA 1998; 95:7933-7938.
29. Shapiro R. Prebiotic cytosine synthesis: a critical analysis and implications for the origin of life. Proc Natl Acad Sci USA 1999; 96:4396-4401.
30. Zubay G. Studies on the lead-catalysed synthesis of aldopentoses. Orig Life Evol Biosph 1998; 28:13-26.
31. Holm NG, Dumont M, Ivarson M et al. Alkaline fluid circulation in ultramafic rocks and formation of nucleotide constituents: a hypothesis. Geochem Trans 2006; 7:7 doi:10.1186/467-4866-7-7.
32. Sacerdote MG, Szostak JW. Semipermeable lipid bilayers exhibit diastereoselectivity favoring ribose. Proc Natl Acad Sci USA 2005; 102:6004-6008.
33. Cooper G, Kimmich N, Belisle W et al. Carbonaceous meteorites as a source of sugar-related organic compounds for the early Earth. Nature 2001; 414:879-883.
34. Sanchez RA, Orgel LE. Studies in prebiotic synthesis. V. Synthesis and photoanomerization of pyrimidine nucleosides. J Mol Biol 1970; 47:531-543.
35. Fuller WD, Sanchez RA, Orgel LE. Studies in prebiotic synthesis: VII. Solid state synthesis of purine nucleosides. J Mol Evol 1972; 1:249-257.
36. Schwartz AW. Phosphorus in prebiotic chemistry. Phil Trans R Soc B 2006; 361:1743-1749.
37. Yamagata Y, Watanabe H, Saitoh M et al. Volcanic production of polyphosphates and its relevance to prebiotic chemistry. Nature 1991; 352:516-519.
38. Beck A, Lohrmann R, Orgel LE. Phosphorylation with inorganic phosphates at moderate temperatures. Science 1967; 157:952.
39. Yamagata Y, Inoue H, Inomata K. Specific effect of magnesium ion on 2'3'-cyclic AMP synthesis from adenosine and trimetaphosphate in aqueous solution. Orig Life Evol Biosph 1995; 25:47-52.
40. Cheng C, Fan C, Wan R et al. Phosphorylation of adenosine with trimetaphosphate under simulated prebiotic conditions. Orig Life Evol Biosph 2002; 32:219-224.
41. Joyce GF, Schwartz AW, Miller SL et al. The case for an ancestral genetic system involving simple analogues of the nucleotides. Proc Natl Acad Sci USA 1987; 84:4398-4402.
42. Bolli M, Micura R, Eschenmoser A. Pyranosyl-RNA: chiroselective self-assembly of base sequences by ligative oligomerization of tetranucleotide-2'3'-cyclophosphates. Chem and Biol 1997; 4:309-320.
43. Eschenmoser A. Chemical etiology of nucleic acid structure. Science 1999; 284:2118-2124.
44. Schoning K, Scholz P, Guntha S et al. Chemical etiology of nucleic acid structure: the α-threofuranosyl-(3' → 2') oligonucleotide system. Science 2000; 290:1347-1351.
45. Egholm M, Buchardt O, Nielson PE et al. Oligonucleotide analogues with an achiral peptide backbone. J Am Chem Soc 1992; 114:1895-1897.
46. Miller SL. Peptide nucleic acids and prebiotic chemistry. Nat Struct Biol 1997; 4:167-169.
47. Sen S, Nilsson L. MD simulations of homomorphous PNA, DNA and RNA single strands: characterization and comparison of conformations and dynamics. J Am Chem Soc 2001; 123:7414-7422.
48. Lazcano A, Guerrero R, Margulis L et al. The evolutionary transition from RNA to DNA in early cells. J Mol Evol 1988; 27:283-290.
49. Freeland SJ, Knight RD, Landweber LF. Do proteins predate DNA? Science 1999; 286:690-692.
50. Deamer DW. Prebiotic amphiphilic compounds. In: Seckbach J, ed. Origins, Genesis, Evolution and Diversity of Life. Kluwer Academic Publishers 2004:77-89.

51. Rushdi AI, Simonet BRT. Lipid formation by aqueous Fischer-Tropsch type synthesis over a temperatrure range of 100-400°C. Orig Life Evol Biosph 2001; 31:103-118.

52. Rao NN, Liu S, Kornberg A. Inorganic polyphosphate in Escherichia coli: the phosphate regulon and the stringent response. J Bacteriol 1998; 180:2186-2193.

53. De Duve C. Blueprint for a cell. Neil Patterson Publishers. 1991:113-116.

54. Liu R, Orgel LE. Oxidative acylation using thioacids. Nature 1997; 389:52-54.

55. Weber AL. Formation of pyrophosphate, tripolyphosphate and phosphorylimidazole with the thioester N,S-diacetyl-cysteamine as the condensing agent. J Mol Evol 1981; 18:24-29.

56. De Zwart II, Meade SJ, Pratt AJ. Biomimetic phosphoryl transfer catalysed by iron(II)-mineral precipitates. Geochimica Cosmochimica Acta. 2004; 20:4093-4098.

57. Russell MJ, Martin W. The rocky roots of the acetyl-CoA pathway. Trends in Biochem Sci 2004; 29:358-363.

58. Wachterhauser G. Before enzymes and templates: theory of surface metabolism. Microbiol Rev 1988; 52:452-484.

59. Wachterhauser G. From volcanic origins of chemoautotrophic life to Bacteria, Archaea and Eukarya. Phil Trans R Soc B 2006; 361:1787-1808.

60. Weber AL. The triose model: glyceraldehyde as a source of energy and monomers for prebiotic condensation reactions. Orig Life Evol Biosph 1987; 17:107-119.

61. Lazcano A, Miller SL. On the origin of metabolic pathways. J Mol Evol 1999; 49:424-431.

62. Keefe AD, Miller SL. Are polyphosphates or phosphate esters prebiotic reagents? J Mol Evol 1995; 41:1432.

63. Carmeli C, Lifshitz Y. Nucleotide transport in Rhodobacter capsulatus. J Bacteriol 1989; 171:6521-6525.

64. Miller SL, Orgel LE. The origins of life on the Earth. Prentice Hall 1974:86-87.

65. Miller SL, Bada JL. Submarine hot springs and the origin of life. Nature 1988; 334:609-611.

66. Islas S, Velasco AM, Becerra A et al. Hyperthermophily and the origin and earliest evolution of life. Int Microbiol 2003; 6:87-94.

67. Ronimus RS, Morgan HW. Distribution and phylogenies of enzymes of the Emden-Meyerhof-Parnas pathway from archaea and hyperthermophilic bacteria support a gluconeogenic origin of metabolism. Archaea 2003; 1:199-221.

68. Lautru S, Challis GL. Substrate recognition by nonribosomal peptide synthetase multi-enzymes. Microbiol 2004; 150:1629-1636.

69. Wong JT. Origin of genetically encoded protein synthesis: a model based on selection for RNA peptidation. Orig Life Evol Biosph 1991; 21:165-176.

70. Stryer L. Biochemistry. WH Freeman 1995; 552.

71. Brack A. The chemistry of life's origins. In: Seckbach J, ed. Origins, Genesis, Evolution and Diversity of Life. Kluwer Academic Publishers 2004:61-73.

72. Bernal JD. The physical basis of life. Routledge and Kagan Paul 1951;

73. Negron-Mendoza A, Ramos-Bernal S. The role of clays in the origin of life. In: Seckbach J, ed. Origins, Genesis, Evolution and Diversity of Life. Kluwer. Academic Publishers 2004:183-194.

74. Nikalje MD, Puhukan P, Sudalai A. Recent advances in clay-catalyzed transformations. Org Prep Proced 2000; 32:1-40.

75. Arrhenius GO. Crystals and life. Helvetica Chim Acta 2003; 86:1569-1586.

76. Ferris JP. Montmorillonite-catalysed formation of RNA oligomers: the possible role of catalysis in the origins of life. Phil Trans R Soc B 2006; 361:1777-1786.

77. Corliss JB, Dymond J, Gordon LI et al. Submarine thermal springs on the Galapagos Rift. Science 1979; 203:1073-1083.

78. Baross JA, Hoffman SE. Submarine hydrothermal vents and associated gradient environments as sites for the origin and evolution of life. Orig Life Evol Biosph 1985; 15:327-345.

79. Van Dover CL. The ecology of deep-sea hydrothermal vents. Princeton University Press 2000; 55.

80. Cockell CS. The origin and emergence of life under impact bombardment. Phil Trans R Soc B 2006; 361:1845-1856.

81. Monard P-A, Kanavarioti A, Deamer DW. Eutectic phase polymerization of activated ribonucleotide mixtures yields quasi-equimolar incorporation of purine and pyrimidine nucleobases. J Am Chem Soc 2003; 125:13734-13740.

CHAPTER 10

The Dawn of the RNA World:
RNA Polymerization from Monoribonucleotides under Prebiotically Plausible Conditions

Pierre-Alain Monnard*

10.1 Introduction

The polymerization of RNA molecules from abiotically synthesized monoribonucleotides is a critical step in the emergence of a metabolism centered on the dual functions of RNA (genetic information repository and catalyst), which is postulated in the RNA-World hypothesis. This chapter presents recent achievements in the area of non-enzymatic RNA polymerization and assesses their significance in the light of the de novo emergence of the RNA World.

10.2. RNA as the Central Molecule for a Prebiotic Metabolism

Extant cellular life is based on well understood principles of molecular biology that comprise the protein-dependent replication of nucleic acids and the nucleic acid-dependent encoding of proteins. Although it is obvious that a simpler system must have preceded today's complex cellular biology, this interdependence of proteins and nucleic acids has long represented a profound puzzle (often depicted as the "chicken and egg" dilemma) for researchers attempting to determine the nature of a primitive "biochemical" metabolism. In fact, a "satisfactory" answer to this dilemma could not be envisioned until the RNA-World concept was first hypothesized in the1960s by F.H.C. Crick[6] and L.E. Orgel[7] among others and later named by W. Gilbert.[8] These authors postulated an autonomous RNA-based "organism" or protocell in which RNA strands could perform several functions presently carried out by proteins, first and foremost that of RNA replication, while acting as genetic information repository. At that time, no catalytic RNAs (so called ribozymes) had been discovered and this idea was rather speculative because it was only supported by the ubiquitous distribution of coenzymes incorporating nucleotides in their structures. Natural selection through replication and mutation was also known to be the sole mechanism for the evolution of complex biochemical systems from simpler ones. The discovery of self-splicing ribozymes in contemporary cells and more recently the demonstration that ubiquitous, ribosomal peptide synthesis is a ribozyme-catalyzed reaction[9] have lent increased plausibility to the idea of an RNA World as the precursor of the current DNA-RNA-Protein World. Although the RNA World seems inevitable at some point during the emergence of cellular Life from inanimate matter, it is not clear whether it represented the first system during evolution towards Life. In fact, the evidence at hand still leaves open questions about the origins and "biochemistry" of the RNA World.

The realization of a functional RNA World requires a series of events (Fig. 10.1) including the abiotic synthesis of RNA monomers and their assembly into oligomers (in the likely presence of metal catalysts) that could be elongated either by monomers or by ligation with another oligomer. These molecules would have had to serve as templates for their own copying or replication and all these events must have initially occurred non-enzymatically. At that stage, a pool of RNAs (a large set of RNA molecules) must have existed from which a set of catalytic RNAs (among them, RNA-based RNA copying molecules) emerged through natural selection that together sustained exponential growth in the prebiotic environment.

(i) Implications of the RNA Activity for RNA Polymerization

Catalysis requires molecular recognition of a target molecule by the catalyst. It is likely that early ribozymes were not as substrate specific as today's protein enzymes, a fact that would have allowed them to take full advantage of the various substrate derivatives synthesized abiotically while still permitting a relatively efficient catalysis. Just as for protein enzymes, the recognition of target molecules by ribozymes is determined by the conformation of the RNA molecules (i.e., their three-dimensional shape) (Fig. 10.2 and Box 10.1).

The assembly of RNA motifs requires relatively long polymers: known synthetic or natural RNA polymers capable of catalysis or molecular recognition tend to be composed of at least 20-30 units with a large nucleobase fraction involved in the formation of structural motifs. In vitro selected RNA ligases and polymerases are formed by even longer strands of at least 50 and 150 monomers, respectively.

Such lengths have implications for the RNA-polymerization process. Indeed, if a 50-mer RNA is needed for a given activity, 4^{50} or 10^{30} different RNA sequences (assuming four different nucleobases) theoretically exist, but 3.5×10^7 kg RNA (the total weight of all possible molecules of that length) cannot likely be made at once. Although a significant number of such sequences could have had the same activity, thereby reducing the above amount, concentration issues must not be overlooked: a threshold concentration of active molecules is always necessary for a given activity and must be taken into account when considering an initiation of the RNA World. This is especially true if several sequences were needed simultaneously in a cooperative role. Thus, the RNA polymerization

*Pierre-Alain Monnard—Los Alamos National Laboratory, Earth and Environmental Science Division (EES-6), Los Alamos, USA.
Current address: Flint Center, University of Southern Denmark, Odense M, Denmark. Email: monnard@ifk.sdu.dk

Prebiotic Evolution and Astrobiology, edited by J. Tze-Fei Wong and Antonio Lazcano. ©2009 Landes Bioscience.

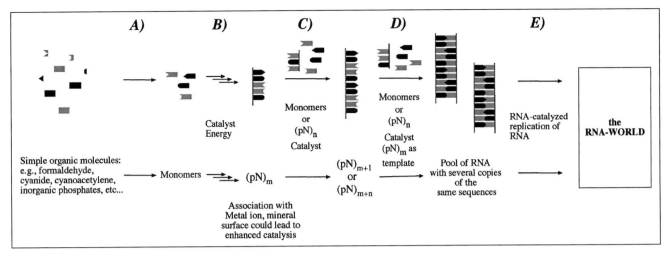

Figure 10.1. Hypothetical route to the de novo RNA World. N stands for any nucleobase, (pN) for a RNA oligomer or polymer. m, m+1 and m+n subscripts refer to the length of a polymer. A) Synthesis of the RNA monomers from simple organic precursors. This phase relies on prebiotic chemistry. B) Oligomerization of the monomers. Monomers are incorporated (polymerized) into oligomers in the likely presence of metal-ion catalysts as no protein enzymes were present to catalyze such a reaction. C) Synthesis of longer polymers. Polymers required for catalytic activity can be obtained either by ligation of previously formed oligomers or further monomer addition. D) Non-enzymatic, template-directed polymerization. To transmit the sequence of catalytic RNA fragments, the ability to non-enzymatically copy these molecules is essential. This process will not only permit an increase of the catalytic activity, but also of the RNA-pool size as well as the emergence of new catalysts because of copying errors. E) The RNA World. When the RNA pool contains molecules that are also catalysts of their own copying, as well as those of any other functional RNAs, a full-fledged RNA World is achieved.

processes must have intrinsically yielded RNA pools containing a large fraction of catalytically active molecules (by what is thus far an unknown process).

(ii) Environmental Conditions

Unfortunately, few clues about the original environmental conditions on primitive Earth remain due to the activity of the biosphere and plate tectonics. However, recent investigations suggest the presence of liquid water and continents as far back as 4.3×10^9 years ago. Higher volcanic activity and lower sunlight levels than today can also be expected. Thus life producing reactions could have occurred over a broad spectrum of conditions in terms of temperature, pressure, ionic strength, pH, etc. and have been located on land, deep within the Earth, in the oceans, or at interfaces between these regions.

10.3 RNA Monomers

Although the synthesis of the different RNA monomers from prebiotic molecules is mostly beyond the scope of this chapter (see ref. 2), a quick summary of research efforts on this topic is necessary. The implication of monomer syntheses (or the lack thereof) for the de novo emergence of an RNA World from prebiotic organic mixtures is essential for our discussion.

(i) Synthesis of the Ribonucleotides

RNA monomers, β-D-ribonucleotides, are rather complex molecules formed by three different types of molecules—a nucleobase, a ribose and a phosphate group (see Box 10.1), each of which needs to be synthesized as an intermediate product (or made available in the case of the phosphate group) before being assembled into the ribonucleotide.

Experiments aiming at prebiotic nucleotide synthesis have generally concentrated on one of these specific moieties:

- They have demonstrated the synthetic pathways for the formation of both purines and pyrimidines of interest among a large number of nonbiological derivatives.
- D-ribose can be generated by the formose reaction from a prebiotic substrate, formaldehyde, but ribose represents only a rather small fraction of all sugar products. In addition, this sugar decomposes with its half-life being between hours and years at pH 7, depending on the temperature. This situation can however be partially remedied either by complexation with borate or derivatization of the ribose.
- Nucleoside synthesis is obtained by coupling a ribose with a nucleobase and currently represents the weakest step in prebiotic nucleotide synthesis. Recent studies however hint at a simplified synthesis when starting with a ribose derivatized by phosphate.[10]
- The nucleoside can react with inorganic phosphate to yield nucleotide 5'-phosphates.

Although these "prebiotically plausible" syntheses are conducted with highly purified reactants, they still deliver low yields for any given product. Multiple steps such as solvent extractions are involved, which are certainly not prebiotic. Finally, they produce a complex mixture of compounds, some of which are isomers (molecules with the same chemical formula and often with the same kinds of bonds between atoms, but in which the atoms are arranged differently) of the intended products and which could easily inhibit polymerization or replication.

The difficulties encountered in the syntheses are so severe that alternatives to the de novo appearance of RNA on the primitive Earth should be seriously considered. For example, (i) the nucle-

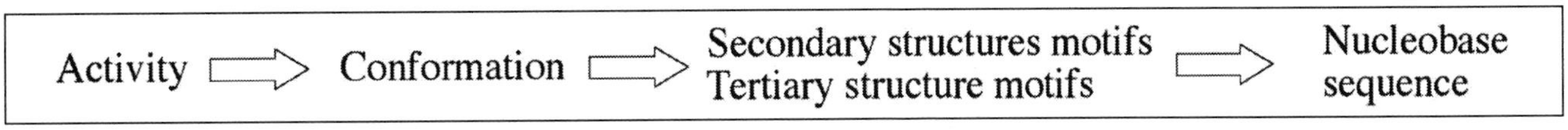

Activity ⇒ Conformation ⇒ Secondary structures motifs / Tertiary structure motifs ⇒ Nucleobase sequence

Figure 10.2. Chain of dependence for RNA activity. The catalytic activity of an RNA depends on its conformation that is obtained from the interactions between secondary and tertiary structure motifs. In turn, the assembly of such motifs is conditioned by the nucleobase sequence.

Box 10.1. Formation of RNA Structures

RNA strands can form a variety of structures that are integral to the in vivo functions of the molecules. RNA structures likely played a significant role in the RNA World as they do for in-vitro selected ribozymes today.

The structures will mainly depend on the RNA molecule sequences and their base-pairing behavior, that is, their capacity to form hydrogen bonds (H-bonds) between two nucleobases, as well as unspecific nucleobase stacking. Other properties such as the torsion angles along the phosphate-ribose backbone (Fig. Box 10.1.1A), the nucleobase and ribose conformations (Fig. Box 10.1.1B) will also contribute to structure formation mainly by constraining the molecular conformations (see ref. 1). The various base pairs are classified into two categories: the common Watson-Crick base pairs (Fig. Box 10.1.1C) and the nonWatson-Crick base pairs (Fig. Box 10.1.1D).

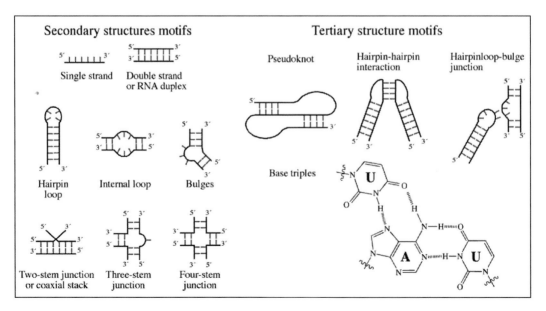

Figure Box 10.1.1. Factors influencing RNA structures. A) RNA polymer is formed of β-D-ribonucleotides linked by phosphodiester linkages. Each nucleotide is composed of a backbone consisting of phosphate (dark highlighted)-ribose (light highlighted) group and a nucleobase, a heterocycle, which is attached to the C1′ of the ribose in a β configuration (both nucleobase and phosphate above the ribose-ring plane). Nucleobases are classified as purines, e.g., Adenine (A) and Guanine (G), or pyrimidines, e.g., Cytosine (C) and Uridine (U). Each single bond (dotted fat lines) along its backbone can rotate. B) The D-ribose can adopt two types of conformations (or puckering forms) C2′- and C3′-endos. C) Watson-Crick base pairs: A•U and G•C interact via two and three hydrogen bonds (dotted lines), respectively. An N-H group acts as hydrogen donor, while a nonbinding electron pair of an oxygen on the nucleobase ring (C = O) or a ring nitrogen are the acceptors. D) Non-Watson-Crick base pairs; on the left, a Wobble base-pair system (G•U); and, on the right, some additional nonWatson-Crick base pairs that occur in RNA.

Structures stabilized by interactions between two strands or by binding of metal ions or proteins on the RNA are classified into categories of secondary and tertiary structure motifs (Fig. Box 10.1.2).

Hairpins are the most common motif in RNA folding. They are composed of a generally short, double-stranded, helical stem and a terminal loop of variable numbers of nucleotides. They clearly illustrate the relation between nucleobase sequence and strand conformation that allows the stabilization of an RNA fold. For example, in the loop sequence UUCG, the conformation of the backbone and the nucleobases allows for the formation of an additional Wobble base pair (G•U) within the loop and the stacking of the remaining loop nucleobases, thereby increasing the thermodynamic stability of the stem.

Figure Box 10.1.2. Examples of secondary and tertiary structure motifs. The backbone is represented by the black line and the nucleobases by the light-gray short lines. Two light-gray lines facing each other represent a base pair. In the base triple (UAU) the A•U is a Watson-Crick base pair whereas U•A is not.

otides might have been delivered by comets and meteorites after having been synthesized under conditions (temperature, pressure, pH) completely different from those we infer for the primitive Earth. This proposition may well be relevant if the amount of carbon compounds delivered by extraterrestrial bodies (over 10^{16} kg during the several first 100 million years following the Earth formation) is considered. (ii) RNA could have been preceded by another more prebiotic molecule, component of a preRNA World that invented the RNA World. (iii) The RNA World might have emerged from a network of catalytic reactions using prebiotic organic molecules to produce more complex molecules. In this case, the system information was intrinsic to the interconnections of the catalytic processes. This idea is often referred to as the "Metabolic World."

The idea of a preRNA World (ii) has led to the investigation of several information polymers with different backbones, deemed simpler to synthesize (Fig. 10.3). It is assumed that this hypothetical RNA ancestor should efficiently form base pairs with itself and RNA monomers in order to obtain information transfer between the two Worlds.

Eschenmoser and his collaborators investigated in depth nucleotide derivatives based on variations of the sugar ring size[11] to try to understand the reasons underlying the choice of ribose and deoxyribose in natural nucleic acids. Others designed new polymers that could solve some of the synthetic/structural issues of RNA: for example, nucleic acid derivatives with a peptide[12] or a glycerol-phosphate[13] backbone. Such derivatives are interesting for gaining insights into RNA properties. Although it is claimed that some of them, such as GNA and PNA, could be prebiotically plausible, no prebiotic syntheses, let alone processes for their non-enzymatic polymerization, have yet been reported. Moreover, some obstacles to the realization of these processes are already well documented, for example, the cyclization of PNA monomers into cyclic dimers.

In summary, there are no satisfactory de novo syntheses for β-D-nucleotides reported yet, which might reflect our poor understanding of prebiotic conditions or may well support the idea that the RNA World was not the first step in the evolution leading to cellular life. However, even if the "Metabolic World" postulate is correct, the catalytic network information would have to undergo transition at some point to an information/catalyst polymer-based form. Thus the study of non-enzymatic RNA polymerization and replication remains essential to an understanding of this process.

(ii) Reactivity of RNA Monomers

To ensure the polymerization of RNA monomers, it is crucial to understand their reactivity. The standard free energy for the synthesis of a phosphodiester bond is approximately +5.3 kcal/mol in aqueous solutions. Thus polymerization will not spontaneously occur unless the monomers are chemically activated (i.e., they are provided with chemical energy stored in the bond between the monomer and an activation group), or external energy (e.g., heat) is supplied. To activate RNA monomers, inorganic polyphosphates (Fig. 10.4) can be used, as they have been shown to promote the polymerization. However, because of their slow polymerization rates, most of the "prebiotic" polymerizations have been and still are, carried out using nucleotides activated as phosphoramidates, usually phosphorimidazoles (Fig. 10.4), even though these molecules might not have been present on early Earth.

Natural RNA is generally linked with 3'-5' linkages, a regioselectivity (Fig. 10.4) that is ensured by polymerase enzymes. Metal-ion mediated, non-enzymatic polymerization tends to yield a mixture of all possible linkages, thus producing RNA analogs. (When we speak of RNA in the remainder of this chapter, we will refer to the mixtures of RNA and its analogs). Their relative frequency depends on the relative reactivity of the nucleophilic hydroxyl groups and also on the type of metal catalysts and the medium used. The heterogeneity of linkages in non-enzymatically synthesized oligomers could prevent RNA function and replication.

10.4 RNA Polymerization or Self-Condensation of RNA Monomers

To qualify as "prebiotically plausible", any RNA polymerization process should likely occur

- in an aqueous environment or at least be supplied with its reacting species from an aqueous medium,
- in the presence of a complex mixture of monomers at low initial concentrations,
- and in the presence of metal-ion catalysts either as dissolved ions or on solid mineral surfaces.
- Furthermore, the polymerization rates must be higher than the decomposition rates of the polymeric products. This

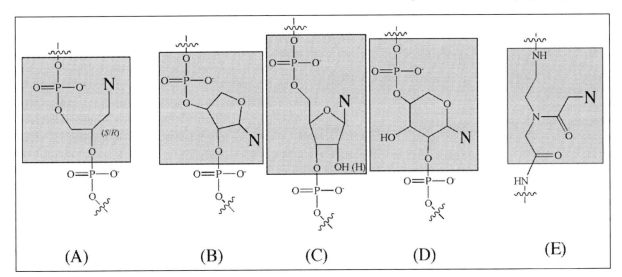

Figure 10.3. Examples of alternative monomers (dark highlight). A) Glycerol nucleic acid, (S/R) GNA; B) Threose nucleic acid, TNA; C) Ribose or deoxyribose nucleic acid depending on the 2' group (OH versus H), RNA and DNA; D) pyranosyl RNA, pRNA and E) Peptide Nucleic Acid, PNA. (A) and (E) backbones are not sugar based. B, D and E can base-pair with themselves. A, B and E can base-pair with C, the natural nucleic acids.

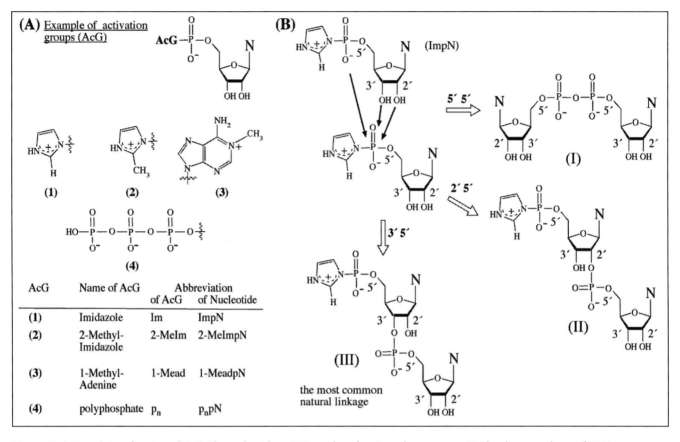

Figure 10.4. Reactivity of activated β-D-ribonucleotides. A) Examples of activated monomers. B) The three products of RNA monomer dimerization have differing regioselectivity: (I) 5′-5′ or pyrophosphate, (II) 2′-5′ and (III) 3′-5′. N stands for any nucleobase.

kinetic aspect is important as polymerization reactions tend to be inhibited in an aqueous environment due to entropic effects.

- Finally, the reaction process must selectively produce a pool of polymers with a high potential for activity.

We will now review the various approaches investigated to perform RNA polymerization from monomers, also called monomer self-condensation, which in part fulfill the requirements listed above.

(i) Self-Condensation in Homogeneous Aqueous Media

It was expected that activated nucleotides in the presence of dissolved divalent metal-ion catalysts (e.g., magnesium, lead or uranyl ions) would form extended stacks (which bring the reacting groups in close proximity) due to the relatively hydrophobic character of

the heterocyclic nucleobases and the interactions of the divalent metal-ions with the negatively charged phosphate and ribose hydroxyl groups (Fig. 10.5).

In principle, the activation groups with the metal-ions should permit one to overcome the energy barrier thereby allowing polymerization to occur. However, at low concentrations of catalyst and monomers, the products are primarily dimers with pyrophosphate and 2′-5′ linkages. Even implementing a scenario such as the tidal pool (this scenario proposes concentrated mixtures of activated monomers and metal ion catalysts in small pools on beaches after the evaporation of the water), reactions only yield few products up to the 4-mers.[14] This is in part due to the hydrolysis of the active species and to the incapacity of the hydrophilic reactive molecules (in particular, U residues) to form stacks. Oligomer elongation either by monomer addition or by ligation of short oligomers is

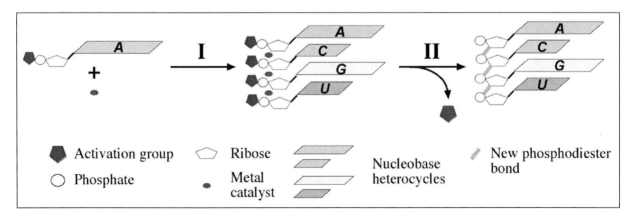

Figure 10.5. Schematic representation of RNA-monomer self-condensation. I) Self-assembly of the stacks. II) Phosphodiester bond formation catalyzed by metal-ion.

equally unsuccessful. That is, an oligomer length that would allow for product activity cannot be reached by non-enzymatic synthesis in homogeneous aqueous solution.

In the light of these unsuccessful attempts, it was surmised that the homogeneous aqueous set-ups do not allow for the ordering of the reacting species by stacking. Thus, a specific environment had to be introduced in the system, which would interact with the monomers and catalysts and concentrate them, thereby reducing water molecule activity and simultaneously ordering them. To date, three environments have been investigated: mineral surfaces, the eutectic phases in water-ice and lipid-bilayer lattices.

(ii) Self-Condensation on Mineral Surfaces, Such as Montmorillonites

The charged nature and regular structure of mineral surfaces can provide a supporting environment for polymerization of RNA. Various inorganic minerals, such as hydroxylapatite ($Ca_5(PO_4)_3OH$) and clays such as montmorillonites (Fig. 10.6), have been used to promote RNA polymerization. Of the various generic clays tested, most showed little or no activity. However, dispersions of chemically treated montmorillonites, called homoionic, demonstrate significant performance.[15]

Activated nucleotides can adsorb either directly on the surface (via hydrophobic interactions or electrostatic interactions between their positively charged activating groups and the negatively charged surface) and/or be held there by metal-ion bridges between the nucleotide phosphates and the surface. The extent of the adsorption will be dependent on the properties of the activating group, the mineral surface and the aqueous solution. Furthermore, polymers adsorb more tightly than individual monomers due to their multiple cohesive interactions.

The montmorillonite-supported ImpA self-condensation in aqueous electrolyte solutions containing $MgCl_2$ (the catalyst) produces RNAs of up to 50 monomeric units in length when small oligomers acting as primers (short deoxyribose oligomers with a last ribonucleotide, $(pdA)_9pA$, or pyrophosphate ribonucleotides dimers, AppA) are present. In such experiments, activated monomers need to be regularly added (once a day for 14 days) to replenish the monomer stocks and prevent the growth-limiting effect of monomer hydrolysis. In contrast, when using monomers with large

heterocycles as their activating group, such as 1-methyladenine, a comparable oligomer length is attained without primers within only 3 days. Moreover, all four nucleotides are incorporated.[16]

This polymerization is selective both at the level of the product nucleobase sequence and the regiochemistry of the linkages (see Box 10.2 for explanation concerning the experimental determination of these parameters). For example, while all possible dimers are formed in ImpA-ImpC mixtures, only a small number of all longer oligomer isomers predicted by the random synthesis are observed.[17] The 3' terminal nucleobase (purine or pyrimidine) of the nascent oligomer and its regiospecificity determine the reactivity for elongation. Because a 3'-5'-linked terminal nucleotide elongates more efficiently, the ratio of 2'-5' or 3'-5' linkages observed on montmorillonite is greatly modified in favor of the natural isomers (3'-5') compared to that in a homogeneous aqueous medium where the 2'-5' linkage is predominant. Finally, since a 3' terminal purine has a higher intrinsic reactivity than that of a 3' terminal pyrimidine, the formation of purine-rich oligomers is favored.

(iii) Eutectic Phase in Water-Ice

Exploration of planetary bodies in recent years has clearly established that water exists elsewhere in the Universe often as ice. On the primitive Earth, the solar irradiation was less intense (approximately 30% weaker) than today; thus water-ice mixtures might have been present.

Water-ice systems represent an interesting environment for RNA monomer self-condensation because such a system can very efficiently concentrate solutes (Fig. 10.7), thereby enabling their organization. Freezing reduces the water activity in the system by dehydrating the sample and owing to the low temperatures it should protect RNA products from thermal decomposition. The exact processes involved (simple concentration effects, interactions with the ice-surface, etc.) are not yet elucidated, but polymerization is only successful if ice nucleation occurs. Supercooled solutions are not conducive to efficient reaction.

When mixtures containing all four imidazole-activated monoribonucleotides in the presence of metal-ions, typically mixtures of Mg(II)/Pb(II), are incubated for up to 36 days at −18.4°C (a temperature slightly below the freezing point), quasi-equimolar, almost-complete incorporation (up to 90%) of all monomers

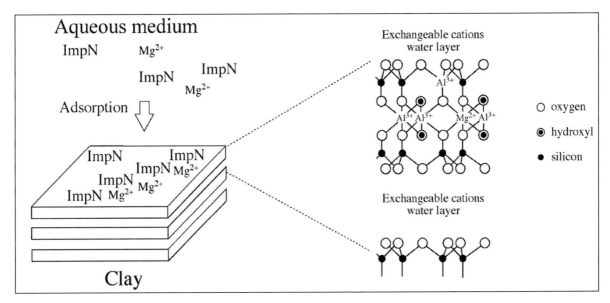

Figure 10.6. Montmorillonite structure and its influence on RNA polymerization (adapted from ref. 16). Montmorillonite is composed of two tetrahedral layers of silicates linked by octahedral aluminates. Si^{4+} and Al^{3+} are partially substituted by alkali metal or alkaline earth metal (e.g., Mg(II)), as well as Fe(III) ions, due to weathering. The homoionic montmorillonite form is obtained when all exchangeable cations are replaced by Na^+ ions. Reactions take place on the surfaces and not on the edges.

Box10.2. RNA Analytics

RNA strands produced by non-enzymatic polymerization can be analyzed in terms of length, nucleobase content and linkage regio-specificity using several methods.

Gel Electrophoresis. This method is used to determine the length of RNA products (greater than 5-mers) by comparison with a RNA ladder that contains RNA molecules of known length. A resolution of one nucleotide can be achieved. Also, the type of linkage for a primer elongated by one nucleotide can be determined.

Chromatography. High pressure liquid chromatography (HPLC) using both reverse-phase and ion-exchange columns is used in RNA analytics. With reverse-phase chromatography, it is possible to separate dimeric products both according to their nucleobase content and their type of linkage (2′-5′, 3′-5′ and pyrophosphate). This type of analysis is performed to analyze the enzymatic decomposition products of RNA polymers and determine the ratio of natural to unnatural linkages. This methodology can also be applied to determine the nucleobase content of RNAs that have been completely decomposed into nucleotides under alkaline conditions or by phosphodiesterases. Ion-exchange chromatography is used to determine the length of short oligomers and their regiospecificity.

Enzymatic Digestion. Ribonuclease or RNase enzymes specifically cleave 3′-5′ linkages and can be sequence-specific both in terms of nucleobase and the RNA structure (single strand or duplex) flanking the cleavage site. These enzymes can help us to determine the ratio of natural to unnatural linkages in conjunction with gel electrophoresis or chromatography.

Mass Spectrometry (MS). In particular, MALDI-TOF MS is used to determine the sequences of short oligomers. This method permits one to analyze the weight of biopolymers without destroying them. The resolution in mass can be as low as 1 proton mass, which allows for the relatively precise determination of sequence identity.

Sequencing. This methodology allows for the determination of nucleobase sequence in natural RNA and DNA. Nowadays, RNA fragments are enzymatically reverse-transcribed into cDNA or complementary DNA. This cDNA is then amplified, inserted in plasmids that are transfected in bacteria for further amplification. The DNA plasmid product is then sequenced. The mixture of linkages in RNA products synthesized non-enzymatically renders the reverse-transcription extremely difficult.

into medium-length mixed oligomers (up to 15- to 30-mers) is observed.[18] Contrary to clays, the reaction occurs in the presence of various heterogeneous mixtures of activation groups and metal catalysts at initial concentrations of reacting species as low as 10^{-6} M. Usually, mixtures of Mg(II)/Pb(II) are used, but Pb(II) ions alone are catalysts for the polymerization at a ratio of catalyst to monomers as low as 1:20.[19] Low concentrations of activated oligomers can effectively be ligated to yield longer polymers or elongated by adding fresh activated monomers. Interestingly, the elongation using monomer addition only proceeds beyond a range of length between 30- to 35-mers if the elongated oligomers are forming secondary structures (Monnard and Szostak, unpublished observations). This fact may indicate a form of selectivity toward structure-forming products, which could lead to a higher number of RNA catalytic species.

Product regiospecificity (3′-5′ linkage) is also improved compared to homogeneous aqueous reactions.[18] Furthermore, the eutectic phase in water-ice has the advantage of reduced hydrolysis

rates. In typical reactions, only 6-10% of an initial ImpU concentration is hydrolyzed to the inactive uridine monophosphate. In contrast, in homogeneous aqueous solutions and on montmorillonite hydrolysis is accelerated by metal ions in direct proportion to the monomer concentration by up to 600-fold.[19] The low hydrolysis rate in the eutectic could have been essential for the generation and selection of long oligomers in an environment likely to have a limited monomer supply.

(iv) Lipid-Bilayer Lattices

The investigation of this particular microenvironment was spurred by the insight that chemical activation of the mononucleotides may not be required if synthesis of phosphodiester bonds could be driven by the chemical potential of fluctuating anhydrous and hydrated conditions, with heat providing activation energy during dehydration. Amphiphile bilayers that compose vesicles or liposomes could represent a model for such a fluctuating environment. Under the right conditions, bilayer structures are formed

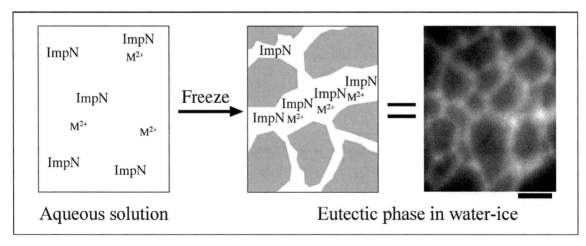

Figure 10.7. Formation of the eutectic phase in water-ice. A dilute solution of activated monomers (ImpN) and metal-ion catalysts (M^{2+}) is frozen below its freezing point, but above the eutectic point (i.e., the temperature at which the whole sample is frozen-solid). During freezing, solutes are concentrated in the liquid phase (eutectic phase) between the pure-ice crystals (see the light micrograph of a reaction mixture containing a fluorescent dye). Note that upon the initiation of freezing, the concentration of the solutes increases which simultaneously lowers the freezing point of the residual solution. Bar = 26.7 μm.

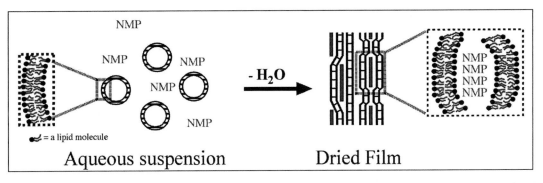

Figure 10.8. Formation of the lipid-bilayer lattices. RNA monomers (NMP, ribonucleotide 5′-monophosphate) are mixed with a liposome suspension (lipid bilayers are the circular and ladder-like columnar structures). Upon dehydration of the mixture, liposomes fuse into multi-layered structures forming a lipid-bilayer lattice that retains some moisture while sequestering monomers in extended stacks (gray line).

when amphiphile molecules are suspended in an aqueous medium. This molecular arrangement is preserved when vesicles are dehydrated. If small molecules, such as RNA monomers, are present in the aqueous medium during the dehydration, they will be trapped between the amphiphile bilayers (Fig. 10.8) and likely ordered into a structure potentially conducive to polymerization.

There are several main differences between this approach[20] and the two previously described ones: (A) The monomers are not chemically activated. (B) No metal-ion is used as a catalyst. As a corollary, the RNA linkage formation must be induced by an external energy source, e.g., heat. (C) The use of amphiphile bilayer lattices ensures that the system remains in a fluid state. (D) A fraction of the RNA products will become compartmentalized at the end of the procedure. We will come back to this latter point.

In experiments conducted by Deamer and collaborators,[20] phosphatidylcholine lipids (these molecules seem unlikely on the prebiotic Earth) dissolved in ethanol were injected into an aqueous solution containing nucleotide monophosphate where they spontaneously formed membranous structures. The suspension was then heated, dried under a flow of carbon dioxide forming lipid-bilayer films and incubated for up to 2 h at temperatures between 70 and 90°C. The samples were then rehydrated and the drying-heating-rehydrating cycles were repeated up to 7 times.

With samples containing only one nucleotide monomer, adenine 5′-monophosphate (AMP) or uridine 5′-monophosphate (UMP), RNA molecules with lengths between 50-100 monomeric units are detected that represent maximally 10% yield of polymers by weight. The maximum yields depend on several experimental parameters: temperature, cycle number, the type of gas used during dehydration of the suspension, the lipid composition and the lipid to nucleotide molar ratio. Mixtures of UMP and AMP produce polymers of reduced length, which match those observed on montmorillonite or in the eutectic phase in water-ice. In the absence of lipids but with the same cycling, no products longer than tetramers can be detected. Finally, the regioselectivity of the RNA products remains unclear as the researchers only stated "products could not be digested by RNase enzymes."[20]

In summary, the results obtained with these three different approaches to the polymerization of RNA from monomers clearly indicate that multiple pathways can exist to yield polymeric molecules with a length compatible with an RNA activity. These polymers still contain mixed linkages, a fact that may well inhibit their catalytic activity and their replication.

10.5 Information Transfer in an RNA World

Replicating an active RNA molecule (Fig. 10.9) is essential to the development of the RNA World because it results in a selective increase in their number. Furthermore, the likely relative inaccuracy in replication early on allowed for a number of small mutations that with selection likely promoted the evolution of new and/or better catalysts. The replication however must have been accurate enough to permit the preservation of the RNA functions. Such a process, also referred to as template-directed polymerization, can only occur if an existing RNA molecule serves as a template on which monomers or short oligomers can assemble due to base-pair formation and then be linked. Note that a catalytical RNA molecule would have to lose its structure, at least partially, to be copied.

Two separate issues impact the ability of monomers and oligomers to ligate during template-directed polymerization: (i) the stability of the hybridized system which improves as the length of complementary chain increases and/or longer oligomers are used; (ii) the likelihood for the neighboring monomers or oligomers on the chain to be ligated to align themselves thereby facilitating polymerization.

The environments investigated for monomer self-condensation are also used for the investigations of the template-directed RNA replication, with the notable exception of the lipid-bilayer lattices that have yet to be studied. In addition to the usual chemicals, the samples contained RNA polymers as templates (Fig. 10.9). Most reactions were conducted in high-ionic strength media and at low temperatures (around or below 0°C), to reduce the repulsion of template and monomers (both negatively charged) and to promote the base-pairing between the reactants, respectively.

(i) Template-Directed Polymerization in Homogeneous Aqueous Solutions

In this environment, an efficient non-enzymatic, template-directed polymerization of purines on their complementary pyrimidine strands has easily been obtained. Because base pairing on nucleic acid templates does not overcome the poor stacking of U-monomers, two consecutive A-residues completely inhibit the reaction (as this implies 2 stacked U's in the complementary strand). Furthermore, it has also been experimentally observed that the template must contain at least 60% of C-residues for efficient "replication" to occur.[21] This observation is likely related to the fact that activated G-residues have the strongest tendency to stack yielding better aligned systems and thereby facilitating polymerization. Although the complementary strand can be produced in the case where the template strand contains ≥60% C-residues, exponential growth however does not occur, nor is the C-residues enriched template itself likely to be reproduced, as its own template would contain ≥60% G-residues and hence ≤40% C-residues. The use of activated oligomers could partly resolve this issue as oligomers will hybridize more stably permitting enhanced polymerization. In fact, the ligation of two RNA oligomer triphosphates[22] or activated pyrimidine pentamers $(Im(pU)_5)$[23]

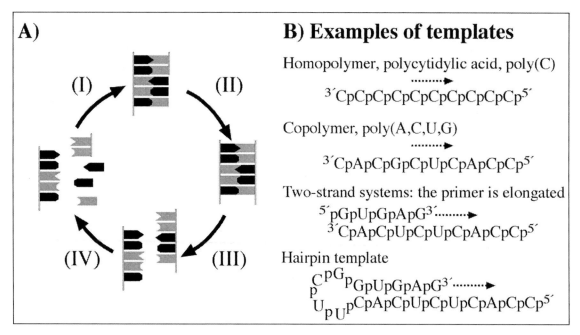

Figure 10.9. Schematic representation of template-directed polymerization. A) Templates, monomers and metal-ion catalysts are mixed and assembled through base-pair recognition in a duplex structure (phase I) that bring the monomers close together. The backbone of the second strand is then synthesized by metal-ion supported catalysis yielding a complete complementary strand (phase II). Double-stranded products are then separated (phase III), becoming templates for their own copying with the complementary strand yielding the original sequence (phase IV). B) Example of templates. The arrows indicate the polymerization direction.

and longer U-oligomers has been demonstrated with predominantly 3′-5′ and 2′-5′ linked products, respectively; the different regioselectivities observed in these two papers result from their differing reaction set-ups.

(ii) Template-Directed Polymerization on Mineral Surfaces

Template-directed RNA polymerization (ImpA on poly(U) and ImpG on poly(C)) can be mediated by several minerals, e.g., by hydroxylapatite and attapulgite, a clay.[24] One may reasonably assume that a template-directed polymerization of recently synthesized pyrimidines with a 1-methyladenine activating group (Fig. 10.4) on montmorillonite will soon be possible. Their intrinsically stronger adsorption should prolong the half-life of the pyrimidine template/monomer complexes on the clay allowing a more effective template-directed polymerization.

(iii) Template-Directed Polymerization in the Eutectic Phase of Water-Ice

The eutectic phase in water-ice seems to be conducive to the template-directed polymerization of ribonucleotides. Stribling and Miller[25] have established that the template-directed oligomerization of ImpG can occur in the eutectic phase in water-ice. Trinks and collaborators[26] have shown that a solution containing poly(U) as templates and 2-MeImpA delivers RNAs with a length longer than 100 monomeric units within a year. Short uridine oligomers (3-5 mer) can also be ligated on poly(A) at −25°C.[23] In fact, we have observed in a low ionic strength medium (Monnard and Szostak, unpublished results) that, for all four nucleobases under noncompetitive conditions (only a single species of activated nucleobase present at a time), the initial elongation rates on RNA hairpin templates (Fig. 10.9) are clearly dependent on the base-pairing of the monomers with the template. They are comparable (within an order of magnitude), yet are at least 1.5 times higher than those observed for nonWatson-Crick systems. This rate enhancement may not be large compared to those observed in solution, but remember that no self-condensation or oligomer elongation occur in homogenous

solutions whereas efficient oligomerization is always observed in self-condensation reactions in the eutectic. Interestingly, whereas monomer self-condensation is strongly inhibited by salt dissolved in this medium,[27] template-directed reactions can occur both in low and high ionic-strength media.

(iv) Linkage Heterogeneity of the Templates

RNA oligomers with mixed linkages have been used as templates for their own replication. It was expected that such templates with mixed linkages would represent an insurmountable obstacle to efficient RNA replication. However, this expectation has turned out to be only partially correct as the successful template-directed polymerization of 2-MeImpG has been carried out with mixtures of 2′-5′ linked oligo(C) templates produced on montmorillonites[28] yielding both 3′-5′ and 2′-5′ linked products. More recently, Sawai and collaborators showed that pure 2′-5′ linked RNA tetramers would still ligate on pure 3′-5′ linked decameric templates into octameric products as pure 3′-5′ tetramers on pure 2′-5′ decameric templates. The ligation efficiency was however lower than that for a uniformly linked (templates and oligomers) system.[29]

In summary, the results of template-directed RNA polymerization provide interesting insights about the conditions needed for an efficient replication of a biopolymer from activated monomers or short oligomers. Note that in the latter case, the accuracy of the replication might be lower than that likely obtained with activated monomers. The results also show that RNAs with mixed linkages may serve as templates although with lower efficiency than their uniformly linked counterparts. Finally, no amplification reactions have yet been carried out.

10.6 RNA Alone as the Motor for Evolution Toward Cellular Life

One can still wonder whether RNA alone was sufficient for the emergence of the RNA World. Assuming a pair of RNA polymerases (though here in the broad sense that includes ligases) existed (one template and one replicator), would they alone have permitted the emergence of RNA life? Indeed, to consider the emergence and evolution of an RNA World alone seems out of context.[30] Life, as we know it, suggests that, very early on, a compartment and metabolic structures would also have been necessary to create a RNA-based protocell and that it only survived because of their synchronous action and reciprocal feedback regulation. For example, a compartment (a structure permitting the segregation of molecules from their environment) might have already played an essential role by permitting RNA polymerization (as shown by the polymerization in lipid-bilayer lattices that, upon final rehydration, will spontaneously reassemble into vesicular compartments partially encapsulating RNA products) and certainly did play an essential part in the accumulation and maintenance of high concentrations of reacting species, as well as the distinct evolution of particular RNAs.

10.7 Conclusion and Further Directions

True RNA amplification by non-enzymatic polymerization, a cornerstone process for the RNA World, remains elusive. However, considerable progress has been made in understanding how RNA fragments could have arisen given a pool of activated monomers and subsequently replicated. Studies have shown that several methodological approaches can support these reactions and yield 30- to 50-mer molecules whose length is compatible with catalytic activity. Furthermore, the replication of an RNA strand can also occur using either monomers or oligomers.

These systems still have weaknesses. First and foremost of which is the purity of the starting compounds. Thus the work to date should be considered as yielding models rather than literal representations. For example, the clay-supported reactions are extremely sensitive to the nature of the activation group on the monomers. The efficiency of the eutectic phase in water-ice seems dependent on the total concentration of solutes, even those not involved in the polymerization. In the bilayer-lattice approach, metal-ions, absent in the reaction mixtures, are known to accelerate RNA decomposition at high temperatures and organic solutes might disrupt bilayer self-assembly.

Impurities or less than optimal conditions will result in slower polymerization. This could be acceptable except for the fact that RNA products can rapidly decompose. Thus the full polymerization of functional RNAs must occur within a time period that is short compared to that for a single hydrolysis reaction in the full-length polymer.

Nonetheless, researchers have already revealed three approaches that start to resolve the issue of RNA polymerization and it seems likely that uncharted solutions still exist. Multiple pathways might have simultaneously promoted RNA polymerization, taking advantage of a broader range of environmental conditions.

The de novo emergence of the RNA World from organic molecules present on the primitive Earth has yet to be demonstrated and inescapable problems remain to be solved. (A) A prebiotic β-D-ribonucleotide synthesis, let alone that of an activated nucleotide, has yet to be discovered. This process should be not only efficient but also selective enough to deliver large amounts of pure monomers from mixtures containing a large library of organic molecules that may interfere with the intended synthesis. (B) The emergence of catalytically active RNAs in a prebiotically synthesized RNA pool, let alone RNAs with replicating properties, is still not demonstrated. (C) The role of other precursors of precellular parts, e.g., the compartment, in the development of a full-fledged RNA-based organism must be elucidated for they could have supported its emergence.

These problems should be considered without forgetting that RNA might not have built the first precellular information and catalytic system. An alternative system based on a self-assembled network of interconnected catalytic reactions without the intervention of RNA or any substitute might well have initiated the series

of evolutionary steps leading to the emergence of living organisms. Nonetheless, the evolution of such a Metabolic World must have ultimately led to the emergence of a heritable information/catalytic-polymer system, likely based on RNA, the precursor of today's DNA-RNA-Protein World. One possible advantage of this scenario could then have been the provision of efficient catalytic pathways for the synthesis of RNA monomers.

Thus, whether or not one believes in the RNA World and further considers RNA as the actual first information/catalytic material, the study of non-enzymatic RNA polymerization and replication offers an exceptional opportunity to explore essential aspects of the emergence of a polymeric information/catalytic system in the absence of a complex metabolism, especially in the current absence of any other plausible, experimental alternative.

Acknowledgement

I wish to thank Dr. C.L. Apel, Dr. M.S. Declue, Dr. H.J. Cleaves, Dr. H.-J. Ziock and H. Fellermann for their comments. This work was supported by the Laboratory-Directed Research and Development program at Los Alamos National Laboratory and in particular through the Protocell Assembly project therein.

References

Further Readings

1. Bloomfield VA, Crothers DM, Tinoco Jr I. Nucleic acids: Structures, properties and functions. Sausalito, California: University Science Books 2000.
   - This book covers the properties of nucleic acids.
2. Orgel LE. Prebiotic chemistry and the origin of the RNA World. Crit Rev Biochem Mol Biol 2004; 39:99-123.
   - The most recent and comprehensive review on the details of the prebiotic chemistry related to the emergence of the RNA World.
3. Cech TR, Atkins JF. RNA-World. 3rd ed. Cold Spring Harbor: Cold Spring Harbor Laboratory Press, 2005.
   - This book contains a collection of articles related to the RNA World and offers a broad overview of all aspects of the research done to understand RNA functions.
4. Müller UF. Re-creating an RNA world. Cell Mol Life Sci 2006; 63:1278-1293.
   - A review on the emergence of essential RNA functions and an RNA-based "organism".
5. Shapiro R. Small molecule interactions were central to the origin of life. Q Rev Biol 2006; 81:105-125.
   - This review presents a summary of the facts that challenge the de novo emergence of the RNA World on the early Earth, as well as a theoretical model of a purely catalytic or "metabolic" World.

Specific References

6. Crick FHC. The origin of the genetic code. J Mol Biol 1968; 38:367-379.
7. Orgel LE. Evolution of the genetic apparatus. J Mol Biol 1968; 38:381-393.
8. Gilbert W. The RNA-world. Nature 1986; 319:618.
9. Steitz TA, Moore PB. RNA, the first macromolecular catalyst: the ribosome is a ribozyme. Trends Biochem Sci 2003; 28:411-418.
10. Zubay G, Mui T. Prebiotic synthesis of nucleotides. Origin Life Evol Biosphere 2001; 31:87-102.
11. Eschenmoser A. Chemical etiology of nucleic acid structure. Science 1999; 284:2118-2124.
12. Böhler C, Nielsen PE, Orgel LE. Template switching between PNA and RNA oligonucleotide. Nature 1995; 376:578-581.
13. Horthota AT, Szostak JW, McLaughlin LW. Glycerol nucleoside triphosphates: Synthesis and polymerase substrate activity. Org Lett 2006; 8:5345-5347.
14. Kanavarioti A. Preference for internucleotide linkages as a function of the number of constituents in a mixture. J Mol Evol 1998; 46:622-632.
15. Ertem G. Montmorillonite, oligonucleotides, RNA and the origin of life. Orig Life Evol Biosph 2004; 34:549-570.
16. Ferris JP. Montmorillonite-catalysed formation of RNA oligomers: the possible role of catalysis in the origins of life. Phil Trans R Soc B 2006; 361:1777-1786.

17. Miyakawa S, Ferris JP. Sequence- and regioselectivity in the montmorillonite-catalyzed synthesis of RNA. J Am Chem Soc 2003; 125:8202-8208.

18. Monnard PA, Kanavarioti A, Deamer DW. Eutectic phase polymerization of activated ribonucleotide mixtures yields quasi-equimolar incorporation of purine and pyrimidine nucleobases. J Am Chem Soc 2003; 125:13734-13740.

19. Kanavarioti A, Monnard PA, Deamer DW. Eutectic phases in ice facilitate non-enzymatic nucleic acid synthesis. Astrobiology 2001; 1:271-281.

20. Rajamani S, Vlassov A, Coombs A et al. Lipid-assisted synthesis of RNA-like polymers from mononucleotides. Origin Life Evol Biosphere 2008; 28:57-74.

21. Joyce GF. Non-enzymatic template-directed synthesis of informational macromolecules. Paper presented at: Symposia on Quantitative Biology, Cold Spring Harbor 1987; 52:41-51.

22. Rohatgi R, Bartel DP, Szostak JW. Nonezymatic, template-directed ligation of oligoribonucleotides is highly regioselective for the formation of 3'-5' phosphodiester bonds. J Am Chem Soc 1996; 118:3340-3344.

23. Sawai H, Wada M. Nonenzymatic template-directed condensation of short-chained oligouridylates on poly(A) template. Origin Life Evol Biosphere 2000; 30:503-511.

24. Schwartz AW, Orgel LE. Template-directed polynucleotide synthesis on mineral surfaces. J Mol Evol 1985; 21:299-300.

25. Stribling R, Miller SL. Template-directed synthesis of oligonucleotides under eutectic conditions. J Mol Evol 1991; 32:289-295.

26. Trinks H, Schroder W, Biebricher CK. Ice and the origin of life. Orig Life Evol Biosph 2005; 35:429-445.

27. Monnard PA, Apel CL, Kanavarioti A et al. Influence of ionic solutes on self-assembly and polymerization processes related to early forms of life: Implications for a prebiotic aqueous medium. Astrobiology 2002; 2:139-152.

28. Ertem G, Ferris JP. Template-directed synthesis using the heterogeneous templates produced by montmorillonite catalysis. A possible bridge between the prebiotic and RNA-worlds. J Am Chem Soc 1997; 119:7197-7201.

29. Sawai H, Wada M, Kouda T et al. Nonenzymatic ligation of short-chained 2'-5'- or 3'-5'-linked oligoribonucleotides on 2'-5'- or 3'-5'-linked complementary templates. Chem Bio Chem 2006; 7:605-611.

30. Szostak JW, Bartel DP, Luisi PL. Synthesizing life. Nature 2001; 409:387-390.

CHAPTER 11

Ribozymes and the Evolution of Metabolism

Randall A. Hughes and Andrew D. Ellington*

11.1 Introduction

In a prebiotic World, organisms as we currently know them did not exist, but the molecules that carried information and performed catalysis would have arisen during this period of biochemical evolution. Comparative analysis of modern organisms suggests that most biological information is stored in the form of nucleic acids (DNA and RNA), while proteins carry out most catalysis. But was this always the case? Obviously, catalysts that enabled the synthesis and decoding of the hereditary material were necessary, while the hereditary material must have encoded the synthesis of these catalysts. This is the chicken-or-egg problem taken back to its earliest roots.

It is difficult to figure out precisely what molecules may have been present at origins. The advent of catalysis would have enriched the chemical diversity of the primordial World by transforming available precursors into products that would themselves be reactants or catalysts for other reactions, ultimately forming autocatalytic cycles. However, the availability of catalysts would have to some extent been limited by the availability of raw materials and the eventual degradation of the catalysts themselves. Maintaining chemical diversity would have been extremely difficult under such a regime, as establishing a cycle and acquiring new catalysts would have been at the mercy of quite random fluctuations in the environment. One solution to this problem would have been the development of a system of information storage in which blueprints for the construction of useful catalysts were encoded. Given that DNA and RNA serve this function in modern biology, that phylogenetic data suggests these compounds are quite ancient and that the building blocks of nucleic acids were likely present in the primordial soup, information storage by nucleic acid polymers and subsequent propagation of this information into useful catalysts would have conferred a vast evolutionary advantage on early genomes and the organisms that harbored these genomes. More importantly, the genetic system could have mutated, been selected and led to adaptation to differing environments.

Early hypotheses on the make-up of the prebiotic World postulated simple catalytic cycles involving information-storing RNAs and catalytic proteins. However, in the early-1980s, Thomas Cech discovered that the Group I intron from *Tetrahymena* prerRNA catalyzes its own excision from the rRNA transcript[4] and Sydney Altman discovered that the RNA portion of ribonuclease P catalyzes the cleavage of precursor tRNA transcripts.[5] These discoveries advanced the notion that nucleic acids could have played a dual role in the prebiotic World and presented a possible solution to the canonical chicken-and-egg problem. Once fleshed out, this idea eventually became known as the 'RNA World' hypothesis.[6] In this view, the prebiotic World may have become populated by simple replicators based upon nucleic acids. Once this link between genotype and phenotype was established in the early RNA World, additional RNA catalysts (ribozymes) could have evolved to catalyze a number of reactions, including early biosynthetic pathways. Eventually these ribozymes would have elaborated to the point where the translation apparatus could evolve and the ribozymes would have been replaced one-by-one by their protein counterparts.

Though few natural ribozymes still exist today, a great deal of metabolism can trace its synthetic heritage to ribozymes that arose during the RNA World. Most modern cofactors, such as ATP, NAD and even folate, appear to be based in some way on ribonucleotides (see Fig. 11.1). White and others have posited that once a critical mass of enzymes/ribozymes evolved to utilize a particular cofactor then the cofactor could not change without the organism going through an extreme decrease in fitness (the so-called 'principle of many users'[7]). To the extent that cofactors became frozen once a quorum of catalysts evolved to depend upon them they can be seen as the most slowly evolving of all biological structures and amongst the oldest metabolic fossils within a cell. Extending this analysis, it can be hypothesized that these cofactors were originally invented and used by RNA-based catalysts and have since been carried over into modern times because it would have been easier to evolve new catalysts than to re-evolve the reactions the catalysts performed.[8] This is certainly true of modern protein enzymes: while protein sequences, structures and substrates can readily change, the reactions performed by and the cofactors utilized by proteins remain relatively fixed. For example, the principle of many users easily explains why dehydrogenases are often related by their adenosine-binding Rossman folds, rather than by their substrate-binding domains.

The question thus becomes what reactions might have been catalyzed by early ribozymes and how these reactions could have contributed to the origin and evolution of life. In this chapter, we will discuss some of the reactions that can be catalyzed by ribozymes generated by directed evolution experiments and their potential roles in expanding our understanding of the chemistry of the prebiotic and RNA Worlds.

11.2 Ribozymes in Replication

The earliest forebears of genomes in a putative RNA World would likely have been simple ribozymes that could have replicated themselves from available materials. The earliest and simplest mechanism for replication would in turn likely have involved self-templated

*Corresponding Author: Andrew D. Ellington—Department of Chemistry and Biochemistry, Center for Systems and Synthetic Biology University of Texas at Austin, Austin, Texas 78712, USA. Email: andy.ellington@mail.utexas.edu

Prebiotic Evolution and Astrobiology, edited by J. Tze- Fei Wong and Antonio Lazcano. ©2009 Landes Bioscience.

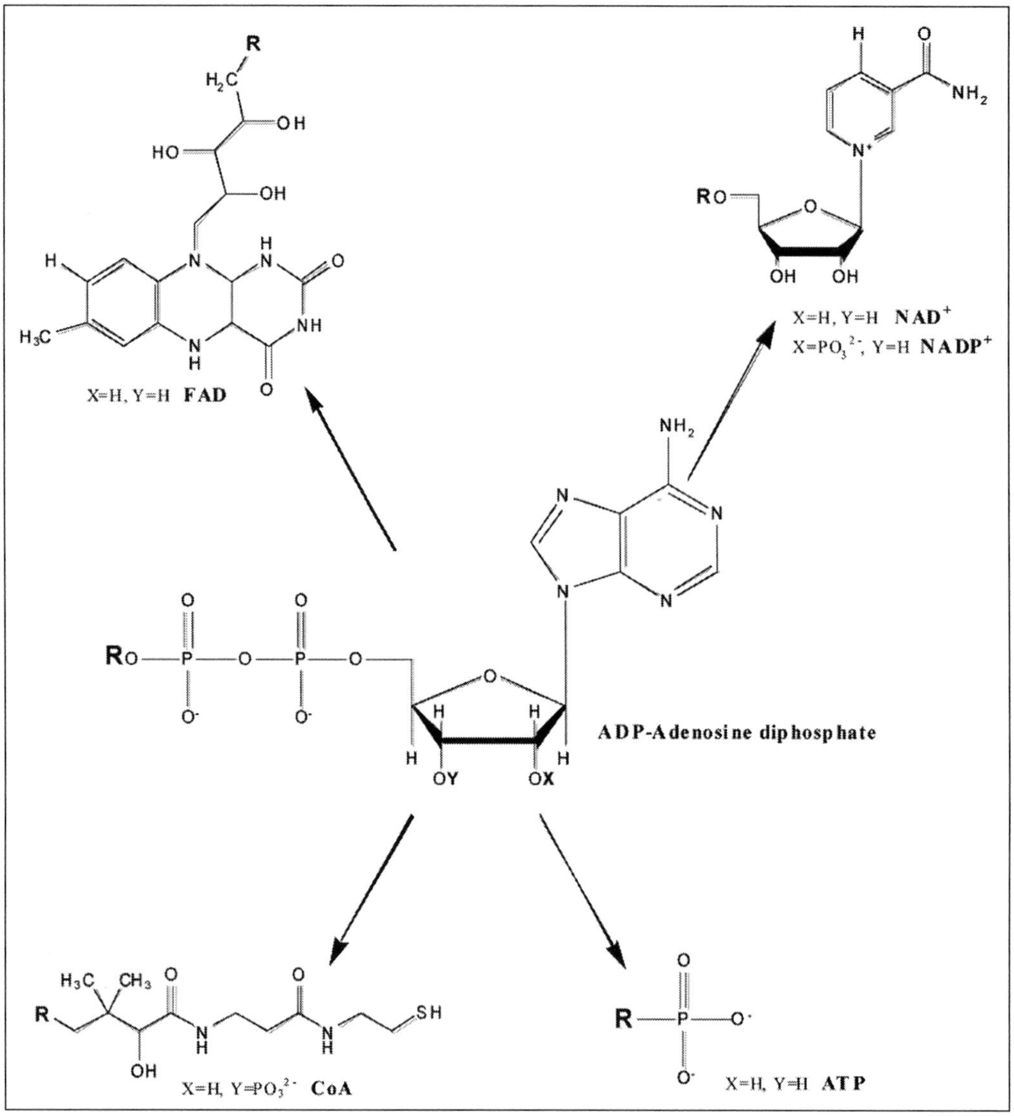

Figure 11.1. Nucleotide-based cofactors. Many cofactors are derived from adenosine nucleotides. R is the attachment point for various cofactor moieties to the ADP skeleton. In addition, many of the cofactors can be phosphorylated on the 3' hydroxyl (Y) or the 2' hydroxyl (X) groups of the ribose sugar.

ligation via Watson-Crick base-pairing. In this regard, the templated replication of short oligonucleotides has actually been observed from trimeric or even shorter substrates.[9,10] Assuming that such mechanisms could have yielded a nascent pool of short oligonucleotide precursors, extension and elaboration of this pool could have yielded longer oligomers with a range of improved catalytic activities. In particular, the formation of more complex RNA structures would have allowed improvements in ligation efficiency beyond simple templating.

As support for these hypotheses, we must go beyond the known natural ribozymes, which are few in number and have by and large evolved to catalyze RNA processing reactions rather than replication. Directed evolution offers an excellent means to recapitulate the RNA World by creating doppelgangers of early catalysts. The utility of directed evolution methods to inquire into origins was powerfully demonstrated by the evolution of a ribozyme that could catalyze the formation of a 3'-5' phosphodiester bond at a ligation junction starting with a 5' triphosphorylated RNA as a substrate.[11] This reaction is very similar to the one catalyzed by protein ligases and polymerases, although the ribozyme is much slower than its protein counterparts. This similarity was further exploited to select for a ribozyme that had polymerase activity using nucleoside triphosphates as substrates.[12] Over four days of incubation this ribozyme could extend an RNA

primer that was annealed to it by six nucleotides, with a fidelity of ninety-two percent. However, the ability of this ribozyme to catalyze its own replication would have been grossly limited by the fact that to complete a second strand the ribozyme would eventually have had to unfold itself. To resolve this issue Bartel and coworkers attempted to evolve a polymerase which could recognize a detached (trans) template:primer complex.[13,14] Through a complicated selection scheme involving a tethered substrate, they were in fact able to select for a variant of the original ribozyme that operated in trans and that demonstrated improved catalysis. This ribozyme was able to extend an RNA primer up to fourteen nucleotides in a twenty-four hour incubation with a fidelity over ninety-eight percent. Unfortunately, low polymerization efficiencies made longer syntheses impossible.

While the Bartel ligase and polymerase stand as avatars of a late RNA World that contained RNA polymerases, it seems unlikely that such complex structures would have arisen in an early RNA World. There must be molecular 'missing links' between the earliest, simplest self-replicating oligonucleotides and late stage polymerases (a 'march of progress' envisioned in Levy and Ellington, 2001;[15] Fig. 11.2). One such missing link might be ribozyme polymerases that utilize oligonucleotides instead of mononucleotides as substrates. A number of small ribozyme ligases have been generated by directed evolution and the crystal structure of one has recently been solved, giving us a glimpse of the chemistries involved in prebiotic replication.[16] The 'march of progress' schema is further supported by computational simulations which showed that the evolution of replicators with greater efficiency and fidelity could be boot-strapped, leading to an increase in early genome size over time.[17] An interesting caveat in these simulations was that the evolution of replicator complexity was directly related to the spatial separation of the replicators themselves. Populations of replicators that were allowed to diffuse freely in the computational grid did not evolve to higher complexity, while those that were spa-

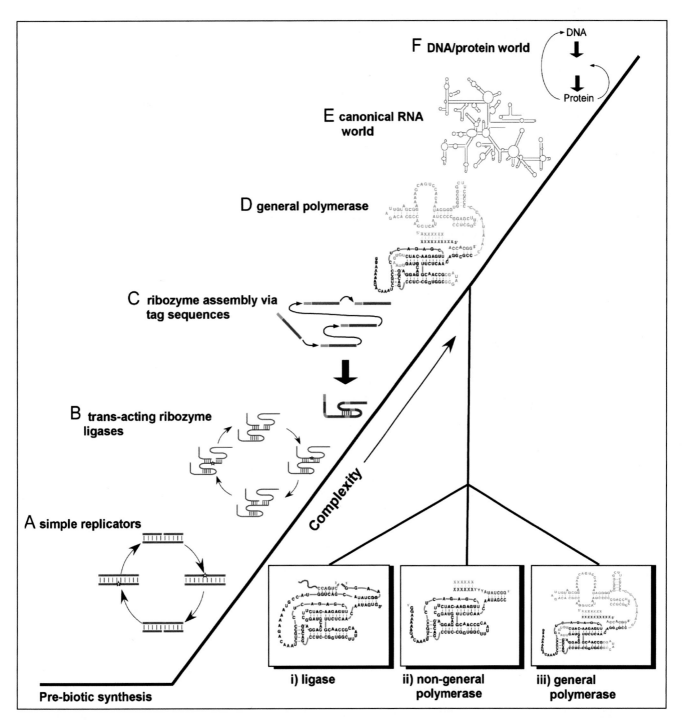

Figure 11.2. 'March of Progress' in ribozyme catalysis. This is one view of how primitive ribozyme replicators might have evolved increasing complexity over time. A) Simple replicators could have emerged from template-directed ligation of oligonucleotides; increasing elaboration would have led to (B) trans-acting ribozymes that catalyzed both the ligation of their own templates and other RNA molecules. C) To avoid parasitism, ribozymes might have taken advantage of short regions of sequence homology ('tags') to generate genomes. D) Long RNA stretches would have provided fodder for the evolution of template-directed polymerase ribozymes capable of rudimentary replication. Better polymerases could have emerged via a pathway similar to the directed evolution of the Bartel polymerase.[10] This polymerase began as a ligase (i), which was then evolved into a nongeneral polymerase with an attached template (ii) and finally into a more general polymerase (iii). The advent of complex polymerases with increased fidelities and processivities would have allowed ribozymes and genomes with increasing sequence complexities to evolve, which could in turn have catalyzed the many biochemistries suspected to have been present in the RNA World (E). Ultimately, the invention of the translation apparatus would have led to the establishment of the modern DNA/protein World (F).

tially separated did. Thus, the evolution of replicators may have been intimately tied to the early cellularization of genomes.

11.3 Ribozymes and Cofactors

One of the acknowledged benefits of using proteins as catalysts is that the available chemical diversity of the amino acid side chains is much greater than that of the four canonical ribonucleotides. Amino acids enable reactions such as proton exchange near neutral pH (histidine), thiol chemistry (cysteine) and nucleophilic attack (serine). Amino acids also have divergent physicochemical properties and include positively charged moieties (histidine, lysine, arginine) and hydrophobic groups (isoleucine, leucine, valine and so forth). These features are largely absent in natural nucleic acids. Nonetheless, even with this seemingly vast advantage many protein catalysts still require cofactors. Interestingly, many of these cofactors are nucleotide-based (Fig. 11.1). As suggested above, this correlation strongly suggests that these cofactors were established in and have been carried over from the RNA World.

The early existence of nucleotide-based cofactors in the RNA World is plausible as both pantetheine (a CoA precursor) and nicotinamide (a NAD precursor) would have been synthesized from materials that should have been prebiotically available.[18,19] There are different models for how cofactors could have been utilized by early ribozymes, but one can imagine that nucleotide-based cofactors could have been covalently appended to nucleic acid catalysts as a consequence of nonspecific ligation reactions. Indeed the Group I intron ribozyme has been shown to be capable of covalently attaching itself to cofactors (NAD+) and cofactor analogs (dephosphorylated CoA) using the same mechanism it normally uses to insert guanosine into RNA during splicing initiation.[20] Similar mechanisms may have also played a role in the establishment of amino acids and translation, as we will discuss below.

As an example of how covalently-attached cofactors would have augmented ribozyme functionality, directed evolution experiments have shown that redox active, alcohol oxidase ribozymes that utilize NAD+ can be selected from random sequence pools.[21] The same ribozyme was also shown to catalyze the reverse reaction—a reduction of benzyl aldehyde back to benzyl alcohol.[22] Two other interesting facts emerged from these studies: first, the alcohol dehydrogenase ribozymes selected had an absolute requirement for Zn^{2+}, which is

also a requirement of the protein enzyme. Second, oxidation of the alcohol could occur by generating NAD+ via uncatalyzed hydride transfer from NADH to FAD, indicating that coupled redox reactions similar to those commonly observed in modern biochemistry could have also arisen in the earliest metabolic pathways.

Directed evolution has also been used to generate ribozymes that can utilize noncovalently attached cofactors. An adenosine-binding RNA domain (the RNA equivalent of the Rossman fold found in proteins) was appended to a random sequence pool and kinase ribozymes that could catalyze the transfer of the thiophosphate from soluble ATP-γS to the 5′-hydroxyl of the ribozyme were selected by capturing self-kinases on a thiol column.[23] One of the selected ribozymes was able to catalyze multiple turnover phosphorylation of an oligonucleotide identical to the 5′ end of the ribozyme, a reaction akin to the one catalyzed by the protein enzyme polynucleotide kinase.

The availability of small amounts of premade cofactors in the prebiotic soup might have led not only to the evolution of cofactor-dependent ribozymes, but also to the evolution of biosynthetic pathways to reproduce the cofactors. Satisfyingly, directed evolution experiments have demonstrated that ribozymes can catalyze the formation of cofactors from precursors. A pyrophosphate transferase ('cappase') ribozyme originally evolved by Yarus and coworkers has been shown to have very loose substrate specificity.[24] Virtually any molecule that contained a phosphate can add itself to the 5′ triphosphate of the ribozyme, displacing pyrophosphate. This ribozyme was initially shown to be able to attach cofactors to itself, similar to the aforementioned experiments with the Group I intron. Other ribozyme variants joined precursors and the 5′ adenosine of the ribozyme to form common cofactors containing canonical disphosphate linkages[25] (Fig. 11.3). For example, the ribozyme could synthesize CoA, NAD and FAD from their 4′-phosphopantetheine, nicotinamide mononucleotide (NMN) and flavin mononucleotide (FMN) precursors, respectively. These experiments again support the hypothesis that cofactors could have been covalently attached to early RNA catalysts, possibly at their termini. As metabolism expanded and the demand for diffusible cofactors increased the covalently-linked cofactors may have become detached and the ribozymes would evolve to synthesize the cofactors using ATP as a substrate in *trans*. Whether cofactors began as tethered appendages

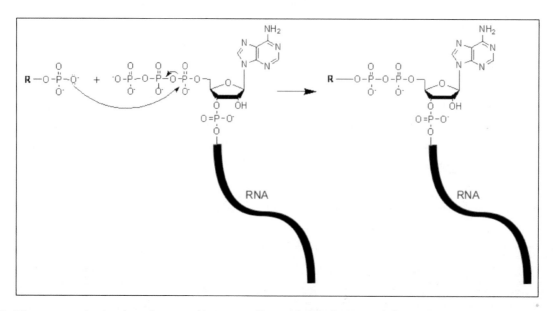

Figure 11.3. Ribozyme mechanism for cofactor synthesis. Virtually any chemical moiety (R) that contains a phosphate can react with the terminal ATP on the 'cappase' ribozyme to attach itself to the ribozyme's 5′ terminus. The R-group could be any of the several cofactor moieties shown in Figure 11.1.

or as diffusible moieties with nucleotide handles is debatable, but what is clear is that the ability to synthesize and utilize cofactors could have been present in the early RNA World.[26]

Directed evolution experiments have been used to extensively explore ribozyme reactions relevant to primordial biosynthesis, many of which require cofactors. For a more detailed review of ribozyme chemistry and cofactors the reader is encouraged to read the references listed in the Further Reading Section.

11.4 Ribozymes, Amino Acids, and the Advent of Translation

The advent of an efficient translational apparatus would have been the pinnacle of biochemical evolution in the RNA World and also its downfall. Efficient translation would have led to longer peptides which could have acted as powerful catalysts. Since protein catalysts have many advantages over ribozymes, once they began to be translated from heritable, genomic RNAs they would have also begun to evolve and to replace their less efficient RNA counterparts. Nonetheless, at the outset highly evolved ribozymes would still have been more efficient than nascent and unevolved proteins. It is therefore likely that the some of the earliest products of translation would have been short peptides that would have acted as cofactors to supply chemical functionalities otherwise unavailable to RNAs (and in turn generating the earliest ribonucleoproteins).

These insights may speak to one of the enduring mysteries of molecular biology: why did the translation apparatus evolve in the first place? From one vantage, translation looks like a textbook example of that well-known oxymoron, anticipatory evolution (i.e., ribozymes were the primary catalysts in a hypothetical early cell and yet somehow knew to create an apparatus to replace themselves). Even if the first translated peptides augmented ribozyme-mediated catalysis, there must have been some simple and obvious evolutionary advantage to augmenting catalysis prior to inventing a complex apparatus to do so. One possible explanation is that since amino acids would have been available in the prebiotic soup, ribozymes could have originally used these simple and available compounds as cofactors. Further increases in catalytic precision and complexity would have driven the conjugation and polymerization of amino acids in series and chains. As an example of how translated peptides could have functioned with RNA catalysts, peptide dependant ribozymes have been selected that activate ligation by >18,000 fold.[27] These experiments also demonstrate that in addition to acting as potential cofactors, peptides could potentially have acted as allosteric effectors regulating metabolism in the RNA World.

These musings in turn beg the question of how amino acids would have been utilized by RNA catalysts. Our discussion of cofactors above provides a possible answer: amino acids could have been either noncovalently bound or covalently appended to ribozymes themselves. Both possibilities can be supported experimentally. Roth and Breaker selected a deoxyribozyme that utilized a noncovalently bound histidine as a cofactor to catalyze the cleavage of an RNA substrate.[28] The choice of histidine in these experiments was particularly interesting as histidine is the only amino acid that can function in general acid or base catalysis at physiological pH values (~7.0). The use of the imidazole sidechain of hisidine for this function is of course widespread in proteins and would have been especially useful for nucleic acid catalysts as they lack facile proton transfer at neutral pH. Other amino acid cofactors might also have been utilized, since directed evolution experiments have shown that RNA aptamers can be evolved to bind specifically to a number of amino acids (see, for example, Majerfeld et al 2005[29] and references therein).

However, the binding constants for amino acid:RNA complexes have proven to be quite weak, suggesting that covalent coupling might have had a greater impact on catalytic efficiency. Amino acids could have become directly attached to nucleotides through relatively simple chemistries[30] and random insertion of these nucleotides into catalysts would have led to at least some variants to show enhanced catalytic abilities. This would also have helped establish the selective pressures that would have led to the development of primordial amino acid biosynthetic pathways, translation and the inception of the genetic code, a hypothesis which has been proposed by Wong.[31] However, in order to make sure this catalytic enhancement was inherited in each generation it would have been necessary to evolve some mechanism to attach a particular amino acid (or sets of closely related amino acids) to particular sites in a ribozyme. In this regard, Szathmary[32] has suggested that amino acids with nucleotide handles attached to them could have been used. This idea is analogous to the binding and use of nucleotide-based cofactors by catalysts. The nucleotide handle or adaptor would have provided a more convenient mechanism for the enzyme to specifically grab onto the amino acid cofactor. Indeed, with an oligonucleotide handle a cofactor could be directed to hybridize to a particular site within a ribozyme. In some ways, oligonucleotide-appended amino acids might have resembled the anti-codon stem loop of modern tRNAs and could have been the evolutionary step that led to encoded peptidation and the advent of the early genetic code.[32]

Assuming that there were sufficient and cogent evolutionary pressures for the invention of translation, then there is still the problem of how a huge, complex ribozyme like the ribosome could have been built up from simpler reaction mechanisms (as hypothesized by Woese as far back as the 1970s). Again, directed evolution can potentially provide experimental instantiation of what may have happened in the distant past, since relatively simple ribozymes can catalyze all of the chemistries needed for protein synthesis.

In modern translation, amino acids are polymerized into proteins by first making high-energy intermediates that then serve as substrates for encoding and ultimately peptide bond formation. High energy aminoacyl adenylates intermediates are synthesized from ATP by a class of enzymes called aminoacyl-tRNA synthetases (aaRS) and their cognate tRNAs (Fig. 11.4, Reaction 1). The aaRS then transfers the amino acid to its cognate tRNA via the formation of an ester between the carboxyl group on the amino acid and the 3' OH of its cognate tRNA to form an acylated tRNA (Fig. 11.4, Reaction 2). The acylated tRNA can then be used in protein synthesis via the peptidyl transferase activity of the ribosome (Fig. 11.4, Reaction 3). Since the ester linkage between the amino acid and the tRNA is a higher energy bond than the subsequent peptide bond this process is energetically favorable and should occur spontaneously when the two substrates are brought into proximity within the ribosome.

Satisfyingly, directed evolution experiments have shown that ribozymes are capable of catalyzing all of these steps. Work from Kumar and Yarus demonstrated that a ribozyme could be selected that could catalyze the formation of an acyl-adenylate from ATP and carboxylic acids[33] (Fig. 11.4, Reaction 1). The subsequent aminoacylation reaction has also been shown to be catalyzed by in vitro selected ribozymes (Fig. 11.4, Reaction 2). Yarus and coworkers again showed that an RNA could be selected that aminoacylated itself with phenylalanine (Phe) via phenylalanine adenylate as the starting material.[34] Incredibly, one of the selected variants, RNA 77, could self-aminoacylate phenylalanine with a speed and specificity that exceeded tRNA aminoacylation by modern protein enzymes.[35] Other directed evolution experiments have shown that selected ribozymes can also catalyze the transfer of other activated amino acids to their cognate tRNAs, a reaction identical to the second reaction catalyzed by the protein based aaRSs (see, for example, Lee et al[36]). The core catalytic activity of the self-aminoacylating RNA 77 eventually could be embodied in a minimal enzyme that was only 29 nucleotides long, though at reduced

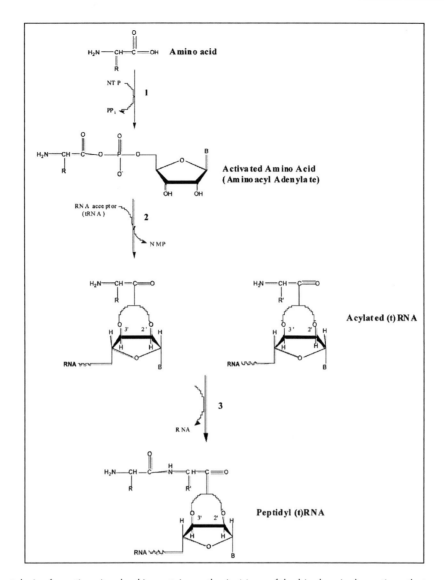

Figure 11.4. Ribozyme catalysis of reactions involved in protein synthesis. Many of the biochemical reactions that occur during translation can also be catalyzed by ribozymes. 1) The activation of an amino acid to form a high energy 'activated' aminoacyl adenylate. 2) The transfer of the activated amino acid on to a hydroxyl on a RNA acceptor. The acceptor RNA in modern translation is tRNA, but in the RNA World could have been a ribozyme with a function similar to modern amino acyl tRNA synthetases. The attachment is indicated to be at either the 2' or 3' hydroxyl. 3) Two acylated amino acids can react to form a peptide bond. The peptide remains attached to the RNA, just as at the P site of the ribosome. Templating of this step has not been experimentally demonstrated, but such templating may have been the precursor of the modern genetic code. B is any nucleobase.

rates compared to the longer variants.[37] These results are significant in that they conclusively show that RNA is able to catalyze two of the three primary steps in translation with a catalyst that is a fraction of complexity of its modern counterparts, hinting that the molecular origins of translation could have arisen from relatively simple RNA catalysts and then evolved to greater complexity over time. The third step of protein synthesis can also be catalyzed by ribozymes (Fig. 11.4, Reaction 3). For example, the minimized RNA 77 could also catalyze the formation of a RNA-Phe-Phe dipeptide (Fig. 11.4, Reaction 3). Zhang and Cech have also selected ribozymes that have peptidyl transferase activity and can form peptides irrespective of the amino acids used.[38] The next step in establishing an experimentally evolved translation apparatus will be to show sequence-specific decoding, something that should be relatively straightforward given that acylated amino acids can be tethered to oligonucleotides, as postulated by Szathmary.[32] Nonetheless, future experiments will be needed to flesh out our understanding on this stage of evolution, perhaps using ribozymes that themselves template peptide formation, similar to the trans-acylation experiments already carried out by Lee et al.[36]

11.5 Conclusions

Since the discovery of catalytic RNAs in the early 1980s, directed evolution experiments have provided experimental validations of theoretical constructs of early evolution. Evolved ribozymes have been shown to catalyze many of the reactions that exist in modern metabolism, ultimately lending support to the existence of RNA as an template and catalyst during the transition from prebiotic chemistries to cellular biochemistries. Most importantly, directed evolution experiments demonstrate that much of the modern translation system could have emerged from relatively simple catalysts in the RNA World. Once established, an efficient translation system would have opened the door for the evolution of modern biochemistry. Future studies of ribozymes should further delimit what processes may have led to the advent of life itself and should help to expand our understanding of the period of biochemical evolution leading up to and encompassing the last common ancestor of modern life, the progenote.

References

Further Readings

1. Wilson DS, Szostak JW. In vitro selection of functional nucleic Acids Annu Rev Biochem 1999; 68:611-647.
   - This is a review of how in vitro selection techniques can be used to generate nucleic acids with desired functions.
2. Chen X, Li N, Ellington AD. Ribozyme catalysis of metabolism in the RNA world. Chem Biodiv 2007; 4:633-655.
   - This is an extensive review of ribozyme catalysis and models for the RNA World.
3. Joyce GF. The antiquity of RNA-based evolution. Nature 2002; 418:214-221.
   - This is an excellent short review of models for the RNA World from prebiotic chemistry through the advent of translation.

Specific References

4. Garriga G, Lambowitz AM, Inoue T et al. Mechanism of recognition of the 5' splice site in self-splicing group I introns. Nature 1986; 322:86-89.
5. Guerrier Takada C, Altman S. Catalytic activity of an RNA molecule prepared by transcription in vitro. Science 1984; 223:285-86.
6. Gilbert W. The RNA world. Nature 1986; 319:618.
7. White HB III. The pyridine nucleotide coenzymes, eds. Everse J, Anderson K, Yu, KS. Academic Press. New York 1982:1-17.
8. White HB III. Coenzymes as fossils of an earlier metabolic state. J Mol Evol 1976; 7:101-104.
9. Sievers D, von Kiedrowski G. Self-replication of complementary nucleotide-based oligomers. Nature 1994; 369:221-24.
10. Zielinski WS, Orgel LE. Autocatalytic synthesis of a tetranucleotide analogue. Nature 1987; 327:346-47.
11. Bartel DP, Szostak JW. Isolation of new ribozymes from a large pool of random sequences. Science 1993; 261:1411-18.
12. Ekland EH, Bartel DP. RNA-catalysed RNA polymerization using nucleoside triphosphates. Nature 1996; 382:373-76.
13. Johnston WK, Unrau PJ, Lawrence MS et al. RNA-catalyzed RNA polymerization: accurate and general RNA-templated primer extension. Science 2001; 292:1319-25.
14. Lawrence MS, Bartel DP. New ligase-derived RNA polymerase ribozymes. RNA 2005; 11:1173-80.
15. Levy M, Ellington AD. The descent of polymerization. Nat Struc Biol 2001; 8:580-82.
16. Robertson MP, Scott WG. The structural basis of ribozyme-catalyzed RNA assembly. Science 2007; 315:1549-53.
17. Szabo S, Scheuring I, Czaran T et al. In silico simulations reveal that replicators with limited dispersal evolve towards higher efficiency and fidelity. Nature 2002; 420:340-43.
18. Dowler MJ, Fuller WD, Orgel LE et al. Prebiotic synthesis of propiolaldehyde and nicotinamide. Science 1970; 169:1320-21.
19. Keefe AD, Newton GL, Miller SL. A possible prebiotic synthesis of pantetheine, a precursor to coenzyme A. Nature 1995; 373:683-5.
20. Breaker RR, Joyce GF. Self-incorporation of coenzymes by ribozymes. J Mol Evol 1995; 40:551-558.
21. Tsukiji S, Pattnaik SB, Suga H. An alcohol dehydrogenase ribozyme. Nat Struct Biol 2003; 10:713-17.
22. Tsukiji S, Pattnaik SB, Suga H. Reduction of an aldehyde by a NADH/Zn^{2+} dependant redox active ribozyme. J Am Chem Soc 2004; 126:5044-45.
23. Lorsch JR, Szostak JW. In vitro evolution of new ribozymes with polynucleotide kinase activity. Nature 1994; 371:31-36.
24. Huang F, Yarus M. Versatile 5' phosphoryl coupling of small and large molecules to an RNA. Proc Natl Acad Sci USA 1997; 94:8965-8969.
25. Huang F, Bugg CW, Yarus M. RNA-catalyzed CoA, NAD and FAD synthesis from phosphopantetheine, NMN and FMN. Biochem 2000; 39:15548-15555.
26. Jadhav VR, Yarus M. Coenzymes as coribozymes. Biochimie 2002; 84:877-88.
27. Robertson MP, Knudsen SM, Ellington AD. In vitro selection of ribozymes dependent on peptides for activity. RNA 2004; 10:114-127.
28. Roth A, Breaker RR. An amino acid as a cofactor for a catalytic polynucleotide. Proc Nat Acad Sci USA 1998; 95:6027-6031.
29. Majerfeld I, Puthenvedu D, Yarus M. RNA affinity for molecular L-histidine: genetic code origins. J Mol Evol 2005; 61:226-235.
30. Robertson MP, Miller SL. Prebiotic synthesis of 5-substituted uracils: A bridge between the RNA world and the DNA-protein world. Science 1995; 268:702-705.
31. Wong JT. Origin of genetically encoded protein synthesis: a model based on selection for RNA peptidation. Ori Life Evol Biol 1991; 21:165-176.
32. Szathmary E. The origins of the genetic code: amino acids as cofactors in the RNA World. Trends in Genetics 1999; 15:223-229.
33. Kumar RK, Yarus M. RNA-catalyzed amino acid activation. Biochemistry a 2001; 40:6998-7004.
34. Illangasekare M, Sanchez G, Nickles T et al. Aminoacyl-RNA synthesis catalyzed by an RNA. Science 1995; 267:643-647.
35. Illangasekare M, Yarus M. Specific, rapid synthesis of phe-RNA by RNA. Biochemistry 1999; 96:5470-75.
36. Lee N, Bessho Y, Wei K et al. Ribozyme-catalyzed tRNA aminoacylation. Nat Struc Biol 2000; 7:28-33.
37. Illangasekare M, Yarus M. A tiny RNA that catalyzes both aminoacyl-RNA and peptidyl-RNA synthesis. RNA 1999; 5:1482-1489.
38. Zhang B, Cech TR. Peptide bond formation by in vitro selected ribozymes. Nature 1997; 390:96-100.

Precellular Evolution:
Vesicles and Protocells

Pasquale Stano and Pier Luigi Luisi*

12.1 Introduction

Living organisms are all composed of cells. Multicellular organisms are composed of several kinds of specialized cells, spatially organized and interacting with each other in complex ways; unicellular organisms, on the other hand, are composed of just one cell that must accomplish all the necessary functions for its maintenance and reproduction. The cellular nature of all forms of life and the evidence that no life exists at hierarchical levels below the cell, can be seen as the most striking property of living organisms.

When we look to an isolated cell, e.g., a bacterium, the most obvious fact lies in front of our eyes: the living can be appreciated as a single entity which is well differentiated from its environment. In other words, the cell has a defined boundary that separates the inside from the outside.

And looking at this interaction between these two regions, we are also able to capture the blue print of cellular life—on the basis of a phenomenological approach. First of all, cells are alive, but "life" cannot be found in any of its individual components—rather, "life" is a collective property that stems from the self-organization of the cellular components and processes—it is a distributed, emergent property. From the historical point of view, several authors recognized that the very turning point of the transition from nonlife to life is the formation of a cell or, in more general term, of a compartment. Harold J. Morowitz, in his book Beginning of Cellular Life,[1] firmly advocates this idea, showing that the universal nature of living compartments is irreducible, since a simpler form of life cannot be found; moreover, he emphasizes the universality of cellular architecture, containing a plasma membrane that separates the cell from its environment.

Before going on with the issue of "compartmentation", a short note on the current theories on the origins of life is called for. The main assumption held by scientists studying origins of life is that life originates from inanimate matter through a spontaneous and gradual increase of molecular complexity, from simple molecules to bio-monomers (as amino acids, sugars, aromatic bases, lipids), to macromolecules and thereby to activity (the so-called Oparin-Haldane hypothesis). Within this framework, two kinds of view have dominated the scientific debate in the last century, namely (i) the metabolism-first approach and (ii) the gene-first approach. Although it is often difficult to sharply categorize researchers in one or the other school of thought, authors such as Alexander I. Oparin, Sidney Fox, Freeman Dyson, Stuart Kauffman, Harold J. Morowitz and Günter Wächtershäuser can be broadly defined as supporters of the first view. Other authors such as John B. S.

Haldane, Leonard Troland, Manfred Eigen, Leslie Orgel, Francis Crick, Carl Woese, Walter Gilbert and Gerard Joyce can be rather considered as adhering to the gene-first hypothesis. An historical discussion of the alternation among different theories, their derivations and evolution has been presented by Iris Fry in her recent book on the origin of life.[6]

Without going into details, this dichotomy ultimately derives from the co-occurrence—in modern cells—of two kind of biopolymers: proteins (enzymes) and nucleic acids (DNA, RNAs), that produce each other in a cyclic manner. Different theories try to explain which of the two was the first agent, the initiator of the cycle.

The "compartment" approach can be considered as being independent of the above-mentioned cases, with the favorite theme of compartmentalized reaction networks. It does not directly resolve the metabolism/genes dilemma, but emphasizes instead the role of compartments in the origin of life. Some of the authors mentioned above have also integrated in their views in one form or another the notion of compartment, such as Dyson or Morowitz. The approach is generally based on enclosed reaction (also genetic) networks in a self-organized bounded system that determines a series of emergent properties such as selective permeability, establishment of electrochemical gradients, sustenance of out-of-equilibrium processes and above all the emergence of cellular individuality.

A basic model for compartments is provided by lipid vesicles (liposomes). Liposomes as well as other kind of vesicles form spontaneously when certain amphiphilic molecules are dispersed in water, (Fig. 12.1). The lipid molecules self-assemble to form a bilayer semi-permeable membrane that encloses an inner aqueous phase and presents a hydrophobic barrier between the environment and the vesicle. What is important from the point of view of the origin of life is the biogenesis of liposome monomers (i.e., the membrane-forming molecules) under prebiotic conditions, as will be discussed in the following paragraphs.

This compartment approach is embodied in the lines of research carried out by David Deamer, Doron Lancet, Yoiko Nakatani, the late Guy Ourisson, Tetsuya Yomo and our group. Over the last 30 years, a considerable number of investigations have been devoted to the study of vesicles composed of phospholipids and other important surfactants such as fatty acids, giving rise to several remarkable findings regarding vesicle properties and reactivity.

In this chapter, we shall illustrate some of the basic concepts underlying cellular and precellular evolution. Toward this aim, we shall discuss first what may be defined as the blue print of cellular life;

*Corresponding author: Pier Luigi Luisi—Biology Department, University of Rome "RomaTre" Viale Guglielmo Marconi, 446; 00146 Rome, Italy. Email: luisi@mat.ethz.ch

Prebiotic Evolution and Astrobiology, edited by J. Tze-Fei Wong and Antonio Lazcano. ©2009 Landes Bioscience.

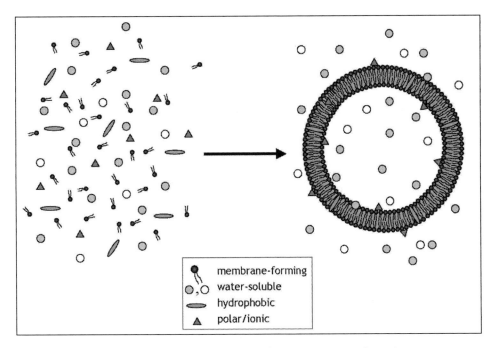

Figure 12.1. The spontaneous self-assembly of membrane-forming surfactants into a vesicle, with an inner water pool that can host water-soluble molecules. Any hydrophobic molecules present will be positioned in the membrane; if ionic surfactants are involved in vesicle formation, the ionic and/or polar solutes with complementary changes will be adsorbed on the surface. Notice that the formation of vesicles yields a three-phase microheterogeneous system (inside/boundary/outside), making possible the establishment of chemical gradients, segregation of macromolecules, selective permeability and above all the emergence of cell-like individuality. Double-chain surfactants as well as some classes of single-chain surfactants can form vesicles under a variety of conditions. In addition to bilayers, mono-layer membranes can be formed by so-called "bola" amphiphiles. Passive entrapment of solutes within vesicles is a process that does not require additional energy; and the self-assembly of surfactants proceeds spontaneously as well ($\Delta G < 0$).

the most powerful theory in this regard is the theory of autopoiesis, which will next be briefly sketched. After that we will describe some of the most basic features of liposomes as models for early cells and give a short review on biochemical reactions occurring in liposomes as a model system for simple cells. We shall also examine a striking reactive pathway of vesicles, namely that of vesicle self-reproduction and the possible integration of metabolism-first and/or gene-first approaches within the framework of the compartment approach. Finally, we shall provide an introduction to the notion and research of "minimal cells"—tangible objects synthesized in the laboratory—that might display minimal cellular functions. This paves the way to an understanding of the structure and functions of the early cell and its evolution to the complex cells of to-day.

12.2 Autopoieis and the Logic of Cellular Life

Cell's life is the starting point for the development of autopoiesis (from Greek auto = self, poiesis = production), introduced in the Seventies by the two Chilean biologists Humberto R. Maturana and Francisco J. Varela.[2,3,7] Autopoiesis does not concern the origins of life, but answers instead to the question "what is life?", analyzing the living organism as it is here and now. In this sense, autopoiesis is a descriptive theory based on phenomenology and starts from the recognition that the main activity of the cell is the maintenance of its own individuality: this is so despite the large number of transformations taking place inside its boundary. This is possible thanks to the fact that the cell regenerates from the inside all components that are being transformed away (boundary molecules included!). Thus, very simply, the blue print for the life of a cell resides in a series of processes that produce all the cell's components that in turn produce the processes that produce such components. This occurs of course at the expense of energy and nutrients from the outside, so that the living cell is thermodynamically speaking an open system, yet characterized by its own operational closure, i.e.,

it contains all the information to organize and reproduce itself from inside the boundary.

This is actually all very simple. Autopoiesis recognizes in the working of the living cell the complementarity between two levels: organization and structure. Organization is the invariant property of all living cells and is based on the production of components that self-organize into an organized reaction network of processes that makes the components and thus perpetuates a circular logic (Fig. 12.2). In contrast, the structure may vary from cell to cell and also in the course of the evolution of a cell.

Several important concepts are associated with this cellular view, such as biological autonomy, social autopoiesis, second order autopoiesis and third order autopoiesis (depending on the complexity of the living organism), etc. The reader is referred to an introduction to autopoiesis in Luisi, 2006[2]), which also contains information about the historical background and considerations on why the autopoiesis theory has not generated a large impact on contemporary biochemistry. Here it may be mentioned that in order to have a complete picture of the living state, the notion autopoiesis needs to be complemented by that of cognition, also developed by Maturana and Varela[7]—namely the selective interaction with the environment.[8,9]

Autopoiesis is closely connected with the definition of life at the cellular level. In fact, the property of being alive belongs to the cell as a whole and not to its individual parts. In this regard, autopoiesis is related to the compartment approach described in the Introduction, not because it explains how compartments are generated, but because it emphasizes that investigations on the origin of living entities must be focused on compartmentalized systems and not on single molecules.

In the following paragraphs, when the self-reproduction of compartments will be discussed, it will be shown how autopoiesis inspired these studies and how a simple autopoietic chemical system was realized in the laboratory.

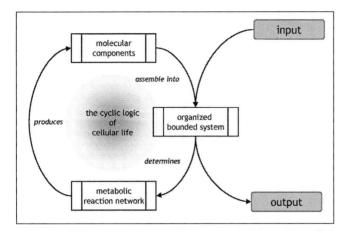

Figure 12.2. The cyclic logic of cellular life. The cell, being an autopoietic unit, is a self-organized bounded system that determines a network of reactions, which in turn produces molecular components that assemble into the organized system, that determines the reactions network, that…

12.3 Vesicles and Other Compartments

The notion that compartments are basic structures of life existed in the literature long before the work of Maturana and Varela. Historically, the first models of compartments were the "coacervates",[10] that are microscopic droplets of proteins and sugars formed by self-assembly, which since they can adsorb small molecules and transform them in a sort of primitive metabolism were supposed to be the precursor of cells. A second attempt to model primitive compartments was made by Fox[11] based on "proteinoid" microspheres formed by the polymerization products of amino acids. These two systems have only historical significance, as most of the studies to-day on protocells are carried out with vesicles, which resemble biological cells more and permit a high degree of biochemical performance. In addition to vesicles, micelles and reverse micelles have been used to construct cellular models especially with regard to self-reproduction (see ref. 2).

Vesicles are spherical compartments ranging on average 50-1000 nm in radius and are formed when appropriate surfactants are dispersed in water. They are formed by a spherically closed semi-permeable boundary, the bilayer membrane, which is self-assembled by the surfactant molecules orienting their hydrophobic tails toward the inner side of the membrane and exposing only their hydrophilic heads to the aqueous solvent (Fig. 12.3). Phospholipids are the most typical vesicle-forming compounds and vesicles formed by lipids are usually called liposomes. Not all phospholipids, however, yield lipid vesicles. For example, the nature of the headgroup can affect the self-assembly properties of the molecule by altering the "surfactant parameter", an empirical value that can be used to envision the most stable assembly pattern of surfactants. Bilayers are formed when the surfactant molecules take on a cylinder-like shape with the value of $v/(a_0 \cdot l)$ about 1.

Phosphatidylcholine vesicles as well as other phospholipid vesicles are very well known and biophysical studies have been carried out to determine their formation, stability, permeability, rigidity, molecular dynamics (such as molecular diffusion in the bilayer, vibrations out-of-the-plane of the membrane, rotational diffusion and flip-flop movements). To have a quantitative idea about vesicle samples, consider a suspension of 10 mM POPC (1-palmitoyl-2-oleoyl-sn-glycero-3-phosphocholine) unilamellar vesicles having a mean radius of 50 nm. On average, a single vesicle is composed of about 90,000 lipid molecules, self-assembled into a spherical bilayer membrane. One milliliter of such suspension contains ca. 10^{14} vesicles, each enclosing an internal volume of ca. 0.0005 fL and the collective internal volume is about 36 μl (3.6%). In contrast, the overall (internal and external) surface is about one square meter. Sub-micrometric lipid vesicles are therefore characterized by a high surface-to-volume ratio and many of their properties are common to those of colloids. The

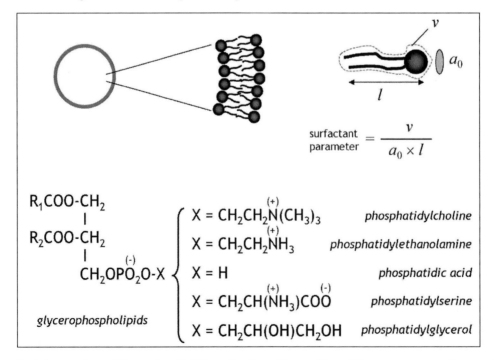

Figure 12.3. Detail of the membrane bilayer. Amphiphilic molecules self-assemble in bilayer structures, where the hydrophilic, polar headgroup is exposed to aqueous solvent. The hydrophobic tails are kept together and away from the aqueous solvent. On the top-right, the so-called "surfactant parameter" and its pictorial representation in terms of basic geometrical elements (v is the molecular volume of the amphiphile, a_0 the "effective" headgroup surface and l is the length). The main glycerophospholipids are shown at the bottom. The surfactant parameter predicts that in most cases they can form bilayers, but some factors as the presence of salts, or insaturations on the acyl chains can change their behavior. Properties of amphiphiles mixtures are difficult to predict.

chemical nature of the lipid bilayer (in terms of polarity) allows the emergence of selective permeability. Water, glycerin, tryptophan, a generic protein and sodium ions have relative permeability in the ratios of $10^9:10^6:10^2:1:1$ for a phosphatidylcholine membrane; thus phospholipids bilayers are impermeable to large molecules and charged ions, whereas small uncharged molecules have higher permeabilities. It is clear therefore that a semi-permeable membrane establishes a selective regime for the uptake/release of molecular components, whereas polymers such as functional macromolecules, when formed inside a vesicle, are well stored and can exert their functions inside a vesicle.

Lipid vesicles, although very useful in constructing cellular models, might have little relevance to origins of life studies. In fact, the complexity of the phospholipid molecular structure makes it difficult to consider such molecules as plausible candidates for the formation of the first protocells even though in some cases the synthesis of phospholipids has been achieved under prebiotic conditions.[2]

In contrast, it has been established that fatty acids were prebiotically available[4] and that fatty acids can spontaneously self-assemble into vesicles.[12,13] Fatty acids have been synthesized under a variety of allegedly prebiotic conditions (see for example ref. 14) and have also been found in the Murchison meteorite.[15] When compared to phospholipid vesicles, fatty acid vesicles show different properties that are strongly dependent on the length of the acyl chain and the number of double bonds present in the chain. Moreover, fatty acids self-assemble into vesicles or micelles according to the ionization degree of their head groups (Fig. 12.4), a behavior that can be rationalized on the basis of the surfactant parameter. Long-chain carboxylates aggregate as micelles when completely ionized (high pH), whereas fatty acids separate out from aqueous solution as oil droplets when completely un-ionized (low pH). At intermediate pH values (generally around 7.5-9.5), fatty acids and their conjugate base (the carboxylate form) are present in equivalent amounts and bilayers can be formed; this observation has been explained on the basis of the hydrogen-bonded dimer shown in Figure 12.4. This intermolecular interaction to form dimers brings about a drastic change in pK: in supramolecular membrane assemblies the pK of long-chain fatty acids is around 4-5 pH units above the pK of short-chain water-soluble monomeric carboxylic acids (e.g., acetic acid).

From these considerations and from the relatively high solubility of fatty acids (~ mM), it follows that fatty acids have a complex phase diagram and are in general more reactive than their phospholipid counterparts. This can be seen as a favorable property for fatty acids being the candidates for early protocells. In the past decades, an increasing number of reports have appeared in specialized journals and vesicles from fatty acid vesicles have been investigated using different techniques, focusing on their structure, morphology, stability, physicochemical properties and reactivity. The interested reader should refer to original papers or to recent reviews regarding these investigations.[4,5,16]

Among the different fatty acids, oleic acid (i.e., (Z)-octadec-9-enoic acid, Fig. 12.4) has received special attention as a model system to explore fatty acid vesicles. In the following section studies on the reactivity and the transformations of oleic acid/oleate vesicles—often referred as oleate vesicles—will be examined with respect to key pathways, that could have relevance to the origin and development of protocells.

In addition to fatty acids, the alkyl and oligoprenyl phosphates, fatty alcohols and monoacylglycerols are also considered to be prebiotically relevant.[5]

12.4 Reactivity and Transformation of Vesicles

Vesicles as such are only models of the shell of biological membranes. In order to obtain a more meaningful biological model for the cell, one has to have biochemicals inside the membranes as well as reactions that are of biological interest. This has been achieved and here we will review some fundamentals of reactions in vesicles, focusing on fatty acid vesicles.

In particular, two aspects of the use of fatty acid vesicles as model of early cells reveal how such synthetic compartments have been investigated in origins of life studies. These are: (A) the realization of fundamental biochemical reactions inside fatty acid vesicles and (B) the self-reproduction of fatty acid vesicles. These two important features of the vesicles' reactivity and transformation will be discussed at first separately. Later it will be shown how a route can be found to functionally couple them together. It is important to realize that the model systems illustrated here are inspired by the autopoietic theory and represent the first attempts to concretely understand the basic process of living cells by constructing them.

(i) Compartmentalized Reactions

The main idea is that simple and complex reactions of biochemical significance can be carried out in the inner cavity of vesicles. A preliminary concept relating to the permeability of vesicles was introduced in the previous section. Lipid vesicles and fatty acid vesicles exhibit selective permeability toward different molecules. Large and charged molecules have low permeability and therefore do not cross the hydrophobic membrane easily. In contrast, small molecules, even if charged, can permeate the bilayer barrier at a measurable rate. Hydrophobic and amphiphilic molecules, on the other hand, accumulate in/on the membrane. This means that it is possible in some cases to feed vesicles through external addition of precursors. The classical application is the entrapment of an enzyme inside vesicles will external addition of substrates. The substrates permeate into the vesicles, where they react with the entrapped enzyme and the reaction product may be released or not from the vesicle, depending on its chemical nature.

Following this scheme (with various modifications), several enzymes have been studied inside vesicles, such as carbonic anhydrase, α-chymotrypsin, β-galactosidase, etc. (for a recent review, see ref. 17). But how can an enzyme be inserted inside the vesicles? The

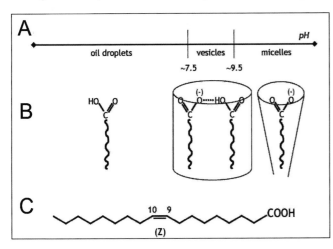

Figure 12.4. Self-assembly structures of fatty acids. A) In general, fatty acids separates out from aqueous solution as "oil" at low pH values, form bilayers (vesicles) at intermediate pH values and form micelles at high pH values (the interval 7.5-9.5 is just an example of typical behavior). B) The different self-assembly patterns at various pH have been explained on the basis of the different protonation states of the molecule, i.e., protonated at low pH, fully ionized at high pH and forming a 1:1 acid:base dimer at intermediate pH near the pK. C) Chemical structure of oleic acid.

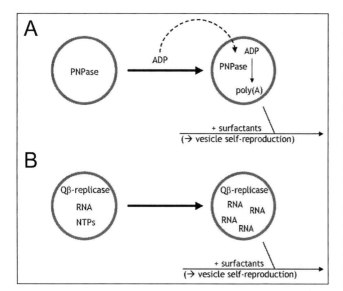

Figure 12.5. Two important reactions carried out inside fatty acid vesicles. Polynucleotide phosphorilase (PNPase) is entrapped inside oleate vesicles, then ADP is added externally; it slowly permeates into the vesicles by crossing the membrane and it is polymerized inside (to give poly(A)). As indicated in the arrow, the process is carried out in self-reproducing vesicles (A). A template RNA, Qβ-replicase and all low molecular weight compounds are trapped inside oleate vesicles and the replication of RNA is followed, simultaneously to the self-reproduction of vesicles (B).

entrapment is generally carried out by preparing the vesicles in the solution where the enzyme is solubilized. Although enzymes are supposed to be passively entrapped during the formation of the vesicles, in some cases, e.g., with basic enzymes and negatively charged surfactants, electrostatic forces can play a large role determining the entrapment yield and the location of the enzyme in the vesicle. After vesicle formation, the enzyme left in the external medium can be removed by gel filtration, centrifugation, dialysis or ultrafiltration. In the case of very large vesicles (>10 μm), the enzyme can also be directly injected into the vesicles by microinjection techniques.

One of the first important experiments on biochemical reactions inside vesicles, with the aim of constructing processes relating to a minimal cellular compartment, was carried out in as early as 1990 by Schmidli et al:[18] the four enzymes responsible for the synthesis of lecithin were entrapped inside lecithin liposomes, with the idea that lecithin would be produced inside the liposomes, leading to the growth and possibly division of the liposomes from the inside—a typical autopoiesis experiment. The experiment was partly successful, but it could not be completed as intended when the enzymes were withdrawn from commercial availability.

In order to simulate two very important processes for the origins of early protocells, the synthesis of poly(adenilic acid)[19,20] and the self-replication of RNA[21] have been carried out in fatty acid vesicles (Fig. 12.5). In the first case, polynucleotide phosphorylase (PNPase) was entrapped inside oleate vesicles (at pH 9) and adenoside diphosphate (ADP) was externally added (Fig. 12.5A). Thanks to the nonzero permeability of the nucleotide, poly(A) was isolated at the end of the reaction from inside the vesicles. Once polymerized, the product in fact cannot escape from the inner vesicle cavity. This system shows how a combination of polymerization and semi-permeability produces an "emergent" outcome, i.e., the formation of protocells containing a macromolecule. This is a model of the early formation of functional vesicles containing a functional biopolymer. In ancient times, a small catalyst might have played the role of PNPase. Notice that, in principle the vesicle membrane

may also assist the polymerization reaction, or reaction inside the vesicle may lead to the formation of some molecule that later acts as a catalyst for polymerization.

In the second case (Fig. 12.5B), an RNA strand was replicated by means of the enzyme Qβ-replicase.[21] The enzyme and the substrates for the reaction (RNA and nucleotides) were trapped simultaneously inside oleate vesicles and the replication of RNA was achieved. Enzymatic RNA replication inside vesicles is of course pertinent to the origin of functional protocells: it is a model for early compartmentalized RNA replication catalyzed either by RNA itself, as in the RNA world hypothesis, or by other catalysts such as proto-enzymes.

As it will be discussed later, the relevance of these investigations (Fig. 12.5A,B) is also related to the simultaneous self-reproduction of the vesicles, which was allowed to occur together with the compartmentalized reactions, so that a system of self-reproducing vesicles with internalized polymerization or internalized self-replication was realized. Therefore these models, even if they do not explain how the first protocells originate, provide a solid experimental approach to the understanding of compartmentalized reactions.

It has also been possible to carry out the PCR reaction in liposomes;[22] and later on ribosomes also could be entrapped in the inner core of vesicles, thus permitting for the first time the synthesis of a polypeptide inside a vesicle.[23] This last experiment opened the gate to a series of subsrquent experiments aimed at expressing protein synthesis in liposomes.

It is well accepted today that proteins can be expressed inside vesicles, by entrapping simultaneously a gene and the entire biochemical machinery needed to carry out the transcription-translation processes. Green fluorescent proteins, T7 RNA polymerase and α-hemolysin have been successfully expressed inside phospholipid vesicles. Again, the interested reader should consult recent reviews for details.[2,24] Some experiments developing the notion of minimal cell are described as follows.

(ii) Self-Reproduction of Vesicles

We have mentioned before the capability of vesicles to undergo "self-reproduction" processes. This is of course very important as it permits a closer modeling of the biological cells. We would like to sketch here the basic principles of such a mechanism and the reader may referr to specialized literature for further details.[2]

Supramolecular structures such as vesicles, micelles and reverse micelles can undergo a "self-reproduction" process. The terminology here is important, since the term "self-replication" i.e., an exact complementary replica of the template molecule, is properly applicable to small molecules as well as nucleic acids. In contrast to self-replication, self-reproduction deals with the formation of new structures via growth-division processes, which largely do not proceed with stringent molecular-level controls as in the case of nucleic acid replication. Molecules self-replicate, whereas cells self-reproduce. As a typical example, a vesicle can grow and divide into two daughter vesicles having different sizes, so that the initial structure (the vesicle) reproduces to form two new vesicles that are similar but not identical.

Pioneering investigations on self-reproduction of supramolecular structures were carried out by Luisi and coworkers in the 1990s, giving rise to the discoveries—in chronological order—of reverse micelles, micelles and vesicle self-reproduction.[25]

Vesicle self-reproduction was achieved in 1994, using fatty acid vesicles and a fatty acid precursor (Fig. 12.6). The experimental strategy is clearly inspired by the autopoietic approach, namely the achievement of a dynamical system that works in the following manner: a bounded particle (the vesicle) takes up the "nutrients" needed for its growth, transforms them—within its boundary—

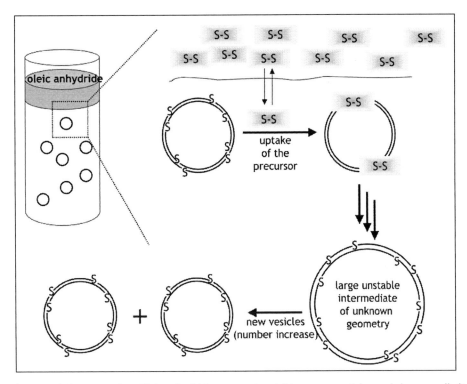

Figure 12.6. Self-reproduction of oleate vesicles. Oleic anhydride, a water-insoluble precursor, is layered above an alkaline solution containing preformed oleate vesicles. Oleate vesicles take up anhydride molecules by dissolving them in the hydrophobic bilayer and increasing the surface where hydrolysis can take place. The products of the reaction, two oleate molecules, increase the vesicle surface and bring about vesicle growth (non necessarily spherical), with a consequent destabilization of the structure that collapses into two or more small vesicles. The overall process leads to an autocatalytic number growth of the vesicles. If preformed oleate vesicles are omitted, they form spontaneously after the initial stage of anhydride hydrolysis and then catalyze the whole process after a lag phase.

into particle components (membrane-forming molecules) and consequently grows, reaches an unstable state and then divides into two or more new particles. The transformation of precursors into elements of the autopoietic particle must occur within the particle itself, for this is the key interior that distinguishes the autopoietic self-reproduction of vesicles from a spontaneous formation of vesicles.

Experimentally, this process can be realized by stratifying oleic anhydride (the precursor) on a basic solution where oleate vesicles are present. Oleic anhydride is slowly hydrolyzed at the interface with the basic solution. Simultaneously to this spontaneous process, other oleic anhydride molecules are taken up by preformed oleate vesicles, since they can be solubilized within the bilayer membrane. Once in the membrane, oleic anhydride can be hydrolyzed and therefore transformed into two oleate molecules, with a corresponding increase of membrane surface. The oleate vesicle is therefore growing by incorporating membrane precursors, which are transformed into building-blocks of the membrane itself, with consequent growth and the eventual production of new vesicles (with a mechanism which is still poorly understood).

The stoichiometry of a growth-division process can be schematically summarized as follows:

$$V + nS \rightarrow 2V$$

where V represent a vesicle and S the membrane-forming compound (for the sake of simplicity, S can represent the precursor as well). Accordingly, the process is autocatalytic, since more vesicles are produced, more precursor can be taken up, more vesicles will be produced and so on. Experimentally, autocatalytic processes are characterized by sigmoidal profiles and this was actually demonstrated in the self-reproduction of oleate vesicles (Fig. 12.7). Notice also that if preformed vesicles are not included in the experimental setup, they can be formed after the initial stage of spontaneous hydrolysis

of the anhydride, with the result that the overall process becomes a spontaneous pathway from surfactant precursors to vesicles without the need of preformed structures. The driving force for entry into the self-reproduction cycle comes from the thermodynamics of anhydride hydrolysis and the self-assembly of fatty acids.

The second example of oleate vesicle self-reproduction exploits the pH-dependent assembly properties of fatty acids. In fact, since at high pH oleate molecules self-assemble as micelles, whereas at intermediate pH (8.5) they form bilayers and therefore vesicles, the addition of oleate micelles into a buffered solution of preformed oleate vesicle also represents a way of feeding vesicles with the aim of observing their growth and division, i.e., self-reproduction. Actually this strategy is currently adopted by many researchers, for it does not have the drawback of the oleic anhydride approach in involving a biphasic system.

Oleate micelles and oleate vesicles are both macroscopically single-phase systems and therefore the process of vesicle self-reproduction can be easily followed by spectroscopic techniques. When oleate micelles (in equilibrium with the monomeric form) are added to a pH 8.5 buffered solution they can form vesicles spontaneously by rearranging their packing modes. However, when preformed oleate vesicles are present in such solution, an additional pathway might be the uptake of oleate (in the form of monomers or micelles) from preformed vesicles, exactly as in the previous case of oleate vesicles and oleic anhydride. In this second pathway, the preformed vesicles grow and divide, starting up self-reproduction dynamics. The two competitive processes of de novo vesicle formation vs uptake-growth-division can be distinguished by labeling the preformed vesicles with a water-soluble marker such as ferritin molecules (Fig. 12.8). By measuring the distribution of ferritin molecules inside the vesicles before and after the addition of oleate micelles, it has been shown that the vesicles can actually grow and a small but significant quantity of ferritin-containing vesicles are

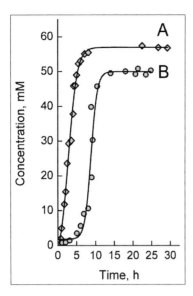

Figure 12.7. Kinetic profiles of the system illustrated in Figure 12.6. When preformed oleate vesicles are present in the aqueous phase, oleic anhydride is quickly hydrolyzed (curve A). When preformed oleate vesicles are not present at the start, they form after the initial alkaline hydrolysis of oleic anhydride, soon after the critical aggregation concentration is reached. The S-shaped profile (curve B) clearly indicates an autocatalytic mechanism, i.e., vesicles that catalyze the formation of other vesicles.

formed by the division of the grown vesicles, as indicated by the reduced number of ferritin molecules per vesicle caused by the distribution of ferritin molecules into "daughter" vesicles.

Modern studies on fatty acid self-reproduction have revealed a series of interesting features that cannot be discussed in detail here, such as the existence of a "matrix" effect, a sort of "templating" effect exerted by preformed vesicles, resulting in the formation of a vesicle population with conserved sizes;[26] the competition between vesicles of different sizes for uptake of oleate micelles;[27] or the competition between osmotically "stressed" and "relaxed" oleate vesicles.[28]

An original contribution combining vesicle self-reproduction and inheritable "compositional" information made by Doron Lancet is shown in Box 12.1.

(iii) Strategies for Functional Coupling of Core and Shell Transformations

We have seen that: (i) simple and complex biochemical reactions (including self-replication) can be carried out with vesicles; (ii) vesicles can self-reproduce spontaneously upon the external addition of a suitable surfactant precursor. Starting from these considerations, which have come from experiments, it might be asked if it is possible to further couple these two processes to create a simple model of an autopoietic self-reproducing cell.

From the theoretical viewpoint, the solution can easily be envisioned: a metabolic network that reproduces itself should be encapsulated within a vesicle and some of the processes in the network should be capable of producing the membrane-forming component from suitable precursors. The system can be simple in construction at the beginning and become more sophisticated at a later stage of development (Fig. 12.9). The vesicle would grow either in terms of internalized molecules based on cyclic networks that produce more and more molecules of the members of the cycles as the reactions proceed, or in terms of membrane surface. As in the case of vesicle self-reproduction, collapse from an unstable state may be expected

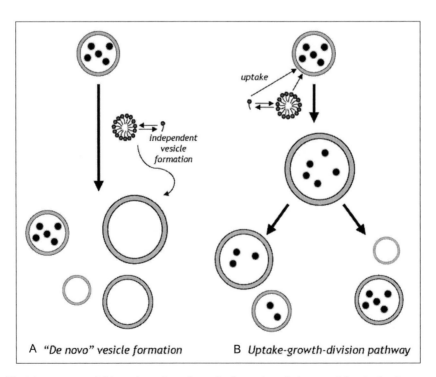

A *"De novo" vesicle formation*   B *Uptake-growth-division pathway*

Figure 12.8. The use of ferritin as water-soluble probe to investigate the formation of oleate vesicles. In the first case (A), oleate micelles form oleate vesicles independently from the preformed vesicles (the experiment was performed using phosphatidylcholine vesicles as preformed vesicles). The mechanism involves the de novo formation of oleate vesicles, with no interaction between oleate micelles—or oleate monomers—and preformed vesicles; consequently, the label is not diluted. In the second case (B) oleate micelles—or oleate monomers—are taken up by preformed phosphatidylcholine vesicles that grow as new membrane-forming molecules are incorporated into the membrane. Following this initial step, the vesicle becomes unstable (the growth might also be nonspherical) and divide into two or more "daughter" vesicles. Accordingly, the probe may or may not be statistically distributed among the new vesicles. The reduction of the average number of ferritin molecules per vesicle suggests the existence of a growth-division pathway (Vesicles, micelles, fatty acids and ferritin molecules are not drawn to scale).

Box 12.1. Lipid "Composomes" and Their Role as Prebiotic Replicators

The role that supramolecular aggregates, as lipid and fatty acid vesicles, can have in the origin of life is not only related to their property of providing a compartment for the development of biochemical reactions. The Graded Autocatalysis Replication Domain (GARD), developed by Doron Lancet and coworkers at the Weizmann Institute, Rehovot, Israel, provides a possible—and unorthodox—scenario for a gene-free propagation of information.[29,30] In fact, it is proposed that in the very early stages of prebiotic evolution, supramolecular assemblies such as vesicles, composed of several kinds of molecular components, could homeostatically grow and split, generating new assemblies that have the same composition as the parent one. The homeostatic growth is favored by selective adsorption and transformation of compounds from the environment. In this way, a particular composition and organization of an assembly represents "information" transmitted through generations of assemblies. The existence of such "composomes" has been demonstrated by detailed computer simulations, that have also shown also the possibility of transition from one composome form to another in a simple evolutionary progression.

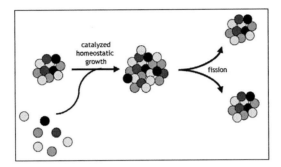

Figure Box 12.1.1. The "composomes", specific compositional states capable of homeostatic growth and information-preserving fission.

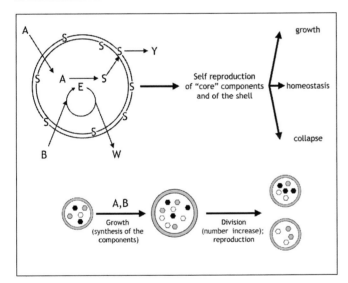

Figure 12.9. Diagram of a self-reproducing autopoietic protocell. A metabolic network is confined in the inner volume of a vesicle. The internal metabolic network "E" (consisting of one or more enzymes) catalyses the formation of the membrane-forming compound S from the precursor A, which is initially present in the environment and permeates through the membrane. S can decay to Y. The whole metabolic network, here indicated as "E", is also reproduced within the vesicle, fed by the set of precursors "B" and producing the waste molecules "W". This would represent a core-and-shell self-production. Notice the production of the internal components and that of the shell component S are functionally linked, because a product of the internal network catalyses the formation of S. Based on the balance between constructive (anabolic) and destructive (catabolic) processes, the protocell can undergo growth, collapse or—when the processes are balanced—enter into a homeostatic regime. When the protocell constructs its components, it grows and divides and the internal components are statistically distributed among the "daughter" cells. Since no control is exerted on vesicle division, some of the new vesicles may be deprived of key components and therefore enter into "death".

to give rise to two or more "daughter" vesicles, each containing part of the internalized components. If the number of internalized components is high, the probability of having all the components within each daughter vesicle is also high and the number of "active" vesicles actually increases. Vice versa, since the process is uncontrolled, the division of internalized molecules may proceed unevenly leading to some daughter vesicles not being able to sustain the metabolic network owing to the lack of some of the components.

The system described in Figure 12.9 is autopoietic and according to early considerations by Maturana and Varela,[7] it might even be called alive; however recent studies have led to a reassessment of this view.[8,31] Going back to experimental approaches, one may ask how the system illustrated in Figure 12.9 can be implemented in the laboratory.

The first attempt in this direction was realized by Zepik et al,[32] who model the system represented in Figure 12.9 by simplifying the network of internal reactions, so that only two reactions are operative: the synthesis and the degradation of S, the membrane-forming molecule. Both reactions, chemical in nature (hydrolysis of anhydride as the S-forming reaction and oxidation of S as the S-depleting reaction), can be regulated in rate so that one can observe the three different states shown on the right hand side of Figure 12.9. In particular, it has been shown that a homeostatic regime can actually be achieved through a the fine regulation of the anabolic and catabolic processes.

From the origin of life point of view, it would be very important to demonstrate that simple chemical systems can evolve, self-assemble and self-organize in the way depicted in Figure 12.9. At the current level of knowledge, however, this objective seems very difficult to achieve, especially if some degree of complexity and/or sophistication is to be built into the system. Moreover, if one wishes to reproduce the pathway leading from simple molecules to the first functional macromolecules (protoenzymes or ribozymes), to the first "metabolic" cycles, to self-replication or self-reproduction, one is faced with a lack of basic knowledge. It can be said—notwithstanding the great efforts devoted to experimental and theoretical research—that almost all the key transitions have simply not be defined.

A possible scenario involves the spontaneous generation of functional macromolecules (one of the greatest challenges in contemporary studies on origins of life), in/on prebiotically plausible self-assembling vesicles, such as the fatty acid vesicles and to give rise to the initiation of confined metabolism. From this stage onward, the self-maintenance and self-replication of the internalized reaction network would have to be attained, followed by the origin of the genetic code and still later by coded self-replication and core-and-shell self-reproduction (Fig. 12.10). In this protocell "evolutionary" pathway, interactions with the environment are fundamental, since the internalized metabolism has to be somehow coupled to the composition of the external medium. Vesicle self-reproduction, mediated by externally supplied or internally synthesized surfactants, would have to fulfill the role of multiplying protocells, at first in a stochastic way, generating diversity and competition between protocells and later on under the control of internal metabolism. Other phenomena, such as fusion, growth-division, competition, selective processes and the entire rich reactive landscape of vesicles could advantageously enable supramolecular systems to escape the world of inanimate matter. This hypothetical scenario, which yet remains largely speculative, focuses on the concept of individuality and population derived from the compartmentation of reaction networks, on their possible evolution and on their stepwise increases in complexity.

In order to understand some of the fundamental transitions depicted in Figure 12.10, several experimental and theoretical investigations have been carried out in recent years. In particular, it would be very important to address the numerous questions and aspects concerning the properties and biological functions of very early cells, those that do not possess a full-fledged arsenal of genes, enzymes, fine metabolic regulations, feedback loops and controlled functions. This "historical" aspect of primeval cells perfectly complements the theoretical issue of the minimal number of components and functions that would make a system "alive". Can we construct such a system?

Within the framework of feasible, concrete and robust—although difficult—experiments, we can "limit" ourselves to the investigation of semi-synthetic constructs that employ extant enzymes/genes and synthetic compartments to build "minimal cells", i.e., cell-like structures that display living properties based on the scheme in Figure 12.9. Of course, the choice of using evolved enzymes and genes stems from the absence of known primitive (and less specific) "enzymes" that would be perfectly fitting for the design and construction of protocells.

The achievement of such a goal would represent, even more than a historical reconstruction of the early living cells, a proof-of-principle concerning the definition and emergence of life as the coordinated, collective property of a network of compartmentalized interacting components/reactions.

Many recent advances (for a review, see ref. 24) indicate that vesicles with internalized reactions that are as complex as protein expression, occurring in single steps or in cascades and observable for periods of a few hours to several days, have become accessible to the experimentalist.

In essence, the key event remains the coupling between the internal metabolism (intended here as the formation and replication of functional macromolecules) and the reproduction of the whole vesicle. The coupling must be functional, i.e., the two processes must be coupled chemically by a reaction or a network of reactions, or by some molecule that is common to the two pathways. Previously, it has been demonstrated that internalized reactions and vesicle self-reproduction can occur simultaneously, as in the case of Qβ-replicase RNA replication and vesicle self-reproduction.

However, the two processes in this instance are just overlaid one on the other, with no functional coupling established between the core- and the shell-reproductions.

A possible approach illustrated in Figure 12.11 consists of a vesicle that synthesizes its membrane components thanks to the presence of a catalyst—an enzyme or a set of enzymes—that is itself produced by an internalized reaction network. In principle this is experimentally feasible by combining in vitro functional protein expression and the above-mentioned processes of vesicle growth-division. Ultimately, all the components required for the production of the enzyme(s) need to be replicated and evenly shared between the daughter vesicles. This construct is designated a semi-synthetic minimal cell.[2,24]

Such minimal cells, far from being perfect, can enrich our understanding of the biophysics and biology of early cells, their requirements in term of substrate permeability, dynamics of internalized reactions, disposal of waste materials, energy requirements, the quest of simultaneous core-and-shell reproduction and interaction with the environment and possibly other cellular individuals. The final goal of realizing the construct depicted in Figures 12.9 and 12.11 has not yet been achieved. We believe, however, insight into several important issues on the origin of cellular life can be gained along the path to this goal, because such "limping" cells, missing in many evolved functions, might represent useful models for the early cells that are open to experimental testing and verification.

12.5 Concluding Remarks

Compartments in general and vesicles in particular, play a fundamental role in the origin of life. We have already seen that, besides providing confined space for the occurrence and persistence of molecular reaction networks, the formation of a compartment establishes the emergence of individuality and, when the autopoietic conditions become fulfilled, the emergence of self-maintenance, self-bounding and homeostasis. In addition, confinement of self-organized reaction networks within a vesicle creates clear-cut distinction between the "system" and its environment. Several authors have recognized from the theoretical viewpoint the relevance of compartments, even though the experimental approach has only been introduced recently. One of the clearest positions is held by Harold J. Morowitz, who states:[1]

> *The formation of closed vesicles was a major event in biogenesis: it represented the origin of supramolecular entities, three-phase systems consisting of a polar interior, a nonpolar membrane core and a polar exterior—the environment. It is difficult to overstress the importance of vesicle closure in cellular genesis. This event established the organism-environment dichotomy in the most general sense. Closure led to a physical separation between outside and inside by a barrier of restricted permeability. Without this barrier, the very idea of a cell is hard to visualize.*

It is not known how the first living cell originated, but what is clear is that no life is possible without a compartment, since life is an emergent and collective property derived from the self-organization of a complex molecular system. Life is not present in single molecules, such as DNA or ribozymes; it arises instead from self-generating autocatalytic and coupled processes that follow a circular logic. Autopoieis offers a powerful descriptive view of life, even if it does not explain how cellular life originated on the Earth.

Experiments have shown that several reactions, including one of great biochemical significance, can be reconstructed inside the tiny aqueous space of lipid- and fatty acid-vesicles and that vesicles can grow and divide through the incorporation of a membranogenic precursor. These encouraging observations are driving current re-

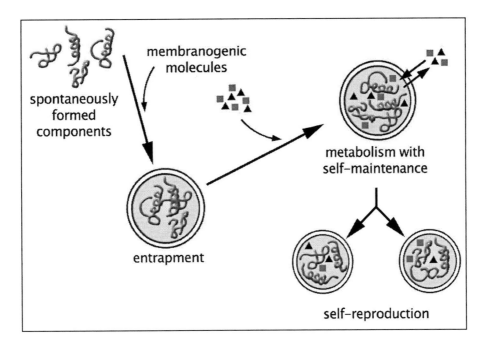

Figure 12.10. A speculative scenario for the emergence of primitive cells. Functional biopolymers and surfactants self-assemble in cell-like structures, i.e., vesicles. Starting from this initial stage, the system must achieve self-maintenance, self-replication and directed membrane synthesis by stepwise chemical evolution. Many steps of this transition are not known, in particular the question of the origin of the first functional macromolecules.

search toward the realization of semi-synthetic constructs where vesicles enclose more complex biochemical pathways, ranging from the expression of genes to the control of vesicle permeability, to internal regulation and to the biosynthesis of lipids, with the aim of synthesizing models of early cells with such minimal and essential functions that they may be called alive. This research area, known as the "minimal cell" area, is attractive also for other disciplines, such as the constructive approach of synthetic biology. Clearly,

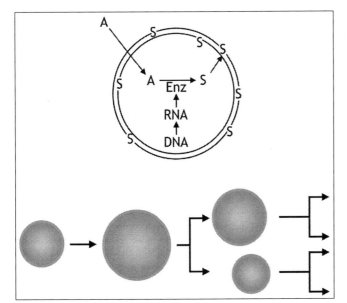

Figure 12.11. An experimental approach to semi-synthetic minimal cells. Enzymes, genes and transcription/translation machineries are entrapped inside vesicles. In situ synthesized enzyme(s) can process suitable precursors to produce the membrane-forming compound S. As a consequence, the vesicles will grow and divide. If the internalized components are not produced, the system will die after a number of divisions owing to an excessive "dilution" of the network elements in some of the vesicles.

semi-synthetic minimal cells do not answer the question of the historical origin of living cells. Nevertheless the efforts involved in their study will bring deeper knowledge (and verification) of the principles underlying cellular life.

Going back to the main topic of this chapter, the origin of protocells, we must admit that very little is known about the mechanisms of their formation. We believe that the formation of compartments is an early step and that various beneficial properties of vesicles such as colocalization of reactants, protection from external molecular parasites and inhibitors, potential surface catalysis, accumulation of hydrophobic pigments in the membrane, maintenance of electroosmotic and chemical gradients, storage of macromolecules, emergence of individuality and ultimately creation of a self-sustained biochemical organized system, all serve to emphasize the significance and need of compartmentation in the transition from the nonliving to living matter. Undoubtedly fatty acid vesicles are currently considered the most coherent model of the early compartments: besides the formation of their fatty acid monomers under plausible prebiotic conditions, the vesicles can form by self-assembly of the monomers, give rise to a semi-permeable boundary, host biochemical reactions, grow and divide spontaneously and therefore self-reproduce.

References

Special Readings

1. Morowitz H. Beginnings of cellular life. Metabolism recapitulates biogenesis. Yale University Press, New Haven, 1992.
   • Essay on the origins of cellular life, with emphasis on the compartment approach and key aspects of metabolic organization, complexity, networks and bioenergetics.
2. Luisi PL. The emergence of life. From chemical origins to synthetic biology. Cambridge University Press, Cambridge, 2006.
   • The book presents a systematic course discussing several theoretical and experimental issues on origins of life studies, as self-organization, emergence, self-replication, autopoiesis, biophysics of compartments, their self-reproduction and use as cellular models.
3. Luisi PL. Autopoiesis: A review and a reappraisal. Naturwissenschaften 2003; 90:49-59.

Box 12.2. Compartments in Other Origins of Life Scenarios

We have seen how the "compartment approach" has been employed as a paradigm for a series of experimental and theoretical investigations. It is clear, however, that compartments alone do not suffice to develop the first cells. The relevance of compartments has been recognized with different personal emphases and biases, by several authors, whose key contributions go beyond a metabolism-first vs gene-first debate. In this paragraph, we will illustrate briefly how some of these authors have integrated the existence and function of compartments into their origin of life scenarios (see also ref. 6).

Manfred Eigen, representing the gene-first viewpoint, has proposed a stable coexistence of several "cooperating" information carrier molecules (RNA) that self-organize into mutually catalytic cycles known as "hypercycles".[33] In this way, several replicators cooperate with, rather than compete against, each other. In its initial formulation, the hypercycle consists of RNA replicators that are translated into replicator enzymes, but a ribozyme hypercycle also has been proposed. Eigen's hypercycles solve the problem of achieving a rich functional set of molecules by overcoming the trap of replication error threshold (Eigen's paradox). One of the problems of hypercycles is that a hypercycle is not an individual in the sense that a bacterium is. Instead it is an ensemble of interacting molecules that is affected by molecular parasites, short-cuts and mutations. To stabilize hypercycles and improve cycle fitness, it has been suggested, although not very passionately, that the hypercycle might be inserted into compartments,[34] creating in this way individuals with inheritable genetic information built on the joint occurrence of hypercycle processes and compartment reproduction. Moreover, the compartment protects the hypercycle from parasites and alternative cycles and brings in the benefits of favorable mutations.[35,36] Similarly, compartmentalized hypercycles can generate fruitful competition between protocells, thus giving rise by natural selection to more effective enzymes and consequently also longer genes.[6]

Freeman Dyson asserts,[37] on the other hand, that a network of organic reactions catalyzed by protein-like molecules preceded the appearance and replication of nucleic acids. His theory has strong roots in the Oparin view of a metabolism-first/coacervates origin. According to Dyson, compartments and protocells containing a metabolic system based on protein-like catalysts are first and essential. Such primitive "cells", mostly inspired by Oparin's coacervates, could grow by absorbing building blocks from the environment and reproduce by statistical division, so that a basic form of compositional (and functional) inheritance become possible. Nucleic acids originated later as by products of such metabolism (or as cell invaders) and were parasites of the metabolism at first, then symbionts and finally fully integrated components of the cell. The compartment in the Dyson model is a necessary element that encloses a specific set of components which give rise to proto-metabolism. What is inherited by the daughter structures from the parent is the system of metabolites, protoenzymes and metabolic paths, namely the organization of the metabolism and the architecture of the whole system. Although Dyson's model lacks specific chemical considerations, especially those defining the nature of the bounded particle, it makes clear reference to compartments and compartmentalized reactions as a means to achieve individuality and conceives of inheritance routes without referring to the nucleic acids.

Moreover, it focuses on homeostasis as the essential character of living cells.

Another important theoretical approach was formulated some years ago by Stuart Kauffman.[38] It follows the paradigm of complex system self-organization, suggesting that some systems can spontaneously reach a state of "order" despite their initial "disorder" state. In order to accomplish this disorder to order transition, the system must be thermodynamically open, so that it can reduce its entropy at the expense of increased entropy in the environment; this way the system can maintain its far-from-equilibrium state without violating the second law of thermodynamics. The roots of this view stem from the dissipative systems of Ilya Prigogine.[39] Kauffman states that the spontaneous emergence of self-organized systems is one of the key factors for the origin of life. Following this idea, he has developed a model on the onset of an autocatalytic reaction network formed by catalytic, interacting biopolymers. The single molecules cannot self-replicate; the system, i.e., the network of interacting catalysts and reactions, need to reproduce itself as a whole. In this view, collective properties emerge as soon as a critical level of molecular complexity is surpassed. Chemically, the catalytic biopolymers are ribonucleic acids and peptides, but the theory does not involve self-replicating RNAs. Compartments enter into Kauffman's theory—although it is not focused on the compartment approach—in two aspects. The first is the confinement required to let reaction proceed by increasing the local concentration of the biopolymers. The second arises from the need of introducing an evolution/selection pathway. New, occasional mutant polymers may be added to the initial autocatalytic set and this will potentially bring the system to diverge toward different directions. If the autocatalytic set is confined in a sort of protocell (Oparin's coacervates, Fox's microspheres and liposomes were suggested), the division process stochastically brings different polymer sets to two daughter cells, this being especially true for polymers present in low copy number in the parent protocell. This will bring into play the evolution of autocatalytic sets within a population of dividing protocells. Kauffman also suggests that liposomes in particular can take on, in addition to their confinement tasks, a key role in facilitating the formation of biopolymers on the basis of osmotic and entropic effects.[38]

With the "minority control" approach, Kunihiko Kaneko offers a new look at the two main problems of the gene-first or metabolism-first approaches, namely the lack of stability against parasites in the former and problematic inheritance in the second. In his theory, the existence of a compartment (a protocell) is a foremost requirement; and two levels of reproduction, of molecules as well as protocells, are assumed, with the view that compartmentation is important to the origin of genetic information. The minority molecules are low in concentration on account of their slow synthesis. However, because they participate in the mutual catalysis required for the reproduction of the protocell as a whole, they actually control the dynamics of the protocells. The minority molecules must be preserved; and their number fluctuations, having a strong influence on the division time of the protocell, must be dampened. The detailed concepts of the theory of minority control can be found by the interested reader in Kaneko's monograph.[40]

- Basic autopoiesis is introduced and discussed, including a historical viewpoint; its implication in understanding the logic of cellular life is clearly explained. Principles of emergence and biological autonomy are also briefly discussed.
4. Monnard PA, Deamer DW. Membrane self-assembly processes: Steps toward the first cellular life. Anatomical Records 2002; 268:196-207.

- A recent review on the role of vesicles (from phopsholipids and fatty acids) in the origin of life. Structure, stability, permeability and their use in experimental investigations are well illustrated.
5. Walde P. Surfactant assemblies and their various possible roles for the origin(s) of life. Orig Life Evol Biosph 2006; 36:109-150.
- In this review, a large collection of data on the various possible roles (not only compartmentation) of surfactant assemblies as micelles, reverse micelles and vesicles is well organized and clearly explained.

Specific References

6. Fry I. The emergence of life on Earth. London:Free Association Books, 1999.
7. Maturana HR, Varela FJ. Autopoiesis and cognition: The realization of the living. Dordrecht:Reidel, 1980.
8. Bitbol M, Luisi PL. Autopoiesis with or without cognition: defining life at its edge. JR Soc Interface 2004; 1:99-107.
9. Damiano L, Unità in dialogo. Mondadori Milano, 2007.
10. Oparin AI. Origin of life on Earth. Edinburgh:Oliver and Boyd, 1957.
11. Fox SW. Proteinoid experiments and evolutionary theory. In: Ho MW, Saunders PT, eds., Beyond Neo-Darwinism. New York:Academic Press 1985:15-60.
12. Gebicki JM, Hicks M. Ufasomes are stable particles surrounded by unsaturated fatty acid membranes. Nature 1973; 243:232-234.
13. Gebicki JM, Hicks M. Preparation and properties of vesicles enclosed by fatty acid membranes. Chem Phys Lipids 1976; 16:142-160.
14. Rushdi AI, Simoneit BRT. Lipid formation by aqueous fischer-tropsch-type synthesis over a temperature range of 100-400°C. Orig Life Evol Biosph 2001; 31:103-118.
15. Lawless JG, Yuen GU. Quantitation of monocarboxylic acids in the murchison carbonaceous meteorite. Nature 1979; 282:431-454.
16. Walde P, Namani T, Morigaki K et al. Formation and properties of fatty acid vesicles (liposomes). In: Gregoriadis G ed., Liposome Technology, 3rd edition, Vol. I. New York:Informa Healthcare, 2006:1-19.
17. Walde P, Ishikawa S. Enzymes inside lipid vesicles: preparation, reactivity and applications. Biomol Bioengin 2001; 18:143-177.
18. Schmidli PK, Schurtenberger P, Luisi PL. Liposome-mediated enzymatic synthesis of phosphatidylcholine as an approach to self-replicating liposomes, J Am Chem Soc 1991; 113:8127-8130.
19. Chakrabarti AC, Breaker RR, Joyce GF et al. Production of RNA by a polymerase protein encapsulated within phospholipid vesicles. J Mol Evol 1994; 39:555-559.
20. Walde P, Goto A, Monnard PA et al. Oparin's reactions revisited: enzymatic synthesis of poly (adenylic acid) in micelles and self-reproducing vesicles. J Am Chem Soc 1994; 116:7541-7544.
21. Oberholzer T, Wick R, Luisi PL et al. Enzymatic RNA replication in self-reproducing vesicles: an approach to a minimal cell. Biochem Biophys Res Commun 1995; 207:250-257.
22. Oberholzer T, Albrizio M, Luisi PL. Polymerase chain reaction in liposomes. Chem and Biol 1995; 2:677-682.
23. Oberholzer T, Nierhaus KH, Luisi PL. Protein expression in liposomes. Biochem Biophys Res Comm 1999; 261:238-241.
24. Luisi PL, Ferri F, Stano P. Approaches to semi-synthetic minimal cells: a review. Naturwissenschaften 2006; 93:1-13.
25. Luisi PL. Self-reproduction of micelles and vesicles: Models for the mechanisms of life from the perspective of compartmented chemistry. Advances Chem Physics XCII, J Wiley and Sons Inc 1996:425-438.
26. Blochliger E, Blocher M, Walde P et al. Matrix effect in the size distribution of fatty acid vesicles. J Phys Chem 1998; 102:10383-10390.
27. Cheng Z, Luisi PL. Coexistence and mutual competition of vesicles with different size distributions. J Phys Chem B 2003; 107:10940-10945.
28. Chen IA, Roberts RW, Szostak JW. The emergence of competition between model protocells. Science 2004; 305:1474-1476.
29. Segre D, Ben Eli D, Lancet D. Compositional genomes: prebiotic information transfer in mutually catalytic noncovalent assemblies. Proc Natl Acad Sci USA 2000; 97:4112-4117.
30. Segre D, Ben Eli D, Deamer D et al. The lipid world. Orig Life Evol Biosph 2001; 31:119-145.
31. Bourgine P, Stewart J. Autopoiesis and cognition. Artificial Life 2004; 10:327-345.
32. Zepik HH, Bloechliger E, Luisi PL. A chemical model of homeostasis. Angew Chemie Int Ed 2001; 40:199-202.
33. Eigen M, Schuster P. The hypercycle: A principle of natural self-organization. Part A. emergence of the hypercycle. Naturwissenschaften 1977; 64:541-565.
34. Eigen M, Schuster P, Gardiner W et al. The origin of genetic information. Scientific American 1981; 244:78-94.
35. Eigen M. Stufen zum Leben: die frühe evolution in visier der molekularbiologie. München:Piper, 1987.
36. Maynard Smith J, Szathmáry E. The major transitions in evolution. Oxford:Oxford University Press, 1995.
37. Dyson FJ. The origins of life. Cambridge:Cambridge University Press, 1985.
38. Kauffman SA. The origins of order. Self-organization and selection in evolution. New York:Oxford University Press, 1993.
39. Nicolis G, Prigogine I. Self-organization in non-equilibrium systems. New York:John Wiley, 1977.
40. Kaneko K. Life: An introduction to complex systems biology. Berlin:Springer, 2006.

CHAPTER 13

Split Genes, Ancestral Genes

Massimo Di Giulio*

13.1 Introduction: The Origin of Genes

If we think about the origin of genes, we almost inevitably imagine the primitive genetic system passing through an evolutionary stage characterised by genes just long enough to encode a few dozen amino acids, before the genes went on to reach greater lengths. This viewpoint is in keeping with the exon theory of genes, which suggests that the discontinuous structure of eukaryotic genes, consisting of exons alternated with introns, reflects the way in which genes have evolved: the exons might actually be minigenes specifying primarily modular structures that were assembled, i.e., joined, by introns to form the discontinuous genes found in the genome of eukaryotes. According to this theory, every exon in eukaryotic gene would correspond to an ancestral minigene and the introns would constitute points of suture between these minigenes.[3] There is a great deal of evidence in favor of this theory.[4]

The identification of ancestral genes, i.e., minigenes postulated by the exon theory of genes, has been made difficult by the long time spans that have elapsed since the origin of these ancestral genes, wiping away all detectable traces of them. However, conceptual developments and the complete sequences of many genomes are now allowing progress toward detection of these very old genes from ancient evolutionary stages. The present chapter analyses the conceptual basis and the empirical results that have been achieved in identifying ancestral genes and thereby clarifying the nature of the Last Universal Common Ancestor (LUCA).[1]

13.2 Theories Formulated to Explain the Origin of Introns

Three theories have been suggested to explain the origin of introns:[4-6]

A. The "introns late" theory suggests that introns and the spliceosome, the cell organelle that splices out the intron from the gene transcript to yield the mature mRNA, first originated in the eukaryotic lineages and, since then, have accumulated in the genomes of eukaryotes.[2,5] Sources of new introns include "reverse splicing" and the insertion of transposable elements.[5]

B. The "introns early" theory proposes that the intron-exon structure of genes was present in LUCA and possibly even earlier. According to this theory, the domains of primordial proteins were shuffled to facilitate their diversification and evolution.[3-5] Later on the introns were lost from the domains of Archaea and Bacteria. The molecular mechanism that allows intron loss is the recombination of spliced cDNA;

primitive reverse transcriptases might have mediated this loss mechanism.[5] In the "intron invasion" version of "introns early", which provides a compromise between "introns early" and "introns late", it is suggested that introns appeared early but were not widespread before the emergence of eukaryotes. They became prominent only as a result of the invasion of eukaryotic genes by numerous introns at the onset of eukaryotic evolution. This invasion triggered such pivotal events in the eukaryotes as the appearance of the spliceosome, the nucleus, the linear chromosomes, the telomerase and the ubiquitin signaling system. Subsequent intron gain has been limited.[7]

C. The "introns first" theory, which is similar to but reaches even further back than the "introns early" theory, suggests that the introns and spliceosome are relics of the RNA world.[5,6] This model is compatible with the observation that putatively ancient snoRNA genes are often specified by introns. Since RNAs were the only available catalysts for assembling an RNA protoribosome prior to the advent of proteins, the snoRNAs would have been used for the assembly of the protoribosome.[5,6] Therefore, according to this theory, snoRNAs predated the protein encoding exons. The processing of prerRNAs and pretRNAs by RNase P might serve as examples of how RNA processing might have taken place before proteins evolved.[5,6] Both the "introns early" and "introns first" theories call upon known classes of enzymic reactions to handle the introns.

13.3 Split Genes and Other Characteristics of *Nanoarchaeum equitans*

Nanoarchaeum equitans, a particularly unusual hyperthermophilic archaeon, was isolated from a submarine hot vent and subsequently characterised. It is a parasite of another archaeon and displays a number of truly exceptional characteristics, e.g., its ribosomal RNA sequence does not contain the typical oligomeric sequences found in other Archaea. On the basis of the uniqueness of its rRNA sequence, *N. equitans* has been classified as the first representative of a new phylum of Archaea, the Nanoarchaeota.[8]

The sequence of the *N. equitans* genome shows that this organism possesses the following truly unique traits: (i) *N. equitans* has an unusually high number of split genes, at least ten encoding proteins and 6 encoding tRNA molecules (Fig. 13.1).[9,10] The ten protein encoding genes are completely split into two pieces, one for the N-terminal portion and the other for the C-terminal portion of the

*Massimo Di Giulio—Laboratory of Molecular Evolution, Institute of Genetics and Biophysics 'Adriano Buzzati Traverso', CNR, Via P. Castellino, 111, 80131 Naples, Napoli, Italy. Email: digiulio@igb.cnr.it

Prebiotic Evolution and Astrobiology, edited by J. Tze-Fei Wong and Antonio Lazcano. ©2009 Landes Bioscience.

```
5' tRNATrp(CCA)

5' gaatctc gggccggtagctcagcctggttagagcggcggtggccat ccccctttaaattt 3'


3' tRNATrp(CCA)

5' ggatggccgccgaggctctt aacccgcaggtccggggttcgaatccccgccggccc gtgg 3'
```

Figure 13.1. An example of tRNA half genes in the *N. equitans* genome. Position-1 in the boxed sequence of the 5' half tRNATrp(CCA) corresponds to base 308,963 on the plus DNA strand and that of the 3' half corresponds to base 380,364 on the minus DNA strand.[10]

protein. These genes occupy noncontiguous regions in the genome. The same is true for the six genes for the tRNAs whose 5' and 3' halves are codified on completely different noncontiguous genes. Such split genes for tRNA half molecules have been identified so far only in *N. equitans*,[9,10] while split genes for proteins have been observed also in other Archaea but never in such high numbers.[9] (ii) Another unusual characteristic of the *N. equitans* genome is the lack of operons. The operons conserved in all other Archaea are completely absent from *N. equitans*; the corresponding proteins are present in nearly full complements but are encoded by scattered genes.[9] Even the super-operon of ribosomal proteins, the largest gene array that is conserved to varying extents in all Archaea and Bacteria, is almost completely absent except for a few minor instances.

Phylogenetic analyses have placed *N. equitans* as the deepest branch in the Archaea domain on the SSU RNA tree,[9] in the *Euryarchaeota* on the SSU RNA tree,[11] or as the deepest branch of the *Crenarchaeota* on the tRNA tree.[12]

13.4 A Theory for the Origin of Genes and Introns

It has been proposed that DNA evolved late only after the evolutionary stage of the Last Universal Common Ancestor (LUCA), i.e., only during the establishment of the main phyletic lines.[13,14] Based on this proposal, LUCA would have an RNA genome. One of the hypotheses compatible with this evolutionary scenario maintains that LUCA possessed a genome made up of pieces of RNA and that one of the major mechanisms that led to the creation of gene products was trans-splicing. In other words, all the mRNAs in LUCA were built up by means of trans-splicing and genes as physically continuous entities did not exist.[1]

In this evolutionary scenario, it can be suggested that the formation of DNA genomes (taking place after the LUCA stage) transformed LUCA's fragmented RNA genome into a continuous DNA genome in which every RNA fragment corresponded to a minigene. This transformation converted all the RNA fragments in the LUCA genome into small scattered "genes", i.e., individual small interrelated genes that would later form parts of the same gene would be scattered within the DNA genome. In this genome model consisting of scattered minigenes, every minigene corresponded to a future "exon" in the eukaryotic genome. Therefore the earliest "exons" were completely scattered inside the DNA genome. At this point the model calls for independent origin of different classes of introns in the three domains of life, the appearance of which greatly accelerated the joining together of individual minigenes to form for example the typical discontinuous but same-strand genes of eukaryotes. According to this model, in the "first" DNA genome stage, there would not be any introns involved in the assembly of genes. Only later on did introns play a fundamental and very active role in assembling the genes that we see today. Contrary to the exon theory of genes[3] which envisages the presence of introns in LUCA,[5] this model perceives introns arising only after the LUCA stage, or at least not partcipating in the construction of the first DNA genomes, because the first genes were in fact minigenes encoding modular

structures directly descended from the fragmented RNA genome of LUCA. Clearly, this model for the origin of genes and introns would find immediate support if the genome of *N. equitans* were in fact a fossil of ancestral genomes.[1,15]

13.5 A Model of the Origin of the tRNA Molecule and Its Evolution

There is a simple way to construct a tRNA molecule (Fig. 13.2). Starting from a hairpin RNA structure and directly duplicating the hairpin, we would obtain a cruciform structure, the subsequent evolution of which might have created the actual tRNA molecule (Fig. 13.2). This model of tRNA origin has been extensively discussed.[1,15-19]

Some of the strongest evidence in favour of this model is historical in nature and pertains to the position of introns in tRNA genes. In compliance with the exon theory of genes,[3] this model of tRNA origin envisages that any intron present in tRNA genes must have been located in the anticodon loop where the intron would split the tRNA molecule into two halves that correspond to the two original single hairpin structures. In accordance with the exon theory of genes, this model envisages that minigenes coding for the hairpin structures existed and their subsequent evolution must have led to the formation of modern tRNA genes.[16-19] Moreover, the position of the intron could be traced to the way in which the tRNA genes were first assembled. On this basis, the presence of introns in the anticodon loops of tRNA genes in all three biological domains furnishes strong historical evidence in favour of the model (Fig. 13.2; refs. 1, 15-19).

This model of tRNA origin lends itself to further extension.[18] Given that hairpin RNAs encoded by minigenes were the structures based on which the complex process of protein synthesis might have been achieved, it follows that the genetic code might have been organized initially using such simple hairpin RNAs structures instead of complete cloverleaf tRNAs, which might have evolved only subsequent to the LUCA stage.[18] Consequently, the origin of the tRNA molecule might well be nonmonophyletic in nature, i.e., the tRNA molecule specific for a given amino acid might have been built from different hairpin RNAs in different biological lineages.[1,18] On this basis, the genes for cloverleaf tRNAs should have originated late in evolution either during or after the establishment of the three separate domains of life.[1,15,18]

In conclusion, this theory of tRNA origin suggests that minigenes encoding hairpin RNAs appeared very early in evolution.

13.6 The Split *Nanoarchaeum equitans* tRNA Genes Are the Ancestral Genes from Which Modern tRNA Genes Evolved

It has been convincingly shown that, based on the following arguments, the split genes coding for the 5' and 3' halves of the tRNA molecule are the plesiomorphic (ancestral) forms of modern tRNA genes:[1,15]

    i. The introns of tRNA genes are normally located 3' to position-37 on the tRNA, which is the same position where

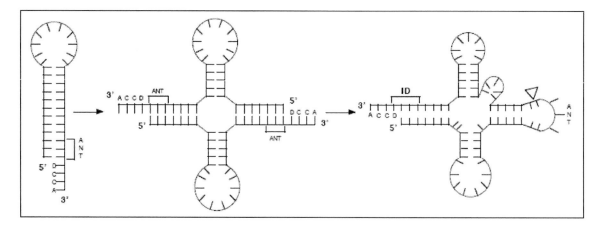

Figure 13.2. Model for the origin of the tRNA molecule. For details see Di Giulio.[15-17] ANT stands for anticodon and ID the nucleotides determining the identity of the tRNA in its recognition by its cognate aminoacyl-tRNA synthetase. The triangle indicates the normal position of intron in tRNA genes.[16,19]

the *N. equitans* half genes for the 5′ and 3′ halves of tRNA are interrupted.[1,10,15] This implies that the tRNA half genes and the tRNA genes carrying an intronic insertion must represent different stages of the same evolutionary sequence (Fig. 13.3A; ref. 1, 15), because the probability of both of these scissions occurring 3′ to position-37 on account of chance is extremely low. If the tRNA half genes and the tRNA genes with introns represent different stages of the same evolutionary sequence, then continuous tRNA genes without any intron must be the final stage of this sequence of tRNA evolution. Thus the sequence of tRNA evolution went through the distinct Ancestral, Intermediate and Current stages shown in Figure 13.3A. This sequence in time, by bringing the two scattered split tRNA genes of the Ancestral Stage at first close to one another separated only by an intron as in the Intermediate Stage and finally continuous with one another as in the Current Stage, has progressively reduced the likelihood of committing any errors during the construction of the cloverleaf tRNA molecule.[1,15] In contrast, based on the alternative view depicted in Figure 13.3B, the evolutionary sequence, leading from continuous tRNA genes to tRNA genes with introns and finally to scattered split genes, would result in error enhancement rather than error reduction in the course of evolution, which is unreasonable.

ii. The absolute gene frequencies found among extant organisms for these three stages also strongly support the evolutionary sequence in Figure 13.3A over that in Figure 13.3B. There are many more species of organisms with continuous tRNA genes than tRNA genes with introns, which in turn

far exceed the number of organisms (so far only *N. equitans*) with split tRNA genes. This is entirely consistent with the relic status assigned to split tRNA genes in Figure 13.3A. In contrast, Figure 13.3B would regard continuous tRNA genes, the majority of tRNA genes in the living world, as evolutionary relics. The fact that relics are as a rule very rare therefore contradicts continuous ancestral tRNA genes and supports split ancestral tRNA genes.[1,15]

Consequently, the half genes encoding the 5′ and 3′ halves of tRNA molecules in *N. equitans* represent the ancestral form of tRNA genes from which modern tRNA genes might have evolved,[1,15] They are also the minigenes predicted by the exon theory of genes.

13.7 Theorem of the Polyphyletic Origin of tRNA Genes

The arguments regarding the tRNA split genes of *N. equitans* can be summarized in a theorem on the nonmonophyletic origin of tRNA genes:

> *"If the half genes of* N. equitans *tRNAs represent the plesiomorphic form of tRNA genes, then the mere observation of the existence of such genes in an organism implies that the hypothesis of the monophyletic origin of tRNA is false and the polyphyletic origin of tRNA genes is true."*

The proof of this theorem is straightforward. If the monophyletic hypothesis is true, from the evolutionary stage of LUCA onward we should be able to observe, by definition, only complete tRNA genes and no half genes, because a monophyletic origin would not allow the existence of two kinds of tRNA genes, continuous and split, at

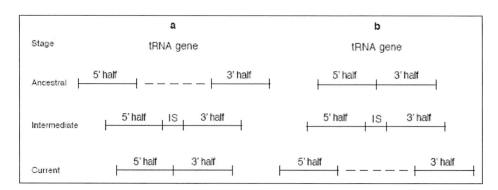

Figure 13.3. Two alternative evolutionary sequences for tRNA genes (IS = intron). a) The hypothesis that evolution progressed from split to unsplit genes and b) The alternative possibility that evolution progressed from unsplit to split genes.[1,15]

any time after the LUCA stage. Therefore the observation of tRNA half genes after the LUCA stage and the evidence that these half genes represent the ancestral form, imply that the monophyletic hypothesis is false.[1,15] It follows that tRNA genes are polyphyletic in origin and that LUCA must have contained half tRNA genes.

13.8 The Split Genes Encoding Some *Nanoarchaeum equitans* Proteins Are Probably Ancestral Genes

It would be truly surprising if there were no relationship between the split genes for the 5′ and 3′ halves of some of the tRNAs in *N. equitans* and the split genes for ten different proteins in this organism.[1,9] The idea is that the split genes encoding some of the *N. equitans* proteins are also the plesiomorphic forms of these genes. An observation that would strongly support this view is as follows. If the two protein fragments encoded by the noncontiguous split genes of *N. equitans* are aligned with the homologous continuous proteins from eukaryotes, we should be able to predict where the introns should be in these eukaryotic genes. In other words, the region in the multiple alignment where the N-terminal fragment from *N. equitans* ends and the C-terminal fragment begins should house the introns within the eukaryotic genes if, as predicted by the intron theory of genes, introns played an important role in assembling genes. In fact, for all six protein genes for which this prediction can be tested, the prediction has been verified by the occurrence of introns at the expected sites (manuscript in preparation). These observations therefore strengthen our belief that at the LUCA evolutionary stage there were only completely split genes, which were later joined, as testified by the presence of introns at the junction sites. Accordingly we must conclude that the noncontiguous split genes encoding some of the *N. equitans* proteins are the plesiomorphic forms of these genes, in accord with the situation of the split *N. equitans* tRNA half genes.

13.9 The Polyphyletic Origin of Genes for Proteins

The theorem on the polyphyletic origin of tRNA genes indicates the existence of completely separate tRNA genes encoding the 5′ and 3′ halves of the molecule at the LUCA stage.[1,15] This along with the observations on the split genes encoding some of the *N. equitans* proteins point to a more generalized state in which all the genes were fragmented into pieces and the mRNAs had to be assembled. This generalized state might apply to LUCA, in which case all the continuous genes of extant organisms would have to be assembled after the main lines of biological divergence had been established, implying that their origins were also nonmonophyletic.[1,18] Consequently, LUCA's mRNAs might have been assembled by means of trans-splicing for translation into protein products whose genes actually evolved only later on, i.e., after the domains of life were established. There is a great deal of evidence that appears to corroborate this nonmonophyletic origin of genes.[1,18] In this regard it is noteworthy that the replication apparatus does not seem to be homologous in the three domains of life.[13] This and other observations suggest that DNA appeared only after the main phyletic lines had been established, i.e., after the LUCA stage.[13,14] This is in perfect agreement with the polyphyletic hypothesis of the origin of genes supported here: a late origin of DNA might also correspond to a polyphyletic, that is to say a late, origin of genes. The late arrival of DNA only after the LUCA stage, within each of the main phyletic lines, would lead to the independent, 'definitive' origins of genes within these phyletic lines. Therefore it seems that we must consider the possibility of a polyphyletic origin for all genes very carefully.

13.10 Conclusions and Prospects

It appears very clear that the split genes of *Nanoarchaeum equitans*, particularly those encoding the 5′ and 3′ halves of some of its tRNA molecules, are the oldest forms of genes known to us. This indicates that organisms are guardians of secrets regarding the early phases of the evolution of life, exemplified by the split tRNA genes being minigenes, i.e., ancestral genes characteristic of the very ancient phases of the origin of tRNAs.[1,15] This renders possible the accurate reconstruction of primordial events such as the origin and its timing, of tRNAs. The fascinating prospect is therefore opened up that, with the development of new theories and the sequencing of new genomes, ancient events long regarded as being scientifically problematic to solve, such as the origin of protein synthesis, will become open to investigation. More generally, we can be confident that the earliest phases of the evolution of life will be solved and that the origin of life itself will find intriguing answers from the study of living organisms.

References

Special Readings

1. Di Giulio M. The nonmonophyletic origin of the tRNA molecule and the origin of genes only after the evolutionary stage of the last universal common ancestor (LUCA). J Theor Biol 2006; 240:343-352.
2. De Souza SJ. The emergence of a synthetic theory of intron evolution. Genetica 2003; 118:117-121.
3. Jeffares DC, Mourier T, Penny D. The biology of intron gain and loss. Trends Genet 2006; 22:16-22.

Specific References

4. Gilbert W. Why genes in pieces? Nature 1978; 271:501.
5. Roy SW. Recent evidence for the exon theory of genes. Genetica 2003; 118:251-266.
6. Poole AM, Jeffares DC, Penny D. The path from the RNA world. J Mol Evol 1998; 46:1-17.
7. Koonin EV. The origin of introns and their role in eukaryogenesis: a compromise solution the introns-early versus introns-late debate? Biol Direct 2006; 1:22 doi: 10.1186/1745_6150-1-22.
8. Huber H, Hohn MJ, Rachel R et al. A new phylum of Archaea represented by nanosized hyperthermophilic symbiont. Nature 2002; 417:63-67.
9. Waters E, Hohn MJ, Ahel I et al. The genome of Nanoarchaeum equitans: insights into early archaeal evolution and derived parasitism. Proc Natl Acad Sci USA 2003; 100:12984-12988.
10. Randau L, Munch R, Hohn M et al. Nanoarchaeum equitans creates functional tRNAs from separate genes for their 5′- and 3′-halves. Nature 2005; 433:537-541.
11. Brochier C, Gribaldo S, Zivanovic Y et al. Nanoarchaea: representatives of a novel archaeal phylum or a fast-evolving euryarchaeal lineage related to Thermococcales? Genome Biol 2005; 6, R42 doi:10.1186/gb-2005-6-5-r42.
12. Wong JT, Chen J, Mat WK et al. Polyphasic evidence delineating the root of life and roots of biological domains. Gene 2007; 403:39-52.
13. Forterre P. Three RNA cells for ribosomal lineages and three DNA virus to replicate their genomes: A hypothesis for the origin of cellular domain. Proc Natl Acad Sci USA 2006; 103:3669-3674.
14. Poole AM, Logan DT. Modern mRNA proofreading and repair: clues that the last universal ancestor possessed an RNA genome? Mol Biol Evol 2005; 22:1444-1455.
15. Di Giulio M. Nanoarchaeum equitans is a living fossil. J Theor Biol 2006; 242:257-260.
16. Di Giulio M. On the origin of the transfer RNA molecule. J Theor Biol 1992; 159:199-214.
17. Di Giulio M. Was it an ancient gene codifyng for a hairpin RNA that, by means of direct duplication, gave rise to the primitive tRNA molecule? J Theor Biol 1995; 177:95-101.
18. Di Giulio M. The nonmonophyletic origin of tRNA molecule. J Theor Biol 1999; 197:403-414.
19. Di Giulio M. The origin of the tRNA molecule: implications for the origin of protein synthesis. J Theor Biol 2004; 226:89-93.

Genetic Code

J. Tze-Fei Wong*

14.1. Introduction

Living matter is rich in information content. The information is carried above all in the polymeric sequence of DNA, which may be transferred to RNA and protein through transcription and translation. These polymers consist of more than one kind of monomer building blocks and the information is embedded in the sequence in which the different monomers are arranged and its length. Comparable heteropolymer information systems are also employed in the sequence of letters and words in languages and sequence of acoustic notes in music. The maximum information content I_{max} of a sequence is given in bits by:

$$I_{max} = N \, Log_2 \, (M)$$

where M is the number of kinds of monomers and N the sequence length. Complex chemical compounds such as chlorophyll and heme store information in the spatial arrangement of their atoms and bonds, but these compounds are difficult to replicate and constrained by modest upper limits of information. In contrast, polymeric information can be made practically unlimited by lengthening the sequence. The information content of human genomic DNA, with four different monomers and a length of three billion basepairs, is:

$$I_{max} = 3 \times 10^9 \, Log_2 \, (4) = 6 \times 10^9 \text{ bits}$$

This amount of information is equivalent to a 400-volume manual for the construction of a human, 600 pages per volume and 5,000 printed characters, each carrying 5 bits, per page. It is not even near the upper limit for DNA. The lung fish genome contains far more DNA than human.

Because DNA (T,C,A,G) and RNA (U,C,A,G) both employ a 4-letter alphabet, DNA letters are transcribable into RNA letters on a one-to-one basis. In translation, as George Gamow foresaw,[6] there has to be correspondence between each amino acid and a combination of nucleotides. On this basis, the genetic code translates the RNA language into protein language by distributing 64 triplet codons to amino acids and termination signals. In telegraphic transmission, the Morse Code encodes letters in the alphabet by up to four 'dots' and 'dashes'. It is an entirely arbitrary code and can be replaced easily by another code; it is deliberately abandoned in favor of other codes in military transmission for the sake of secrecy. Since the genetic code is no less arbitrary, its basically universal usage by all living organisms suggests that the code was established prior to the earliest branchings of present day organisms approximately 3.6 billion years ago (Fig. 14.1). It is as old if not older than the oldest rocks.

The Rosetta Stone, discovered in the Egyptian harbor Rosetta, was created in 196 BC. It bears the engraved translations of a single passage in two Egyptian languages, hieroglyphic and Demotic and in classical Greek. These translations of the passage, a decree from Ptolemy V regarding wine from the vineyards and other vital matters of state, enabled Jean-Francois Champollion to decipher the principles of hieroglyphic writing and uncover the history of ancient Egypt. Rarely have governmental decrees brought such lasting utility. The universal genetic code likewise bears a translation of different languages, those of nucleic acids and proteins (Fig. 14. 2). With the expectation that decipherment of the structure of the code will bring unique insight into the origin of protein language, there has been no research focus of biochemical archaeology that compares in sustained intensity with the decades of analysis devoted to this code based on a diversity of approaches.

14.2 Birth of the Code

(i) The Sidechain Imperative

Ribozymes, RNA aptamers, enzymes and binding proteins all rely on an optimized sequence giving rise to a three dimensional folding that positions critical sidechains around a substrate or ligand. Accordingly the acquisition of an ensemble of effective sidechains represents a key evolutionary imperative. In the RNA World,[7] RNA molecules had to serve the dual functions of informational replicators and ribozymes/aptamers. In replication through complementary basepairing, two complementary purine-pyrimidine basepairs are optimal. Having only a single basepair abolishes information content. Having three or more basepairs could increase replication errors from misreadings. Yet two kinds of basepairs bring only four letters into the RNA alphabet, which severely limits sidechain versatility for ribozyme or aptamer activities. One solution to the conflicting demands of replication and function was to limit the number of bases to keep replication errors low, but introduce postreplication modifications to enhance sidechain variety.[8] As long as the modified nucleotides did not participate in replication, they would not cause replicative errors. To-day, 95 kinds of post-transcriptional modifications are found on tRNAs, rRNAs, snRNAs and mRNAs and across all three domains of life.[9] Up to 80% of the nucleotide positions on tRNA aside from its 3'-CCA terminus have been subject to modification.[10] As well, the LUCA genome based on the *ancient eight* species contain such RNA modifying enzymes as dimethyladenosine transferase, pseudouridine synthase, queuine/archaeosine tRNA-ribosyltransferase, 2-methylthioadenine synthetase and tRNA(1-methyladenosine)methyl-transferase, attesting to the primordial origin of RNA modifications (Appendix 15.1 and ref. 11). However, every modification called for a modifying ribozyme that had to distinguish its substrate nucleotides from

*J. Tze-Fei Wong—Applied Genomics Center, Fok Ying Tung Graduate School and Department of Biochemistry, Hong Kong University of Science and Technology, Clear Water Bay, Hong Kong, China. Email: bcjtw@ust.hk

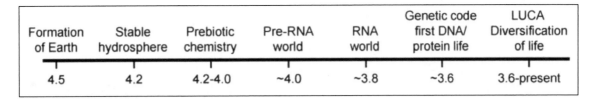

Formation of Earth	Stable hydrosphere	Prebiotic chemistry	Pre-RNA world	RNA world	Genetic code first DNA/ protein life	LUCA Diversification of life
4.5	4.2	4.2-4.0	~4.0	~3.8	~3.6	3.6-present

Figure 14.1. Timeline of early history of life in billions of years according to Joyce.[3]

other nucleotides.[12] As these RNA modification systems became increasingly cumbersome, the incentive for evolving surrogate biopolymers to serve as catalysts and ligand-binders would mount. Since these surrogate biopolymers had no genetic informational role, the variety of their sidechains could be increased without impacting the fidelity of genetic information. The RNA World likely explored different classes of surrogate biopolymers before opting for the polypeptides. The RNAs would direct the synthesis of the polypeptides and pass on to them the bulk of the catalytic and ligand binding responsibilities.

The choice of polypeptides as surrogate polymer came as no surprise. Amino acids available from the prebiotic environment carry superlative sidechains for catalysis and ligand binding. The prowess of proteins in ligand binding is exemplified by the variable region of antibodies where 216 amino acid residues, half from the L chain and half from the H chain of IgG, suffice to generate a huge variety of antibodies with distinct antigen specificities. The prowess of proteins in catalysis is unsurpassed by any other class of substance so far encountered in catalysis science. Catalysis by either ribozymes or enzymes (C) is described by the formation of a complex between catalyst and substrate S, followed by transformation of substrate to reaction products with regeneration of free catalyst:

$$C + S \rightleftharpoons CS \rightarrow C + \text{reaction products}$$

The velocity v of the catalysed reaction is described by the Michaelis-Menten Equation:

$$v = k_{cat} \cdot [C_o][S]/([S] + K_m)$$

The catalytic rate constant k_{cat} determines the maximum rate at which S can be converted into products at total catalyst concentration $[C_o]$ and the Michaelis constant K_m determines the S concentration needed to half-saturate the binding sites on the catalyst. RNAs through their many hydrogen bonding groups excel in folding up to a rich variety of configurations to provide high-affinity binding sites for all sorts of ligands and ribozymes usually display a low K_m. However, having only four kinds of sidechains, they often cannot achieve a high k_{cat}. For example, the enzymes RNase A, RNase T1 and RNase T2 hydrolyze their substrates with a k_{cat} of 5,000 to 180,000 min^{-1}, whereas ribozymes that cut RNA, including both Group I ribozymes and hammerhead ribozymes, typically display a k_{cat} of 100 m^{-1} or less.[13-15] The RNA World lasted possibly much less than a billion years (Fig. 14.1), whereas the DNA-Protein World is over three billion years old and still going strong. So there must be some underlying disadvantage(s) of the RNA World relative to the DNA-Protein World. Difficulty in attaining high k_{cat} could be a key element in this regard.

(ii) RNA Peptidation

While the evolutionary advantage of switching from ribozymes to enzymes could be straightforward, the switch had to be implemented in compliance with two basic requirements. First, the formation of proteins must strictly obey instructions from the RNA genes, so that the wisdom already accumulated in the RNA sequences through evolution would not be wasted. The system could not afford to twice invent the genes. It followed that a dictionary, or

code, had to be developed to translate RNA language into protein language. There were alternative approaches to the process. The coding units, or codons, might contain overlapping or non-overlapping nucleotides and each codon might comprise one, two, three or more nucleotides. Likely most if not all of these alternatives were explored by the ribo-organisms and those found wanting were eliminated, leaving the non-overlapping triplet code at the end as an optimal balance accomodating enough amino acid variety and not being overly cumbersome. Secondly, planning ahead is a trait that does not surface in biological systems until the vertebrate stage. Accordingly every step of genetic code development had to be accompanied by some immediate advantage for the system. One development mechanism that meets both requirements is the two-stage RNA Peptidation mechanism[16] (Fig. 14.3).

The starting point (left, Fig. 14.3) was a ribozyme segment unassisted by any postreplication modification. Later, modifications would be recruited to add extra sidechains to the ribozyme. Since amino acids were available in the prebiotic environment, they would be included in some of the added modifications. The incorporation of aminoacyl- or peptidyl-sidechains to the RNAs finds support in the range of peptide-containing nucleotide type molecules that are utilized by organisms to-day (Table 14.1) and the discoveries of ribozymic aminoacylation of RNA[17-19] and amino acid and peptide activation of ribozyme.[20-21] A ribozymic nonribosomal peptide synthetase (NRPS) system, the enzymic version of which is used nowadays to synthesize peptide antibiotics,[22] could be employed to energize peptide-bond formation on RNA. RNA nucleotide sequence directed the sequence in which amino acids were incorporated into a peptide on the RNA, matching different amino acids to cognate three-nucleotide patches, to yield a first-stage peptide-enhanced ribozyme (middle, Fig. 14.3). In flavoproteins, the flavin cofactor is usually firmly bound to the enzyme; in some instances, e.g., D-amino acid oxidase, it may also dissociate reversibly from the enzyme. Likewise, peptides and amino acids may be attached to the RNA as in the case of other posttranscriptional modifications, or they may bind to the RNA reversibly.[23] With covalent attachment, unintended binding of a peptide factor to other ribozymes could be prevented and construction of a constellation of multiple sidechains around a catalytic site, as in the case of most enzyme active sites, facilitated.

Even short peptides, e.g., 14 amino acids in length, are known to be endowed with catalytic activity and 10^6-fold fewer peptide sequences than RNA sequences need to be searched to obtain an effective catalyst.[8] Thus the catalytic versatility of the peptide sidechains on the RNA very soon and unmistakably manifested itself. Thereupon they were detached from the RNA to function on their own as primitive enzymes, replacing the ribozymes. In this second, triplet-coding stage, the cognate three-nucleotide RNA patches that directed amino acid incorporation acted as triplet codons (right, Fig. 14.3). Different adaptor RNAs would become aminoacylated with different amino acids, leading to a family of precursors to modern aminoacyl-tRNAs.[3] These aminoacylatable adaptor RNAs could be simple stem-loop minihelices,[24,25] or elaborate aminoacylating ribozymes.[3] Still later, when DNA took over from RNA as genes, the RNA molecules are confined to their present day roles associated

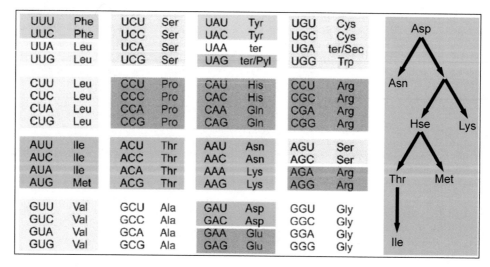

Figure 14.2. Universal genetic code. Codons for different biosynthetic amino acid families are color coded. Pyrrolysine (Pyl) only has part use of UAG and selenocysteine (Sec) part use of UGA. The Asp-family biosynthetic pathways are shown on the right.

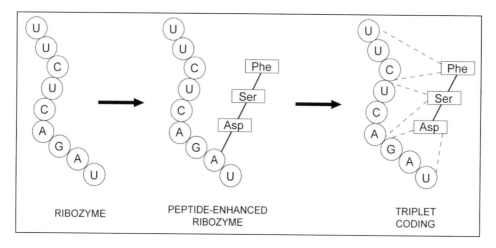

Figure 14.3. Development of triplet coding by the RNA Peptidation mechanism.

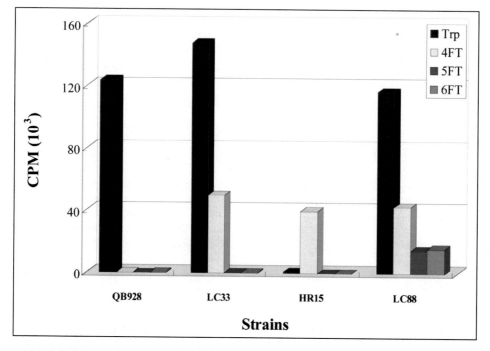

Figure 14.4. Growth of *B. subtilis* strains on Trp and 4-, 5- or 6-fluoro-Trp measured by [^{33}P]-phosphate incorporation into colonies on agar.[61]

Table 14.1. Compounds containing peptide or amino acid moieties linked to nucleotidyl or organic bases[16]

Compound	Base-Amino Acid Link	Function
Glycyl-tRNA$_1$	O-peptidylnucleoside	Bacterial cell wall synthesis
UDP-N-acetylmuramyl-pentapeptide	UDP-GNAc-lactyl-peptide	Bacterial cell wall synthesis
Coenzyme A	ADP-pantothenyl-cysteamine	Acyl transfer
Folic acid	Pteroyl-poly-glutamate	One-carbon transfer
Factor 420	Flavinoid-linked (Glu)$_2$	One-carbon transfer
Carbon dioxide reduction factor	Flavinoid-linked (Glu)$_2$	One-carbon transfer
Actinomycin	Phenoxazone-dicyclo-pentapeptide	Antibiotic
Adenylated protein	Ribose-phospho-tyrosine	Protein regulation
ADP-ribosylated protein	N- and C- glycosides	Protein regulation
Viral RNA-protein	5'-Protein-pUUAAAACAG-for polio virus	Primer in RNA synthesis
N-peptidyl-tRNA	Purine-6-carbamoyl-threonyl-amido group	Transfer RNA
O-peptidyl-tRNA	O-peptidyl-nucleoside	Intermediate in protein synthesis

with protein synthesis: as rRNA, tRNA, mRNA and ribozymes catalyzing ribosomal peptide bond formation and structural RNA processing.

(iii) Code Expansion

Once the genetic code was established, the sidechain imperative that drove the formation of polypeptide surrogates began to press for increased amino acid variety in the genetic code beyond the Phase 1 amino acids available from the prebiotic environment. The only avenue in this direction was development of novel biosynthetic pathways within the cell to produce the Phase 2 amino acids for the code[26-28] (Table 1.1). These Phase 2 amino acids added to the protein sidechains with phenyl, indole, imidazole, sulfhydryl, amide and cationic groups. The catalytic adeptness of imidazole, for instance, is such that a simple Ser-His dipeptide is capable of catalyzing DNA, protein and ester cleavages.[29]

As Box 14.1 shows, all alphabetic languages require a competent collection of letters to represent the human voice. Once competence is attained, the alphabet freezes. Likewise, by trying out novel Phase 2 amino acids, selecting those best suited and finding them satisfactory, the genetic code froze for the next three billion years. The prefreeze code expansion was in effect a search for sidechain excellence for the proteins. Whenever a novel amino acid was added to the code, it brought with it new variety but also noise, e.g., when Gln was introduced into the code by pretran synthesis (Fig. 1.4) and took over the erstwhile CAA-CAG codons from Glu, the code benefited from the gain of a new amide amino acid, but replacing all CAA-CAG encoded Glu residues with Gln created considerable noise. Just as one hesitates to talk loudly in a library but not in a market place, noise is more acceptable against a noisy background than against a quiet one. Likewise, the noise created by a novel amino acid carried less evolutionary penalty when background translation error/noise was higher. Thus low error/noise determined when code expansion adding novel amino acids was to cease.[28] Because the translation error rate depended on the competence of proteins in the translation machinery, only excellence of performance by all of

---

Box 14.1. Evolution of Alphabets

In human languages, the letters of an alphabet represent different sounds of the human voice. Because the human voice is capable of making 40 different basic sounds, one needs up to 40 letters in an effective alphabet. Since a combination of letters can sometimes be used instead of a new letter to represent a sound, e.g., 'sh', 'ch', 'oe' and so on, different alphabets can vary in the number of their constituent letters:

Cyrillic	33
English	26
Hebrew	22
Hungarian runan (archaic)	39
International Phonetic	40
Latin	23

Two characteristics of alphabet evolution are particularly relevant to the protein alphabet. First, the number of letters has to be adequate, for too few letters will under-represent the human voice, ending up with grunts rather than language. Secondly, once an alphabet is established, it tends to stay frozen. An alphabet is a cultural heritage, and any attempt to change its letters, e.g., bringing the number of letters in the English alphabet to equal that of the Cyrillic alphabet, or vice versa, will run into considerable resistance.

By the same token, the number and balance of amino acids in the protein alphabet must be adequate to support a high level of performance by the proteins. For example, a protein consisting of only the three amino acids Gly, Ala and Val can generate an enormous amount of information content based on the formula $I_{max} = N \, Log_2(M)$ by making N huge, but the information generated is not very useful. No matter how long the protein is or how much its sequence is permutated, such a protein will be severely limited in the kinds of enzymic reactions it can catalyze. It lacks the sidechain variety required for excellence. Since the protein alphabet is also a cellular heritage as the foundation of the inherited information in the proteome, it tends to stay frozen as well. Changes to this alphabet will introduce a great deal of nonsensical noise into the protein sequences, and therefore elicit substantial resistance.

these proteins could bring about a low enough error rate to freeze the code. Therefore code evolution would never cease until excellence was achieved in the encoded amino acid ensemble. Accordingly, the accomplishments of protein molecules is not an accident or good fortune but the consequence of rigorous control of code expansion by error-feedback.

The sidechain imperative has been one of the most powerful and insatiable driving forces in prebiotic/early biotic evolution. Starting from the RNA World, it has introduced postreplication modifications, peptidyl ribozymes, Phase 1 genetic code, Phase 2 genetic code, last minute insertion of selenocysteine (Sec) and pyrrolysine (Pyl) into the half frozen code (Fig. 14.2) and, even after the code became totally frozen, proliferation of posttranslational modifications (PTM) to add many more amino acid sidechains to proteins (Table 14.2). The PTM glycosylations alone include sugars O-linked to Ser or Thr, or N-linked to Asn, in linear or branched chains and comprising combinations of mannose, galactose, fucose, acetylgalactosamine, acetylglucosamine and sialic acid.

(iv) Code Evolution

The genetic code is a fascinating admixture of the rational in its orderly allocation of the majority of 4-codon boxes to one amino acid or equally to two amino acids and the seemingly irrational such as the split codon domains of Ser. At one time this strange structure of the code and its universality were thought to be the outcome of 'frozen accident'.[31] Since then at least three evolutionary mechanisms shaping the code have been identified.

Error Minimization

A feature of the code is that physically similar amino acids tend to occupy codons that are *neighbors* in the code, neighbors meaning any two codons characterized by only a single-base difference between them. For example, the CUU-CUC-CUA Leu codons, the AUU-AUC-AUA Ile codons and GUU-GUC-GUA Val codons are neighbors and these amino acids all have bulky hydrophobic side chains. As a result, misreading CUU as AUU or GUU generates only minor disturbance of protein structure. Codon allocations that serve to reduce the impact of errors could be favored in evolution, especially in primordial eons when replication, transcription and translation mistakes were frequent. This mode of error minimization holds not only for Leu-Ile-Val, but also for others such as Ser-Thr, Phe-Tyr, Asp-Glu and Lys-Arg.[32-36]

However, there are three areas in the code, Non-Ops I-III, where violation of error minimization is self-evident:

1. *Non-Op I*: The physicochemical difference between any two amino acids can be calculated by the Grantham chemical-difference formula combining the three parameters of composition, polarity and volume. Among the 190 possible pairs formed from the 20 encoded amino acids, Cys-Trp ranks as the least-alike pair: the chemical distance of 215 between Cys and Trp is many times the chemical distance of 5 between Leu and Ile.[37] Yet the Cys UGU-UGC codons share the same UGN box as the Trp UGG codon.

2. *Non-Op II*: The ultimate in error minimization is the zero error incurred when a codon is misread for its synonymous codon. Accordingly, it is expected that the first step toward error minimization is to maximize the placement of synonymous codons in neighboring positions. This is complied to by the majority of codons. However, the Ser AGU-AGC codons are not neighbors to any of the Ser UCN codons, in a clear departure from error minimization.

3. *Non-Op III*: The frequency of total codon usage is 2.48% for Met and 1.39% for Cys in aerobic *Escherichia coli* K12 and 1.66% for Met and 1.31% for Cys in anaerobic

Methanopyrus kandleri (species closest to LUCA: see Section 15.2). Yet Met is allocated only one codon while Cys receives two. Such disproportionate codon allocations suggests non-optimization of the code with respect to any physical guidelines.

It is now known from the coevolution theory (Section 14.3) that Non-Ops I-III do not represent an abandonment or lapse of error minimization. The UGN box used to belong to Ser, forming a contiguous domain with the Ser UCN and AGY codons. However, when Ser produced Cys and Trp through biosynthesis, it ceded its UGN codons to Cys, Trp and termination signal. Still later, Sec, another Ser-derived amino acid, managed to acquire a share of UGA, thereby joining its Ser-derived siblings Cys and Trp in the same box. This ceding of the Ser UGN codons to its biosynthetic products interrupted the contiguity of the Ser codon domain and caused both Non-Op I and Non-Op II. The single codons given to Trp and Met suggests that these amino acids arrived late in code expansion. Accordingly the ceding of Ser UGN codons was completed evidently not too long prior to the rise of LUCA and the freezing of the code, leaving inadequate time for evolution to repair Non-Op I and Non-Op II toward error reduction. Likewise, the late arrivals of Met and Trp imply that there was little time to fine tune the number of codons they receive and hence Non-Op III. For the same reason, there also might not be time for other late changes in the code to be optimally adjusted for error reduction before the freeze.

Besides insufficiency of time for adjustment, error minimization is limited by the fact that every codon has six neighbors, three per base position, even counting only the first two bases. Improvement in error reduction with respect to some of the six neighboring positions could bring deterioration with respect to others. Overall, the extent of error minimization achieved in the code is about 40-45%[38,39] and it contributes a 10^{-6} selection factor toward the emergence of a unique code.[34]

Stereochemical Interaction

When an aminoacyl-tRNA compound is positioned on the ribosome for peptide bond formation, the amino acid attached to its 3'-terminus might be too far away from the codon and anticodon for direct physical interactions with them. Direct interactions could be more easily effected, however, in the case of a primitive tRNA minihelix[25] or where the codon or anticodon sequence is present in the tRNA acceptor stem.[40] As well, the amino acid and its anticodon on the tRNA might both bind to the active site of the aaRS, thereupon interacting with one another either directly on the aaRS or indirectly through the aaRS. For example, a hydrophobic aaRS active site might preferentially bind a hydrophobic amino acid together with a tRNA possessing a hydrophobic anticodon, thereby promoting a general correlation between hydrophobic amino acids and hydrophobic anticodons.[5] So both direct and indirect stereochemical interactions could be significant in bringing about the experimentally observable correlations between the hydrophobicities of amino acids and their codons/anticodons.[32,33,41,42]

RNA aptamers capable of binding Trp, Arg, Val, Ile, Tyr, Phe, His, Trp or Leu have been found to contain in their amino acid binding pockets a cognate codon or anticodon triplet for the bound amino acid.[43,44] These findings add to the suggestion from hydrophobicity correlations that amino acid-codon/anticodon interactions played a significant role in deciding codon assignments.[41-47] This might be especially the case during the initial codon partition between the Phase 1 amino acids, before either translational errors or coevolution with amino acids became important guides. The Stereochemical Interaction mechanism contributes a 0.04%, or 4×10^{-4} selection factor toward the emergence of a unique code.[43]

Amino Acid Biosynthesis

Inspection of the codon locations for various biosynthetic amino acid families reveals correlations between amino acid biosynthesis and codon assignments.[48,49] The mechanisms by which such correlations came to be established in the course of genetic code evolution are examined in the next section.

14.3 Coevolution Theory

The coevolution theory (CET) describes the mechanisms that brought about the entry of biosynthetically-derived Phase 2 amino acids into a code initially encoding only the environmentally-derived Phase 1 amino acids and the transfer of codons from Phase 1 to Phase 2 amino acids:

> *"The structure of the codon system is primarily an imprint of the prebiotic pathways of amino-acid formation, which remain recognizable in the enzymic pathways of amino-acid biosynthesis. Consequently the evolution of the genetic code can be elucidated on the basis of the precursor-product relationships between amino acids in their biosynthesis. The codon domains of most pairs of precursor-product amino acids should be contiguous, i.e., separated by only the minimum separation of a single base change."[26]*

The evidence for CET is widely based.[5,50-52] For the purpose of proof, CET may be examined in terms of its four individually falsifiable tenets.

Tenet 1. The prebiotic environment did not supply all twenty canonical amino acids at life's origin, but had to be complemented by sourcing through inventive biosynthesis.

This tenet may be falsified by either production of all 20 canonical amino acids under acceptably prebiotic conditions, or discovery of all 20 canonical amino acids on meteorites giving evidence for their extraterrestrial synthesis. Over the three decades since the proposal of CET, little progress has been made along these two lines of potential evidence against Tenet 1. On the contrary, the triple convergence of evidence from genetic code structure, atmospheric amino acid synthesis and meteoritic amino acids has brought unambiguous support for the prebiotic availability of only the environmentally derived Phase 1 amino acids and not the biosynthetically derived Phase 2 amino acids (Table 1.1):

Phase 1: Gly, Ala, Ser, Asp, Glu, Val, Leu, Ile, Pro and Thr

Phase 2: Phe, Tyr, Arg, His, Trp, Asn, Gln, Lys, Cys and Met

Furthermore, in the temporal order of amino acid entries into proteins deduced from consensus chronologies, the nine oldest amino acids identified—Gly, Ala, Val, Asp, Pro, Ser, Glu, Leu and Thr are all Phase 1 amino acids,[53] which provides yet further support for the entry of Phase 1 amino acids into the code ahead of Phase 2 amino acids.

Since the instabilities of the Phase 2 Gln and Asn are such that their concentrations in the prebiotic environment could not exceed 3.7 pM and 24 nM respectively, it is futile to look to their availability from the environment (Section 9.2),[50] thereby proving Tenet 1.

Tenet 2. Pretran synthesis provided mechanisms for the initial encoding of some Phase 2 amino acids.

Among organisms, the incorporation of Gln, Asn and Cys into proteins is mediated by one of two alternative pathways: either through GlnRS, AsnRS, or CysRs, or via pretran synthesis of Gln-tRNA from Glu-tRNA (Fig. 1.4), Asn-tRNA from Asp-tRNA, or Cys-tRNA from phosphoSer-tRNA. Tenet 2 may be falsified if the use of aaRS is found to predate use of pretran synthesis in

Table 14.2. Some posttranslation modifications[30]

Acetylation	ADP-ribosylation
Methylation	Flavin attachment
C-terminus amidation	Oxidation
Biotinylation	Palmitoylation
Formylation	Phosphatidylinositol attachment
Gamma-carboxylation	Phosphopantetheinylation
Glutamylation	Phosphorylation
Glycosylation	Pyroglutamate formation
Glycylation	Proline racemization
Heme attachment	Arginylation
Hydroxylation	Sulfation
Iodination	ISG15 protein attachment
Isoprenylation	SUMO protein attachment
Prenylation	Ubiquitination
Myristoylation	Arg conversion to citrulline
Farnesylation	Disulfide bridges
Geranylgeranylation	Proteolytic cleavage

all of these instances, demonstrating that the pretran pathway is merely an evolutionary late development unrelated to preLUCA events. Contrary to such an outcome, multiple lines of evidence have identified a *Methanopyrus*-like LUCA. Since *Methanopyrus* is devoid of GlnRS, AsnRS as well as CysRS[5,11,54] (Appendix 15.1), it follows that pretran synthesis was employed at the LUCA stage, predating the introduction of GlnRS, AsnRS and CysRS. Moreover, comparative phylogenetics indicate that use of pretran synthesis was primordial for Gln-tRNA and Asn-tRNA and coprimordial with CysRS for Cys-tRNA.[55,56] The incorporation of Sec into proteins also depends on pretran synthesis.[57,58] These findings demonstrate that use of pretran synthesis predated the use of aaRS for the incorporation of at least Gln, Asn, Sec and in all likelihood Cys as well into proteins, proving Tenet 2.

The proving of Tenet 2 in turn adds to the proof of Tenet 1. Since Gln, Asn and Sec had to enter into proteins through pretran synthesis, they were not supplied from the environment.

Tenet 3. Biosynthetic relationships between amino acids were an important determinant of codon allocations.

The observable correlations in the code between amino acid biosynthesis and codon allocations include: (a) location of the codons of Ile, Met, Thr, Asn and Lys, members of the Asp biosynthetic family, in the third row of the code; (b) colocation of Cys and Trp, both derived from Ser, in the UGN box despite their physical unlikeness;[37] and (c) allocation of part use of UGA codon to Sec, another amino acid derived from Ser through pretran synthesis, placing it in same UGN box as Cys and Trp (Fig. 14.2). These are strong correlations even though, being statistical in nature, they may not constitute rigorous proof.

A rigorous proof of Tenet 3 follows from the proof of Tenet 2. Gln is allocated CAA and CAG because a Glu-tRNA carrying the anticodon to these codons was converted to Gln-tRNA via biosynthetic pretran synthesis, not because of any stereochemical interaction between Gln and its codon/anticodon, error minimization, or some other factor. Likewise, biosynthesis was solely responsible for the allocation of AAU and AAC to Asn, UGU and UGC to Cys and part use of UGA to Sec. Therefore, biosynthesis was the sole determinant deciding codon allocations to these amino acids, thus proving Tenet 3.

Tenet 4. The amino acid ensemble encoded by the genetic code is mutable, allowing early code expansion to admit the Phase 2 amino acids.

Minor variations are well known among organisms and organelles with respect to the codons assigned to the 20 amino acids, testifying to the postLUCA mutability with respect to codon assignments. What is universal, never mutated during the past three billion years, is the canonical ensemble of 20 amino acids in the code, which raises the question of whether or not this ensemble is mutable at all. Since CET proposes that the code was extensively shaped by the addition of Phase 2 amino acids to the code, it unconditionally requires the code to be freely mutable during its evolution with respect to its encoded amino acids. Tenet 4 is falsified if prolonged efforts should fail to mutate the encoded amino acids. Accordingly experiments were carried out to mutate the genetic code of *Bacillus subtilis* in this regard. The results described below (Section 14.5) showed that 4-fluoro-Trp could replace Trp and even displace Trp from the canonical amino acids,[59-61] proving Tenet 4.

CET is proven through the proof of its four tenets.

14.4 Selection of Unique Code

There are an astronomical number of possible permutations, or alternate codes, differing in the arrangement of the codons for the different amino acids. Although minor variations in codon assignments to amino acids are found in a wide range of nuclear and organellar systems especially metazoan mitochondria,[62] the usage of basically the same unique codon arrangement by all known organisms suggests that there are no more alternate codes from the LUCA stage left in the living world. Permutations of codon packages yield about 2×10^{19} possible alternative codes.[28] Since the universe is only 15×10^9 years, or 4.7×10^{17} seconds old, forty kinds of life forms each bearing an alternate code would have to be competitively eliminated per second if competition between organisms with different codes represents the sole mechanism for removing alternate codes. This is a physical impossibility. It follows that code disallowance and code competition had to work side by side to pave the way to a unique code. The three major evolutionary mechanisms participate in finalizing the unique code differently:

a. Error Minimization steers the genetic code toward enhanced error-impact reduction through the elimination, via competition, of life forms carrying laggard codes with inefficient error-impact reduction. This mechanism contributes a selection factor of 1×10^{-6}, or one in a million selection, toward the emergence of a unique code.[34]

b. Stereochemical Interaction does not rely on competition between life forms to remove alternate codes. Instead, alternate codes deficient in amino acid-codon/anticodon interactions are never formed, viz. such codes are historically disallowed. This mechanism contributes a selection factor of 4×10^{-4}, or a four in ten thousand selection, toward the emergence of a unique code.[43]

c. Coevolution with amino acid biosynthesis, by dividing up the amino acids into biosynthetic groupings during codon allocation, disallows vast tracts of alternate codes, e.g., once the UGN box was assigned to Ser in the code, only amino acids biosynthetically derived from Ser such as Cys, Trp and Sec could receive codons from a subdivision of the UGN box. This mechanism contributes a selection factor of 10^{-11}, or one in a hundred billion selection, toward the emergence of a unique code.[28] This mode of alternate code reduction is analogous to the reduction of seating arrangements at a wedding reception by affinity grouping of the guests (Box 14.2).

Altogether these three mechanisms are capable of a 4×10^{-21} selection, which is sufficient for the selection of a unique code out of 2×10^{19} possible codes. This sufficiency solves the mystery of the remarkable emergence of a universal code. The relative contributions made by the three mechanisms to the emergence are:

Coevolution: Error Minimization: Stereochemical Interaction = 40,000,000: 400:1

The coevolution process is therefore the preeminent factor shaping the finalized structure of the code.

14.5 Genetic Code Mutation

Many amino acid analogues are readily incorporated into prokaryotic and eukaryotic proteins, but they invariably fail to support indefinite cell growth. To test Tenet 4 of CET, experiments were conducted in 1983 to mutate the amino acid code of *B. subtilis* QB928 (a Trp-auxotroph that cannot revert out of its Trp-dependence) in order to replace the use of Trp for long term cell growth by its analogue 4-fluoroTrp (4FT). In two mutational steps, QB928 gave rise to strain LC33 (LC = large colonies on 4FT plates), which can grow indefinitely on 4FT forming colonies on agar (Fig. 14.4). Although two mutational steps might be accompanied by more than two mutations, the number of mutations required to arrive at 4FT utilization were still surprisingly few. When LC33 was further mutated, it gave rise to HR15 in another two mutational steps (HR = high 4FT/Trp growth ratio), which grows well in 4FT but not on Trp. Thus Trp is replaceable by 4FT in LC33. In HR15 and its derivative strains such as faster growing HR23 and Met-independent MR3, Trp has in fact been displaced from the amino acid code: it can no longer support long term cell growth. Growing on 4FT, these strains are inhibited by Trp now acting as an inhibitory analogue. However, the cells can back mutate to enable Trp to regain its lost growth-support capacity (Fig. 14.5). LC33, which grows on Trp and 4FT, has also yielded through mutations the LC88 strain, which grows on Trp, 4FT, 5-fluoroTrp (5FT) and 6-fluoroTrp (6FT) (Fig. 14.4), even though 5FT is a potent inhibitor of bacterial growth.[59-61]

The fall of Trp from its status as a canonical amino acid in HR15 indicates that genetic code evolution was by no means limited to the addition of novel amino acids to the code. Amino acids also could be tried out, found wanting and rejected from the code. This type of turnover would explain the otherwise inexplicable absence

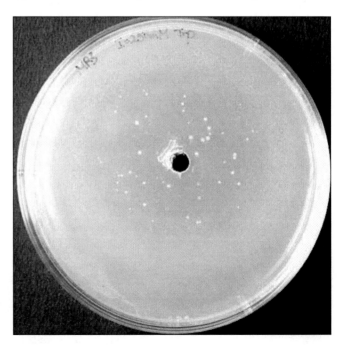

Figure 14.5. Inhibition by Trp placed in center well against *B. subtilis* MR3 growing on 4FT. Revertant colonies that have regained the ability to grow on Trp are visible within the cleared Trp-inhibition zone.[61]

Box 14.2. Seatings at Wedding Reception

The distribution of codons to different amino acids may be compared to the seating of guests at a wedding reception. There are 20 guests, A, B,.....S,T, and there are twenty seats. There will be a total of $p!(q!)^p$ different seating arrangements, where p is the number of seating sections, and q is the number of heads per section. In the first seating approach, all 20 guests are treated as a single group, drawing lots to determine seat assignment, so that p = 1 and q = 20. This yields a total of 2.4×10^{18} different seating arrangements for the guests.

Since there are five affinity groups of guests, four per group --- (1) A-D are bride's relatives, (2) E-H are groom's relatives, (3) I-L are bride's coworkers, (4) M-P are groom's coworkers, and (5) Q-T are neighbors, a second seating approach is to divide the seats into five sections, and randomly draw lots to allocate these sections to groups 1-5. The four seats within each section will be randomly distributed to the four individuals within the same group. In this case, p = 5 and q = 4. This yields a total of only 9.6×10^8 seating arrangements. Thus the constraint imposed in the second approach disallowing all mixed-group seating within the same section reduces the number of possible seating arrangements by a factor of 4×10^{-10}.

Likewise, when the 64 codons in the genetic code are distributed to the 20 amino acids and termination signal without any affinity grouping, the number of possible codes differing in codon allocations is approximately 2×10^{19}. If the 20 amino acids are divided into affinity groups based on biosynthetic relationships, the number of possible codes becomes 2×10^8. This reduction of allowable alternate codes by a factor of 10^{-11} greatly facilitates the selection of a unique code out of all the possible codes.

---

of α-aminobutyric acid from the code. This amino acid is readily produced by atmospheric abiotic synthesis. In the electric spark synthesis in Table 9.1, the amount of α-aminobutyric acid produced was less than Gly and Ala, but more than Glu and Asp. Its absence from the genetic code to-day is likely due to initial try-out by the life forms followed by rejection. Being a 4-carbon hydrophobic amino acid, it might have lost out to amino acids like Val, Leu and Ile which offer a more bulky alkyl sidechain and stronger hydrophobicity.

The relatively small number of mutational steps employed to achieve replacement or displacement of Trp from the code by its fluoro-analogues suggests that there are a small number of Trp residues among the 12,625 Trp residues of the *B. subtilis* proteome where the protein structure around the residue is so restrictive that Trp replacement by 4FT, 5FT or 6FT leads to defective function and therefore growth inhibition. When the structural constraint is relaxed around these critical residues, replacement of Trp by 4FT, 5FT or 6FT is tolerated, as in LC88. However, if the structural constraint is altered in such a manner that it restricts protein function with Trp but not with 4FT, it would cause Trp to lose its growth-support capacity and become an inhibitory analogue.

The introduction of 4FT into the encoded amino acids of *B. subtilis* represents the first known instance of a genetic code mutation altering the ensemble of canonical amino acids. It provides decisive support for Tenet 4 of the coevolution theory and establishes that, in spite of the freezing of the genetic code with respect to its canonical amino acids since the earliest branchings of living species, the code remains intrinsically mutable. The thawing of this once frozen code has ushered in a new era of genetic code mutations. There are two mutational approaches. The top-down or genome wide approach adopted for the encoding of 4FT in *B. subtilis* has been extended to *E. coli* and coliphage.[63-65] In the bottom-up approach, special *orthogonal* aaRS-tRNA pairs are employed to place an unnatural amino acid into selected positions in protein sequences.[66,67] These orthogonal pairs, based on the lack of reactivity between cross domain aaRS-tRNA pairs,[68] interact minimally with other aaRS and tRNAs inside the cell.

Because the amino acid alphabet has been a fundamental, unchanging attribute of all living organisms, the new organisms bearing novel encoded amino acids may be regarded as new forms of life[69] that are transforming the biology of the past into a *sequel* where straitjacketing by the traditional twenty encoded amino acids is relaxed to allow increased dimensions of freedom in evolution based on wider variations in the amino acids.[70] *Bacillus subtilis* strain HR15, which grows on 4FT but not Trp as a canonical encoded amino acid, thus represents the first example of synthetic bilogy.

14.6 Conclusion

The universal genetic code was constructed in two phases. A predecessor code for environmentally derived Phase 1 amino acids underwent expansion to include biosynthetically derived Phase 2 amino acids. In the process the Phase 2 product amino acids received codons from their Phase 1 precursors. This expansion, by breaking up the original codon domains of the precursors, created a strangely structured code with seemingly irrational features such as the unequal allocations of one to six codons to different amino acids, split codon domains for Ser, placement of the codons of physically dissimilar Cys and Trp in the same UGN box and the partial use of a codon by Pyl and Sec. Despite this strange structure, the reigning objective of the expansion, viz. to enhance the capabilities of the encoded ensemble of amino acid sidechains, was superbly accomplished. This ensemble has provided the variety and balance of sidechains for the remarkable performance of proteins manifest in the catalytic perfection of diffusion controlled enzymes where the rate of catalyzed reaction is only limited by diffusion of substrate into the enzyme active site,[71] the development of multicellular life and the advent of human intelligence, arts and sciences. After a deep freeze for three billion years, the code is experimentally thawed since 1983 and its intrinsic mutability is once again making possible code expansion to yield synthetic life forms using synthetic genetic codes.

In evolution, although natural selection based on competition between organisms is of pivotal importance, it has been suggested that developmental constraints could play a significant role jointly with competition in limiting and channeling evolution, e.g., the reduced front legs of *Tyrannosaurus* might have been selected as a by-product of other adaptations such as larger hind legs.[72,73] The emergence of a unique genetic code clearly illustrates the workings of channeling: disallowance of alternate codes by developmental constraints is even more important than elimination of alternate-coded species in achieving the one out of 2×10^{19} selection of the universal code. In Chance and Necessity, Jacques Monod[74] defined three frontiers that represent the foremost challenges of biology: the problem of life's origins, the riddle of the code's origins and the central nervous system. In this regard, the proving of the coevolution theory establishes the subdivision of the codon domains of precursor amino acids to accommodate the influx of product amino acids from biosynthesis as the key to deciphering the riddle of genetic code structure. Just as the map of a country so often tells the story of its history, the universal genetic code is a lasting inscription of the history of its coevolution with the primordial pathways of amino acid biosynthesis.

References

Further Readings

1. Eigen M, Winkler-Oswatitsch R. Steps towards life. Oxford University Press, 1992.
2. Gesteland RF, Cech TR, Atkins JF eds. The RNA world 2nd ed. Cold Spring Harbor Laboratory Press, 1999.
3. Joyce GF. The antiquity of RNA-based evolution. Nature 2002; 418:214-221.
4. Lahav N. Biogenesis. Oxford University Press, 1999.
5. Wong JT. Coevolution theory of the genetic code at age thirty. BioEssays 2005; 27:416-425.

Specific References

6. Gamow G. Possible relation between deoxyribonucleic acid and protein structures. Nature 1954; 173:318.
7. Gilbert W. The RNA World. Nature 1886; 319:618.
8. Benner SA, Burgstaller P, Battersby TR et al. Did the RNA World exploit an expanded genetic alphabet? In: Gesteland RF, Cech TR, Atkins JF eds. The RNA World 2nd ed. Woodbury:Cold Spring Harbor Laboratory Press. 1999; 163-181.
9. Motorin Y, Grosjean H. Appendix 1. Chemical structures and classification of posttranscriptionally modified nucleosides in RNA. In: Grosjean H, Benne R, eds. Modification and Editing of RNA. ASM Press 1998; 543-549.
10. Auffinger P, Westhof E. Appendix 5: Location and distribution of modified nucleotides in tRNA. In: Grosjean H, Benne R, eds. Modification and Editing of RNA. ASM Press 1998; 569-571.
11. Mat WK, Xue H, Wong JT. The genomics of LUCA. Frontiers in Biosc 2008; 13:5605-5613.
12. Lane BG. Historical perspectives on RNA nucleotide modifications. In: Grosjean H, Benne R, eds. Modification and Editing of RNA. ASM Press 1998; 1-20.
13. Cech TR, Herschlag D. Group I ribozymes: substrate recognition, catalytic strategies and comparative mechanistic analysis. In: Eckstein F, Lilley CMJ, eds. Catalytic RNA. Heidelberg:Springer-Verlag, 1996: 1-17.
14. L'Huillier PJ. Efficacy of hammerhead ribozymes targeting a-lactalbumin transcripts: experiments in cells and transgenic mice. In: Eckstein F, Lilley CMJ, eds. Catalytic RNA. Heidelberg:Springer-Verlag ,1996; 284-300.
15. Wong JT, Xue H. Self-perfecting evolution of heteropolymer building blocks and sequences as the basis of life. In: Palyi G, Zucchi C, Caglioti L, eds. Fundamentals of Life. Paris:Elsevier, 2002: 473-494.
16. Wong JT. Origin of genetically encoded protein synthesis: a model based on selection for RNA peptidation. Orig Life Evol Biosph 1991; 21:165-176.
17. Zhang B, Cech TR. Peptide bond formation by in vitro selected ribozymes. Nature 1997; 390:96-100.
18. Illangasekare M, Sanchez G, Nickles T et al. Aminoacyl-RNA synthesis catalyzed by an RNA. Science 1995; 267:643–647.
19. Saito H, Kourouklis D, Suga H. An in vitro evolved precursor tRNA with aminoacylation activity. EMBO J 2001; 20:1797-1806.
20. Robertson MP, Knudsen SM, Ellington AD. In vitro selection of ribozymes dependent on peptides for activity. RNA 2004; 10:114-127.
21. Roth A, Breaker RR. An amino acid as a cofactor for a catalytic polynucleotide. Proc Natl Acad Sci USA 1998; 95:6027-6031.
22. Lautru S, Challis GL. Substrate recognition by nonribosomal peptide synthetase multi-enzymes. Microbiol 2004; 150:1629-1636.
23. Szathmary E. Coding coenzyme handles: a hypothesis for the origin of the genetic code. Proc Natl Acad Sci USA 1993; 90:9916-9920.
24. Schimmel P, Henderson B. Possible role of aminoacyl-RNA complexes in noncoded peptide synthesis and origin of coded synthesis. Proc Natl Acad Sci USA 1994; 91:11283-11286.
25. Schimmel P, Ribas de Pouplana L. Transfer RNA: from minihelix to genetic code. Cell 1995; 81:983–986.
26. Wong JT. A coevolution theory of the genetic code. Proc Natl Acad Sci USA 1975; 72:1909-1912.
27. Wong JT. Co-evolution of the genetic code and amino acid biosynthesis. Trends Biochem Sci 1981; 6:33–36.
28. Wong JT. The evolution of a universal genetic code. Proc Natl Acad Sci USA 1976; 73:2336-2340.
29. Li Y, Zhao Y, Hatfield S et al. Dipeptide seryl-histidine and related oligopeptides cleave DNA, protein and carboxyl ester. Bioorg Med Chem 2000; 8:2675-2680.
30. http://en.wikipedia.org/wiki/Posttranslational_modification.
31. Crick FHC. The origin of the genetic code. J Mol Biol 1968; 38:367-379.
32. Sonneborn TM. Degeneracy of the genetic code, extent, nature and genetic implications. In: Bryson V, Vogel HJ, eds. Evolving genes and proteins. New York: Academic Press 1965: 379-397.
33. Woese CR, Dugre DH, Dugre SA et al. On the fundamental nature and evolution of the genetic code. Cold Spring Harbour Symp Quant Biol 1966; 31:723-736.
34. Freeland SJ, Hurst LD. The genetic code is one in a million. J Mol Evol 1998; 47:238-248.
35. Freeland SJ, Wu T, Keulman N. The case for an error minimizing standard genetic code. Orig Life Evol Biosph 2003; 33:457-477.
36. Ardell DH, Sella G. No accident: genetic codes freeze in error-correcting patterns of the standard genetic code. Phil Trans R Soc London B 2002; 357:1625-1642.
37. Grantham R. Amino acid difference formula to help explain protein evolution. Science 1974; 185:862-864.
38. Wong JT. Role of minimization of chemical distances between amino acids in the evolution of the genetic code. Proc Natl Acad Sci USA 1980; 77:1083-1086.
39. Di Giulio M, Medugno M. The historical factor: the biosynthetic relationships between amino acids and their physicochemical properties in the origin of the genetic code. J Mol Evol 1998; 46:615-621.
40. Rodin S, Rodin A, Ohno S. The presence of codon-anticodon pairs in the acceptor stems of tRNAs. Proc Natl Acad Sci USA 1996; 93:4537-4542.
41. Lacey JC Jr, Wickramasinghe NSMD, Cook GW. Experimental studies on the origin of the genetic code and the process of protein synthesis: a review update. Orig Life Evol Biosp 1992; 22:243-275.
42. Jungck JR. The genetic code as a periodic table. J Mol Evol 1978; 11:211-224.
43. Knight R, Landweber LF, Yarus M. Tests of a stereochemical genetic code. In: Lapointe J, Brakier-Gingras L eds. Translation Mechanisms. Landes Bioscence 2003; 115-12.
44. Legiewicz M, Yarus M. A more complex isoleucine aptamer with a cognate triplet. J Biol Chem 2005; 280:19815-19822.
45. Hendry LB, Bransome ED Jr, Hulson MS et al. First approximation of a stereochemical rationale for the genetic code based on the topography and physicochemical properties of "cavities" constructed from models of DNA. Proc Natl Acad Sci USA 1981; 78:7440-7444.
46. Shimizu M. Molecular basis for the genetic code. J Mol Evol 1982; 18:297-303.
47. Seligman H, Amzallag GN. Chemical interactions between amino acid and RNA: multiplicity of the levels of specificity explains origin of the genetic code. Naturwissen 2002; 89:542-551.
48. Pelc SR. Correlation between coding triplets and amino acids. Nature 1965; 207:597-599.
49. Dillon LS. The origins of the genetic code. Bot Rev 1973; 39:301-345.
50. Wong JT, Bronskill PM. Inadequacy of prebiotic synthesis as origin of proteinous amino acids. J Mol Evol 1979; 13:115-125.
51. Wong JT. Question 6: Coevolution theory of the genetic code: a proven theory. Orig Life Evol Biosph 2007; 37:403-408.
52. Di Giulio M. The coevolution theory of the origin of the genetic code. Physics Life Rev 2004; 1:128-137.
53. Trifonov EN, Gabdank I, Barash D et al. Primordia vita. Deconvolution from modern sequences. Orig Life Evol Biosph 2006; 36:559-565.
54. Wong JT, Chen J, Mat WK et al. Polyphasic evidence delineating the root of life and roots of biological domains. Gene 2007; 403:39-52.
55. O'Donoghue P, Sethi A, Woese CR et al. The evolutionary history of Cys-tRNA^Cys formation. Proc Natl Acad Sci USA 2005; 102:19003-19008.
56. Sauerwald A, Zhu W, Major TA et al. RNA-dependent cysteine biosynthesis in archaea. Science 2005; 307:1969-1972.
57. Commans S, Bock A. Selenocysteine inserting tRNAs: an overview. FEMS Microb Rev 1999; 23:335-351.
58. Feng L, Sheppard K, Namgoong S et al. Aminoacyl-tRNA synthesis by pretranslational amino acid modification. RNA Biology 2004; 1:16-20.
59. Wong JT. Membership mutation of the genetic code: loss of fitness by tryptophan. Proc Natl Acad Sci USA 1983; 80:6303-6306.
60. Bronskill PM, Wong JT. Suppression of fluorescence of tryptophan residues in proteins by replacement with 4-fluorotryptophan. Biochem J 1988; 249:305-308.
61. Mat FWK, Xue H, Wong JT. Genetic encoding of 4-,5-,6-fluorotryptophans: Role of oligogenic barriers. Am Soc Microbiol 104th Meeting 2004; R-029.
62. Abascal F, Posada D, Knight RD et al. Parallel evolution of the genetic code in arthropod mitochondrial genomes. PLoS Biology 2006; 4:e127.
63. Bacher JM, Ellington AD. Selection and characterization of Escherichia coli variants capable of growth on otherwise toxic tryptophan analogues. J Bacteriol 2001; 183:5414-5425.

64. Bacher JM, Ellington AD. The directed evolution of organismic chemistry: unnatural amino acid incorporation. In: Lapointe J, Brakier-Gingras L eds, Translation Mechanisms. Landes Bioscences 2003; 80-94.

65. Bacher JM, Hughes RA, Wong JT et al. Evolving new genetic codes. Trends Ecol Evol 2004; 19:69-75.

66. Xie J, Schultz PG. Adding amino acids to the to the genetic repertoire. Curr Opin Chem Biol 2005; 9:5548-554.

67. Budisa N. Engineering the genetic code. Wiley-VCH. 2006;

68. Kwok Y, Wong JT. Evolutionary relationships between halobacterium cutirubrum and eukaryotes determined by use of aminoacyl-tRNA synthetases as phylogenetic probes. Can J Biochem 1980; 58:213-218.

69. Hesman T. Code breakers. Scientists are altering bacteria in a most fundamental way. Science News 2000; 157:360-362.

70. Cohen P. Life the sequel. New Scientist 2000; 167:32-36.

71. Wong JT. Kinetics of enzyme mechanisms. London:Academic Press, 1975: 200-201.

72. Gould SJ, Lewontin RC. The spandrels of San Marco and the Panglossian paradigm: a critique of the adaptationist programme. Proc Roy Soc London B 1979; 205:581-598.

73. Larson EJ. Evolution. New York:Modern Library, 2004: 280-281.

74. Monod J. Chance and Necessity. New York:Vintage Books, 1972: 138-148.

CHAPTER 15

Root of Life

J. Tze-Fei Wong[*]

15.1 Introduction

The usage of the universal genetic code basically by all living organisms with only minor variations points to the descent of all living species from a Last Universal Common Ancestor, or LUCA, at the root of the tree of life. Knowledge of the nature of LUCA is pivotal to understanding the pathways that gave rise to LUCA, as well as the pathways that led from LUCA to present day life.

An important method for finding the root of life is based on the sequences of protein *paralogs*, namely sister sequences which stem from a duplication of the same gene but end up serving different biochemical functions. These sister sequences were nearly identical at the start. As time went on, they diverged more and more and the extent of their divergence measures how far they have moved away from one another. The first attempts at using mutual rootings of paralogous protein trees to search for LUCA located the root of life in the Bacteria domain of the small subunit ribosomal RNA (SSU rRNA) tree (Section 2.4). However, there were at first very few available protein paralog sequences from all three biological domains and it was also not known at the time that the paralogous rooting method is fraught with artifacts such as long branch attraction, horizontal gene transfer and mutational saturation that could readily invalidate rootings based on just a few species.[5,6] The initial paralogous rooting of the ValRS-IleRS trees,[7] for example, does not contain enough species to detect any of the artifacts now known from an extensive study[8] to heavily beset the IleRS tree. As a result, the early paralogous rootings in the Bacteria domain are unreliable and the unreliability has given rise to pessimism questioning whether the root will ever be found.[9,10]

15.2 Transfer RNA

With the search for LUCA based on protein paralogs troubled by artifacts, other biopolymers have to be looked to for sequence information. DNA and ribosomal RNAs are not useful in this regard because there is no DNA or rRNA paralog in cells. This leaves only the tRNAs. A limitation with tRNA sequences has always been that they are too short, containing only about 75 bases, some of which are only semi-variant. Thus the amount of sequence information from any particular tRNA sequence is small. This limitation may be overcome however by analyzing the entire tRNAomes of species. Since the genomes of free living organisms contain more than thirty tRNA genes, altogether these genes will furnish over two thousand base residues.

(i) Alloacceptor Distances

Based on the coevolution theory of the genetic code (Section 14.3), during the development of the code some of the tRNAs belonging to precursor amino acids were transferred along with their anticodons to product amino acids. That being the case, the kinships between the original tRNAs and their transferred copies could leave behind detectable sequence similarities between same-species tRNAs with different amino acid acceptor specificities, which may be designated as *alloacceptor* tRNAs, in distinction from *isoacceptor* tRNAs that accept the same amino acid. Comparisons of alloacceptors disclose many tRNA pairs exhibiting a high degree of sequence homology, especially among species in the Archaea domain on the the universal tRNA phylogenetic tree. The tRNA^Phe-tRNA^Tyr pair from the archaeon *Aeropyrum pernix* (or Ape—see species abbreviations in Fig. 15.1) strikingly differ from one another at only four base positions. In comparison, there are distinctly more differences between the same pair from either the bacterium Eco or the eukaryote Ecu (Fig. 15.2). These findings suggest that tRNA^Phe-tRNA^Tyr are paralogs derived from a primordial gene duplication. Just as sisters are closer genetically than first-cousins, who are in turn closer than second-cousins and so on, these two paralogous tRNAs gradually diverge more and more from each other with the progress of evolution. On this basis Ape evidently has evolved much less than Ecu or Eco from the root of life.[11]

In the cell, an aminoacyl-tRNA synthetase (aaRS) must recognize its cognate tRNA accurately and charge only that tRNA with its amino acid substrate. If it charges by mistake a noncognate tRNA, a wrong amino acid will be incorporated into proteins. This recognition process depends on identification by the aaRS of special nucleotide residues on the cognate tRNA called indentity elements, which are not found on the noncognate tRNAs. Crystal structures of aaRS-tRNA complexes show the aaRS and its cognate tRNA making close contacts at these positions and usually more than three idenity elements are required for accurate recognition. Accordingly, tRNA^Phe and tRNA^Tyr, from the moment they became paralogs to separately accept Phe and Tyr, would always need to maintain a difference of three or more bases between them as differentiating identity elements. This suggests that the tRNA^Phe and tRNA^Tyr of Ape, with only a 4-base difference, have likely undergone little change for an estimated 3.6 billion years (Fig. 15.2). Such sequence ultra-conservatism is unheard of among proteins.

From vertebrate evolution, it is known that the earliest vertebrates are a group of jawless fishes (*Agnatha*). Many of them are now extinct, but others such as lamprey and hagfish are still alive and well. From the phylogenetic viewpoint, it may be said that lamprey is closer than the rabbit to the ancestral vertebrate. This by no means implies that the lamprey is a more ancient organism than the rabbit. In fact they are both modern animals

*J. Tze-Fei Wong—Applied Genomics Center, Fok Ying Tung Graduate School and Department of Biochemistry, Hong Kong University of Science and Technology Clear Water Bay, Hong Kong, China. Email: bcjtw@ust.hk

Prebiotic Evolution and Astrobiology, edited by J. Tze-Fei Wong and Antonio Lazcano. ©2009 Landes Bioscience.

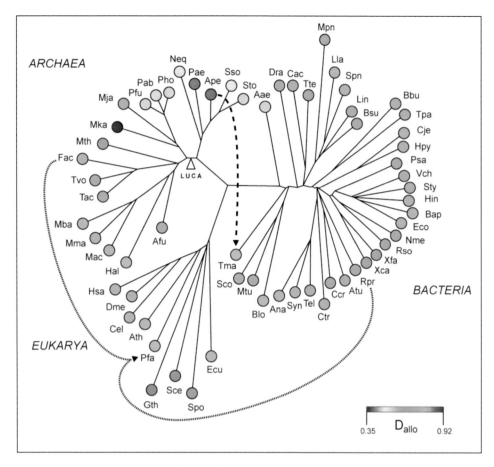

Figure 15.1. Universal tRNA phylogenetic tree with the D_{allo} distances of various species shown in thermal scale.[11,8] Dashed line shows formation of Tma-proximal LBACA from ancient relative of Ape. Dotted lines show formation of Pfa-proximal LECA from endosymbiosis between Fac-related archaeal host and Rpr-related bacterium. Species names: ARCHAEA. Crenarchaeota: Ape *Aeropyrum pernix*, Neq *Nanoarchaeum equitans*, Pae *Pyrobaculum aerophilum*, Sso *Sulfolobus solfataricus*, Sto *Sulfolobus tokodaii*; Euryarchaeota: Afu *Archaeoglobus fulgidus*, Fac *Ferroplasma acidarmanus*, Hal *Halobacterium NRC-1*, Mja *Methanococcus jannaschii*, Mka *Methanopyrus kandleri*, Mac *Methanosarcina acetivorans*, Mba *Methanosarcina barkeri*, Mma *Methanosarcina mazei*, Mth *Methanothermobacter thermautotrophicum*, Neq *Nanoarchaeum equitans*, Pab *Pyrococcus abyssi*, Pfu *Pyrococcus furiosus*, Pho *Pyrococcus horikoshii*, Tac *Thermoplasma acidophilum*, Tvo *Thermoplasma volcanium*; BACTERIA. Aae *Aquifex aeolicus*, Tma *Thermotoga maritima*, Dra *Deinococcus radiodurans*, Ctr *Chlamydia trachomatis*, Bbu *Borrelia burgdorferi*, Tpa *Treponema pallidum*, Blo *Bifidobacterium longum*, Mtu *Mycobacterium tuberculosis*, Sco *Streptomyces coelicolor*, Ana *Anabaena sp*, Syn *Synechocystis 6803*, Tel *Thermosynechococcus elongatus*, Bsu *Bacillus subtilis*, Cac *Clostridium acetobutylicum*, Lla *Lactococcus lactis*, Lin *Listeria innocua*, Mpn *Mycoplasma pneumoniae*, Spn *Streptococcus pneumoniae*, Tte *Thermoanaerobacter tengcongensis*, Atu *Agrobacterium tumefaciens*, Ccr *Caulobacter crescentus*, Rpr *Rickettsia prowazekii*, Nme *Neisseria meningitidis*, Rso *Ralstonia solanacearum*, Bap *Buchnera aphidicola*, Eco *Escherichia coli*, Hin *Haemophilus influenzae*, Psa *Pseudomonas aeruginosa*, Sty *Salmonella typhi*, Vch *Vibrio cholerae*, Xca *Xanthomonas campestris*, Xfa *Xylella fastidios*, Cje *Campylobacter jejuni*, Hpy *Helicobacter pylori*; EUKARYA. Ath *Arabidopsis thaliana*, Cel *Caenorhabditis elegans*, Dme *Drosophila melanogaster*, Ecu *Encephalitozoon cuniculi*, Gth *Guillardia theta*, Hsa *Homo sapiens*, Pfa *Plasmodium falciparum*, Sce *Saccharomyces cerevisiae*, Spo *Schizosaccharomyces pombe*.

and both their lineages are traceable back to the same vertebrate beginning in the late Cambrian period close to 500 million years ago. However, because the lamprey lineage has stayed close to the habitat of the ancestral vertebrate, they have experienced relatively little need to develop for example extensive anatomical and physi-ological modifications. In contrast, the rabbit lineage went on land, changed to air breathing and started to run around on four legs. Not surprisingly the rabbit looks and behaves unlike the early *Agnatha*. Therefore, for the purpose of understanding the respiratory physiology or brain structure of the earliest vertebrates, lamprey makes a

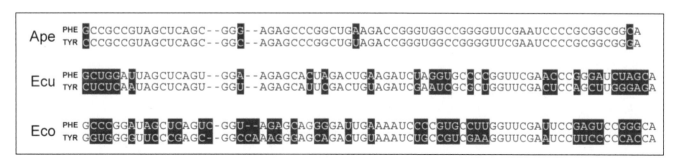

Figure 15.2. Some tRNA^Phe-tRNA^Tyr pairs. Unlike bases in the paired tRNAs are shaded.

far better model than rabbit. By the same token, judging by tRNA sequences, the Ape (archaeal, not primate) lineage for whatever the reasons has undergone much less molecular evolution than Eco or Ecu. It follows that Ape is a closer model for LUCA compared to Eco and Ecu.

There are twenty kinds of alloacceptor tRNAs in the genome of any free living organism and one may determine for any genome the average pairwise genetic distance for its 190 alloacceptor pairs to obtain its alloacceptor distance D_{allo}, which varies from a lowest 0.351 for *Methanopyrus kandleri* (Mka), to the second-lowest 0.402 for Ape, up to 0.760 for humans (Hsa) and the peak value of 0.839 for Sce (namely yeast). These results indicate that the various Mka tRNAs are tightly clustered in sequence space with a high level of resemblance between them. In contrast, the tRNA sequences in Sce are far more dispersed. These results may be interpreted based on the cluster dispersion model of tRNA evolution (Fig. 15.3) where the tRNAs were initially closely packed in sequence space in the center P of the diagram, but evolved outward away from one another into unoccupied sequence space. In the sky, the cosmic Big Bang scatters stars from a point source, with the result that the distances between stars continually increase with time. In the tRNA dispersion brought about by mutations, the distance between any two tRNA lineages mostly increases with time, e.g., for tRNAs A and B, but occasionally it may also decrease, e.g., for C and D. For any genome, D_{allo} is a meaure of its evolved distance from P: the slower evolving species would have smaller D_{allo} and the faster ones larger D_{allo}.

On this basis, LUCA would be located closest to the least evolved genomes identifiable by their minimal D_{allo} distances. Figure 15.1 shows the D_{allo} values of various genomes in thermal scale on the tRNA tree. The genomes with low D_{allo} are not scattered all over the tree but centered at the deep-branching Archaea, which suggests that D_{allo} is not an erratic and therefore useless parameter but a well behaved one. Since Mka in the *Euryarchaeota* and Ape in the *Crenarchaeota*, divisions of the Archaea domain are the two genomes with the lowest D_{allo}, LUCA is located between the branches leading to Mka and Ape in these two divisions.[11] Moreover, because the elongator tRNA[Met] and initiator tRNA[Met] accept the same amino acid but for different functions, they are not treated as alloacceptors for calculating D_{allo}. When the distance between these two tRNAs is estimated for the various genomes, the minimum elongator-initiator distance is again displayed by Mka.[12] Accordingly the D_{allo} distances and the elongator-initiator tRNA[Met] distances contribute Lines 1 and 2 to the array of evidence for locating LUCA proximal to Mka (Table 15.1). Line 1 is further supported by constraint analysis, which shows archaeal tRNAs to be the ancestral group relative to viral, eukaryotic and bacterial tRNAs.[13]

The universal tRNA tree confirms the three-domain structure of the SSU rRNA tree (Fig. 2.3), the deduction of which by Woese constitutes an outstanding accomplishment in molecular evolution.[14,15] However, the tRNA tree shows the Gram-positives to be deeper-branching than cynaobacteria in the Bacteria domain. In this regard it differs from the early SSU rRNA tree[16] which shows the cyanobacteria to be deeper-branching than Gram-positives, but agrees with recent SSU rRNA tree[17] and SSU/LSU rRNA tree[18] that have reversed this branching order in the early SSU rRNA tree. Because SSU rRNA is devoid of paralog and therefore cannot supply a basis for finding the root of life by itself, the SSU rRNA tree was rooted in the Bacteria domain instead of the Archaea domain[15] based on preliminary paralogous rootings of elongation factor and ATPase employing in each case only a single achaeal species, which is grossly inadequate in view of the pitfalls of this type of rootings. Thus the intra-archaeal location of LUCA on the tRNA tree (Fig.

Table 15.1. Lines of evidence locating LUCA close to Methanopyrus[8]

Line	Type of Evidence
1.	Alloacceptor tRNA distances
2.	Initiator-elongator tRNA[Met] distances
3.	Anticodon usages
4.	Aminoacyl-tRNA synthetase distances
5.	Archaeal root of ValRS
6.	Lack of GlnRS in Mka
7.	Lack of AsnRS in Mka
8.	Lack of CysRS in Mka
9.	Lack of cytochromes in Mka
10.	Early Euryarchaea-Crenarchaea separation
11.	Mka as deep-branching archaeon
12.	Primitivity of methanogenesis
13.	Primitivity of anaerobiosis
14.	Primitivity of hyperthermophily
15.	Primitivity of barophily
16.	Primitivity of acidophily
17.	Use of CO_2 as electron acceptor
18.	Chemolithotrophy
19.	Advantage of hydrothermal vents
20.	Minimalist regulations

15.1) departs only from these preliminary paralogous rootings and not from the SSU rRNA tree itself in this regard.

Earlier, tRNA sequence space analysis by statistical geometry has suggested that the process of translation started from a distribution of RNA molecules comprising GC-rich sequences less than 100 nulceotides in length.[19] This suggestion is validated by the 72.5% GC content of the tRNAs of Mka close to LUCA. Figure 15.4 shows the GC-rich consensal Mka tRNA as ancestral tRNA archetype. Even though the majority of Mka tRNAs and accordingly also the consensal Mka tRNA do not possess a long variable arm located 3' to the anticodon stem, it is found that the variable arm is ancient in origin and likely to be present among LUCA tRNAs.[13] Prior to the development of translation, sequences similar to the tRNA-archetype might have played other roles such as the formation of peptidyl-ribozymes (Section 14.2), 3'-aminoacylatable structures on RNA,[20] or service as 'genomic tag' replication-initiation sites at the 3'-terminus of linear RNA genomes.[21] The encoding of primitive tRNA genes is discussed in Section 13.5.

(ii) Anticodon Usages

The collection of tRNA genes in a genome, determined from the complete genomic sequence, reveal the nature of the anticodons employed by the species. These species-specific anticodon usages provide a wealth of interesting information on how genetic coding is implemented in different species. In the genetic code there are thirteen standard 4-codon boxes where the four codons in the box are allocated either all to the same amino acid, e.g., the family box of GUN for Val, or equally to two amino acids, e.g., AAU-AAC for Asn and AAA-AAG for Lys. In most bacterial and eukaryotic species these thirteen boxes are read by varying combinations of anticodons (Fig. 15.5). For example, Sce employs three anticodons bearing a 3'G, U or C to read the UUN, CAN, AAN, GAN, AGN and GGN boxes; three anticodons bearing a 3'A, U or C to read the GUN, UCN and ACN boxes; two anticodons bearing a 3'G or U to read the CUN box; two anticodons bearing a 3'A or U to read the CCN and GCN boxes; and two anticodons bearing a 3'A or C to read the CGN box. In contrast, no archaeon uses more than two combinations. The overall results underline a sharp divide between the complex, mainly multiple-combination anticodon usages

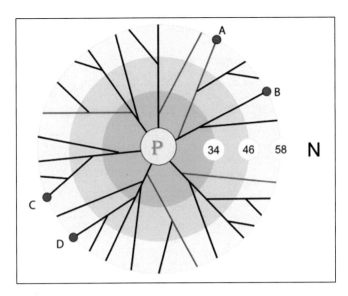

Figure 15.3. Cluster-dispersion model of tRNA evolution. Branchings are gene duplications which generate either new isoacceptors (black lines) or alloacceptors (red lines). The representative numbers of tRNA genes in the tRNAome at different stages are shown in circles.[11]

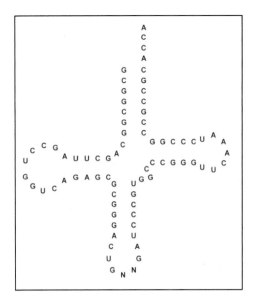

Figure 15.4. Consensal *Methanopyrus* tRNA.

of Bacteria and Eukarya on the one hand and the simple, mainly single-combination anticodon usages of Archaea on the other:[22]

> **Bacteria**, 34 species: 2 species each use five or more combinations, 7 use four combinations, 19 use three combinations, 5 use two combinations and only Tma uses a single GNN+UNN+CNN combination.
>
> **Eukarya**, 7 free living species: 1 species uses five combinations, 1 uses four combinations, 3 use three combinations and 2 use two combinations.
>
> **Archaea**, 18 free living species: 13 species each use a single GNN+UNN+CNN three-anticodon combination. Mka, befitting its proximity to LUCA, displays even greater simplicity in its use of a single GNN+UNN two-anticodon combination. Mja, Mth and Tvo use a transitional mixture of these two kinds of combinations. Fac oddly uses GNN+UNN+CNN for 12/13 of its standard boxes, but ANN+UNN+CNN for its CUN family box for Leu.

Insofar that evolution typically moves from the simple toward the complex, the simple archaeal anticodon usages compared to the complex bacterial and eukaryotic usages favors greater primitivity of Archaea (Line 3, Table 15.1). As well, the suggestion has been made based on ribosome morphology that an 'Eocyte' domain should be split off from the Archaea domain.[23] However, the use of a single GNN+UNN+CNN combination by all four free living crenarchaeons and the great majority of euryarchaeons points strongly to the unity of the Archaea domain.

While Archaea is phylogenetically distinct from Eukarya and Bacteria, hitherto few unique archaeal characteristics unshared by both of the other domains have been identified besides the archaeal use of ether-lipids versus the bacterial and eukaryotic use of ester-lipids. In this regard, the simple versus complex anticodon usages represents a rare divide between Archaea and Bacteria-Eukarya with respect to a fundamental molecular biological characteristic. What is the evolutionary significance of this divide? One plausible answer suggested by Knud Nierhaus (personal communication) relates to the ribosomal elongation fac-

		Mka	Tac*, Tma	Tvo	Fac	Sce	Bsu	Eco			Mka	Tac*, Tma	Tvo	Fac	Sce	Bsu	Eco			Mka	Tac*, Tma	Tvo	Fac	Sce	Bsu	Eco			Mka	Tac*, Tma	Tvo	Fac	Sce	Bsu	Eco
UUU	*F*								UCU	*S*					A																				
UUC	*F*	G	G	G	G	G	G	G	UCC	*S*	G	G	G	G		G	G																		
UUA	*L*	U	U	U	U	U	U	U	UCA	*S*	U	U	U	U	U	U	U																		
UUG	*L*		C	C	C	C	C	C	UCG	*S*		C	C	C	C		C																		
CUU	*L*				A				CCU	*P*					A			CAU	*H*								CGU	*R*					A	A	A
CUC	*L*	G	G	G		G	G	G	CCC	*P*	G	G	G	G			G	CAC	*H*	G	G	G	G	G	G	G	CGC	*R*	G	G	G	G			
CUA	*L*	U	U	U	U	U	U	U	CCA	*P*	U	U	U	U	U	U	U	CAA	*Q*	U	U	U	U	U	U	U	CGA	*R*	U	U	U	U			
CUG	*L*		C	C	C		C	C	CCG	*P*		C	C	C			C	CAG	*Q*		C	C	C	C		C	CGG	*R*		C	C	C	C	C	C
									ACU	*T*					A			AAU	*N*								AGU	*S*							
									ACC	*T*	G	G	G	G		G	G	AAC	*N*	G	G	G	G	G	G	G	AGC	*S*	G	G	G	G	G	G	G
									ACA	*T*	U	U	U	U	U	U	U	AAA	*K*	U	U	U	U	U	U	U	AGA	*R*	U	U	U	U	U	U	U
									ACG	*T*		C	C	C			C	AAG	*K*		C	C	C	C			AGG	*R*		C	C	C	C	C	C
GUU	*V*				A				GCU	*A*					A			GAU	*D*								GGU	*G*							
GUC	*V*	G	G	G	G		G	G	GCC	*A*	G	G	G	G		G	G	GAC	*D*	G	G	G	G	G	G	G	GGC	*G*	G	G	G	G	G	G	G
GUA	*V*	U	U	U	U	U	U	U	GCA	*A*	U	U	U	U	U	U	U	GAA	*E*	U	U	U	U	U	U	U	GGA	*G*	U	U	U	U	U	U	U
GUG	*V*		C	C	C	C			GCG	*A*		C	C	C				GAG	*E*		C	C	C	C			GGG	*G*		C		C	C		C

Figure 15.5. Anticodon usages.[22] (*the Tac usage pattern is shared by the following archaeons: Ape, Pae, Sso, Sto, Afu, Hal, Mac, Mba, Mma, Pab, Pfu and Pho).

tor LepA. This factor, one of the most highly conserved proteins, is present in all bacteria and in nearly all eukaryotes, but not in Archaea. It enables back-translocation during translation, prevents ribosome stalling and enhances ribosomal tolerance to changes in ionic concentrations.[24] Thus Bacteria, by having Lep A, might be more tolerant of internal-milieu variations than Archaea and therefore better equipped to adapt to wide ranging ecologies. To cope with internal-milieu variations, the base-pairing strengths of codon-anticodon pairs might also have to be fine-tuned, thereby accounting for the multiple-combination anticodon usages of Bacteria and Eukarya. Based on this possibility, the formation of the Bacteria domain from Archaea could be the consequence of adaptive advances such as LepA, enabling the Bacteria to enter into new ecological niches far more easily than Archaea, including the human body. This would help to explain the broad ecological distribution of Bacteria compared to the relative confinement of Archaea to extreme environments and why there are so many human infections caused by bacteria and so few if any by archaeons.

15.3 Proteins

Based on fossils, palaeontology has unearthed a magnificent panorama of organisms that populated planet Earth during the fossil-bearing periods starting in the Cambrian period. Further back into the the Precambrian 600 million years ago, the Ediacara fauna discovered near Adelaide in Australia, for example, comprise shell-less specimens of jelly-fish and segmented worms. Still further back in time, fossil microorganisms could be recognized microscopically from their relatively uniform imprints inside rocks and molecular evolution may be traced through protein and nucleic acid sequences. Proteins such as cytochrome c, histones and hemoglobin have yielded invaluable information on biological evolution. However, for tracing events back to LUCA times three billion years ago, proteins tend to be too fast evolving and burdened by horizontal gene transfers (HGT) and other perturbations. For example, the RNA polymerase β and β' subunits tree of Bacteria positions *Aquifex* away from the root of the clade, but moves it close to the root upon removal of two *Mycoplasmas* from the tree.[25] Likewise, the tree of a combined-protein set shows either spirochaetes or thermophiles as the earliest bacterial group depending on whether some species are excluded from the tree.[26]

Two approaches may be called upon to reduce the impact of artifacts in extracting phylogenetic information from protein sequences. Sequence homology of paralogous proteins can be estimated through pairwise comparisons without invoking tree construction and inclusion of a large number of species in tree construction can be utilized to maximize detection of invalidities, as illustrated in the following sections.

(i) aaRS Distances

The BLASTP algorithm may be used to estimate the genetic distance between proteins.[27] Its application to the 190 pairs of the twenty kinds of aaRS within any genome on the tRNA tree generates 190 bitscores. Whenever a bitscore greater than 60 is observed between two aaRS within any genome, the two aaRS might be regarded as potential paralogs derived from gene duplication. On the basis of this criterion, 10 out of the 190 aaRS pairs are potentially paralogous.[12]

Among the genomes from the tRNA tree, the ValRS-IleRS pair reaches the highest maximum bitscore of 506.5 in Mka (Table 15.2), qualifying readily as potentially paralogous. Nine other pairs also achieve a maximum bitscore higher than 60 in one of the genomes, upward from the 66.2 shown by the ThrRS-GlyRS pair. The top bitscore is not always found in the same genome, even though Mka

Table 15.2. Potentially paralogous pairs of aminoacyl-tRNA synthetases[12]

aaRS pair	Max Bitscore	Top Score	2nd Highest	3rd Highest
ValRS-IleRS	506.5	Mka	Mth	Mja
ValRS-LeuRS	232.3	Tma	Hal	Mka
LeuRS-IleRS	202.6	Pab	Pfu	Pho
MetRS-LeuRS	94.4	Tte	Cac	Bsu
ThrRS-ProRS	91.7	Mka	Pab	Sco
MetRS-ValRS	82.4	Cac	Lla	Tte
IleRS-MetRS	75.1	Mka	Lla	Aae
TrpRS-TyrRS	75.1	Mka	Sso	Sto
SerRS-ProRS	67.4	Lla	Spn	Tte
ThrRS-GlyRS	66.2	Mka	Pho	Pfu
Average (Q_{ARS})		Mka 138.5	Mth 119.3	Mja 115.2

turns out to be top scorer in five out of the ten pairs. Averaging the ten bitscores achieved by any genome gives its Q_{ARS} quotient, which measures how closely its aaRS paralogs still resemble one another today. The Q_{ARS} of the various genomes are shown in thermal scale on the tRNA tree in Figure 15.6. The three highest Q_{ARS} of 138.5, 119.3 and 115.2 for Mka, Mth and Mja, respectively, exceed by far for example the 88.2 for Bsu, 60.4 for Eco and 40.9 for Hsa, demonstrating the existence of a clear-cut gradient among the species. Mka is indicated by its highest Q_{ARS} score to be the slowest evolver in aaRS genotypes on the tree in Figure 15.6 (Line 4, Table 15.1), just as it is indicated by its lowest D_{allo} score to be the the slowest evolver in tRNA genotypes on the tree in Figure 15.1. However, there are significant differences between the two trees relating to some of the other species. Mth and Mja are quite fast evolving in tRNA genotypes but slow evolving in aaRS genotypes. On the other hand, the Crenarchaea are slow evolving in tRNA but quite fast evolving in aaRS.

(ii) Archaeal Root of ValRS Tree

Since the ValRS-IleRS pair displays the highest maximum bitscore among all the paralogous aaRS pairs, these two protein sequences could be suitable for paralogous rooting provided there are not excessive perturbations. In order to detect such perturbations, the ValRS and IleRS sequences from the large number of genomes on the tRNA tree are employed to construct the paralog tree in Figure 15.7. The IleRS sequences from Archaea, Bacteria and Eukarya are each split into separate groupings, with the middle cluster on the tree comprising sequences from all three domains. As a result, rooting the IleRS tree using the ValRS sequences as outgroup would be lacking in validity. In comparison, the ValRS sequences are more orderly and largely divided into domain-specific clusters, even though the bacterial sequences of Blo and Rpr are mislocated in the archaeal cluster. Therefore rooting of the ValRS tree using the IleRS sequences as outgroup is comparatively more valid. In this regard, the link from the IleRS outgroup joins the ValRS tree on the red line linking the Archaea ValRS to their junction (point J) with the ValRS from Bacteria and Eukarya, which roots the ValRS tree in the Archaea (Line 5, Table 15.1).[8]

(iii) Missing Genes

In present day organisms, the Phase 2 amino acids Gln, Asn and Cys are incorporated into proteins via GlnRS, AsnRS and CysRS in some species, but via pretran synthesis from Glu-tRNA, Asp-tRNA and o-phospho-Ser-tRNA respectively in other species.

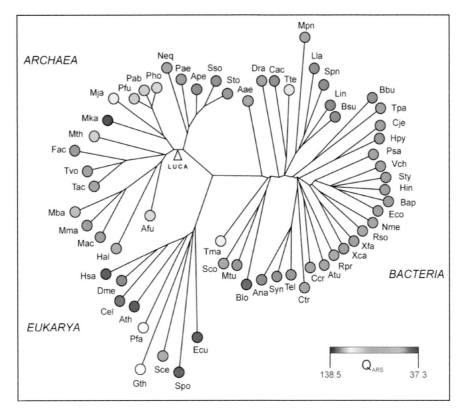

Figure 15.6. Universal tRNA tree with the Q_{ARS} scores of various species shown in thermal scale.[12]

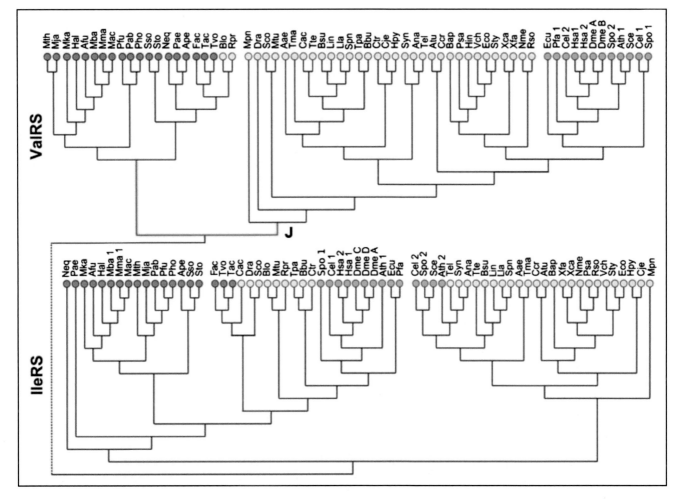

Figure 15.7. Consensus maximum parsimony tree of ValRS-IleRS. (Archaea in red, Bacteria in yellow, Eukarya in blue).[8]

In genetic code evolution, use of pretran synthesis predated the use of aaRS (Section 14.3). Accordingly, using pretran synthesis instead of GlnRS, AsnRS or CysRS to incorporate Gln, Asn or Cys into proteins is a primordial trait and slow-evolving species that have not evolved far from LUCA may be expected to be deficient in one or more of these three aaRS. The fact that the genes for these three aaRS are all missing from the Mka genome adds further evidence to Mka's closeness to LUCA (Lines 6-8, Table 15.1).

Cytochromes are heme-containing proteins that participate widely in electron transport in mitochondria, chloroplasts, sulfate-reducing organisms and even the anaerobic methanogenic *Methanosarcina*. It is therefore a surprise that cytochrome genes are missing from the genomes of Mka, Mth, Mja, Pfu, Pab and Pho, which are clustered together on the tRNA tree. Since Mka, Pfu, Pab and Pho display some of the lowest D_{allo} distances and Mka, Mth and Mja display the three highest Q_{ARS} quotients, this cytochrome-less group represents an *ancient six* group of genomes that are ultra-conservative in molecular evolution. Since Mka, Mth and Mja employ H_2 and CO_2 to make methane, whereas Pfu, Pab and Pho produce H_2 and CO_2 metabolically, the cytochrome deficiency of the group evidently does not stem from metabolic similarity, but from their closeness to a cytochrome-less LUCA. In this light, the cytochrome deficiency of Mka contributes Line 9 to Table 15.1.[8]

(iv) Ancestral Proteins

The composite phylogenetic tree built from 32 proteins, yielding a timescale for protein evolution, points to the Euryarchaea-Crenarchaea separation being the most ancient biological event, Mka as the deepest branching archaeon and advent of methanogenesis as far back as 3.8-4.1 Gya[28] (Lines 10-12, Table 15.1).

The accumulation of atmospheric oxygen from oxygenic photosynthesis brought about extensive proteome adaptations in most organisms. Among the free living organisms, Tma, Mka and Mth exhibit the least post-oxygen proteome adaptations, confirming that Mka is an ultra-conservative organism that has retained the anaerobic character of LUCA[29] and this is also suggested by the amino acid composition of reconstructed ancestral proteins (Line 13).[30] Reconstructed ancestral proteins in combination with the structure of the genetic code futher support LUCA being a hyperthermophile, a barophile living in ocean depths and an acidophile, much like Mka to-day.[31-33] The correlation between hyperthermophily and deep-branching primitivity applies to both Archaea and Bacteria.[34] Moreover, the thermostabilities of ancestral elongation factors resurrected by cloning and expression reveal that ancestral organisms far back in time lived at elevated temperatures.[35,36] These findings, by throwing light on properties common to Mka and LUCA, contribute Lines 14-16 to Table 15.1.

15.4 Metabolism

Large amounts of electron acceptors might not be needed to initiate life, which could start off in a small hospitable incubator site, but they would be essential for the LUCA lineage destined to multiply profusely, overwhelm all competitors and establish universal common ancestry for all future organisms. The abundant electron acceptors on primitive Earth were CO_2 and to a lesser extent sulfate.[37,38] Accordingly, methanogens like Mka that used CO_2 as electron acceptor were well equipped in terms of electron transport to become LUCA (Line 17, Table 15.1). Using CO_2 as electron acceptor in chemolithotrophy, as in the case of methanogenesis, would further favor the large scale proliferation of an organism by freeing it from dependence on organic energy substrates (Line 18).

The hydrothermal vents where Mka lives would make an appropriate home base for LUCA, for methanogenesis at the vents makes possible chemolithoautotrophy, adding the abundance of H_2 as electron donor to the advantage of plentiful CO_2 as electron acceptor. As well, the elevated temperature of the vents would compel the development of hardy biochemical capabilities such as a genetic code with an optimally balanced ensemble of encoded amino acids and a DNA informational machinery (Line 19). The minimalist metabolic regulatory circuits of Mka[39] are also consistent with its primitivity (Line 20).

15.5 Roots of Bacteria and Eukarya

(i) LBACA

Since LUCA was an archaeon, it was also the Last Archaeal Common Ancestor (LACA) that gave rise to the Archaea domain. However, the Last Bacterial Common Ancestor (LBACA) and the Last Eukaryotic Common Ancestor (LECA) need to be identified. Again tRNA sequences are useful for this purpose. Because all bacteria were derived from Archaea through LBACA, the tRNAs of bacteria close to LBACA might yet retain stronger archaeal resemblances than other bacterial tRNAs. To explore this possibility, all archaeal tRNAs were individually compared with all bacterial tRNAs. The comparison revealed, out of the 1.2 M archaeal-bacterial pairs, six hi-sim pairs with high sequence similarity indicated by a genetic distance less than 0.15 (Table 15.3). These six hi-sim pairs include three from Tma and two from Aae on the bacterial side, with the two lowest-distance pairs both coming from Tma. On the archaeal side, Ape contributes to three out of the six hi-sim pairs, including the two lowest-distance pairs. These results suggest that LBACA was closest to Tma and second-closest to Aae on the bacterial side and the archaeal progenitor of LBACA was closest to the crenarchaeon Ape (Fig. 15.1). They also confirm the topology of the tRNA tree, which shows Aae and Tma as the deepest branches of Bacteria and the Bacteria domain arising from the Crenarchaea side of LUCA rather than the Euryarchaea side. In addition, the use by Tma of a single GNN+UNN+CNN anticodon combination to read all its standard codon boxes, just like Ape and the great majority of Archaea (Fig. 15.5), strongly supports its proximity to LBACA. In contrast, Aae employs GNN+UNN+CNN for eight boxes, GNN+UNN for four boxes and ANN+CNN for one box, a three-combination usage that is typically bacterial, not at all archaeal, in character.

Notably, Tma is an anaerobe whereas Ape is an aerobe. So it would be important to examine anaerobic and microaerophilic relatives of Ape, as well as microaerophilic and aerobic relatives of Tma, to deter-

Table 15.3. Hi-sim and ultrahi-sim cross-domain tRNA pairs[8]

Archaea-Bacteria Pairs	Distance
Ape(Ile)cau-Tma(Ile)cau	0.104
Ape(Ile)gau-Tma(Ile)cau	0.104
Mja(Asn)guu-Mtu(Lys)uuu	0.120
Ape(Thr)cgu-Aae(Thr)cgu	0.122
Afu(Ser)gga-Tma(Ser)gga	0.139
Sso(Thr)cgu-Aae(Thr)cgu	0.139

Eukarya-Archaea Pairs	Distance
Pfa(Phe)gaa-Fac(Phe)gaa	0.088
Pfa(Phe)gaa-Tac(Phe)gaa	0.104
Pfa(Phe)gaa-Tvo(Phe)gaa	0.104
Hsa(Phe)gaa-Fac(Phe)gaa	0.137
Pfa(Leu)uaa-Sto(Leu)caa	0.139

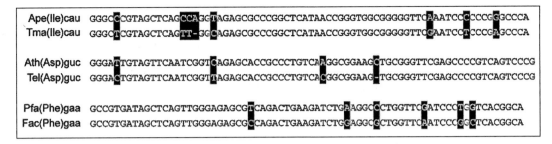

Figure 15.8. Some hi-sim and ultrahi-sim cross-domain tRNA pairs. Unlike bases on the paired tRNAs are shaded.[8]

mine if they might be even closer to the archaeal-bacterial transition than Ape and Tma.

(ii) LECA

When all Eukarya tRNAs were individually compared to all tRNAs from Archaea and Bacteria, the comparisons yielded, out of the 0.98 M eukaryotic-bacterial pairs and 0.58 M eukaryotic-archaeal pairs, not only hi-sim pairs, but also ultrahi-sim pairs with a genetic distance less than 0.10. All eight ultrahi-sim Eukarya-Bacteria pairs detected involve the plant Ath as the eukaryotic partner and the *Cyanobacteria* Tel, Ana and Syn as the bacterial partners.[8] Since *Cyanobacteria* are close to the precursor of chloroplast, these strong resemblances between Ath and cynaobacterial tRNAs most likely arose from the migration of tRNA genes of cyanobacterial origin from the chloroplast genome to the nuclear genome in the plant cells.

The one ultrahi-sim and four hi-sim Eukarya-Archaea pairs detected, consisting of tRNAs for Phe and Leu, are dominated by Pfa on the Eukarya side and the *Thermoplasmatales* Fac (2x), Tac (1x) and Tvo (1x) on the Archaea side (Table 15.3). The sole ultrahi-sim pair is the Pfa(Phe)gaa-Fac(Phe)gaa pair that accept Phe and carry a GAA anticodon (Fig. 15.8). This suggests that LECA likely arose from endosymbiosis of an archaeon host related to *Ferroplasma*, or Fac, which lives in metal sulfide-rich environments[40,41] with Rpr and possibly another bacterium (Fig. 15.1). This is supported by several lines of evidence:

a. The anticodon combination ANN+UNN+CNN is a favorite among eukaryotes, which utilize this combination to read 67% of their single amino acid family codon boxes. In contrast, this combination is not used by any archaeon or bacterium except for the CUN family box in the archaeon Fac (Fig. 15.5) and the bacterium Lla,[22] which suggests that this combination was introduced into Eukarya via endosymbiosis from the Fac-related host.

b. The lack of a cell wall in *Thermoplasma* is well suited to an engulfing endosymbiotic host[42] and this attractive characteristic is shared by *Ferroplasma*.

c. The biochemistry of *Thermoplasma* is advantageous for an archaeal progenitor of Eukarya in an endosymbiotic event in a sulfide-rich environment[43] and the same applies to *Ferroplasma*.

d. An Eukarya-*Thermoplasmatales* relationship is indicated by supertree-based phylogenetic signal stripping.[44]

e. *Thermoplasmas* are thermophiles with an optimal growth temperature of 55-59°C whereas *Ferroplasma* is a mesophile with an optimal growth temperature of 37°C.[40,41] The scarcity of thermophilic eukaryotes and the ready adaptability of eukaryotes to growth at 37°C, traceable eons later to the 37°C human body temperature, favor a mesophilic *Ferroplasma*-like endosymbiotic host over a thermophilic *Thermoplasma*-like endosymbiotic host.

There is substantial evidence that an α-proteobacterium close to *Rickettsia* was the likely precursor of mitochondria through

endosymbiosis. The endosymbiotic event whereby Fac became the progenitor of Eukarya could be the same event involving *Rickettsia* that gave rise to mitochondria. However, the possibility is not ruled out that the Eukarya-generating and mitochondria-generating events might be separate events involving the same or different bacterial partners.[45-47]

Since the Phe-acceptor(gaa) and Leu-acceptor(uaa,caa) tRNAs have retained the highest Eukarya-Archaea sequence resemblance, tracing these tRNA sequences through primitive eukaryotes and *Ferroplasma*-related species could facilitate delineation of the phylogenies of these organisms and determination of their relationships with LECA.

15.6 The LUCA Genome

(i) Non-Minimal Genome

The uncovering of cytochrome deficiency at the LUCA stage based on the *ancient six* genomes of Mka, Mth, Mja, Pfu, Pab and Pho suggests the use of these genomes to define the LUCA genome: genes that are present in all of these six genomes may be regarded at first approximation to be constituent genes of the LUCA genome. However, because the Mka-Mth-Mja group and the Pab-Pfu-Pho group have dissimilar modes of energy metabolism, the genes common to the two groups would lack the specific energy metabolism genes from either group. As LUCA could not survive without any energy metabolism genes, the LUCA gene set has to be supplemented with such genes. Since LUCA was in all likelihood a hyperthermophilic methanogen, the methanogenesis genes common to Mka, Mth and Mja are added to the plausible LUCA genome based on the *ancient six* (Appendix 15.1).[48]

Proteins families are grouped into Clusters of Orthologous Groups, or COGs, which typically serve the same function.[49] The COGs common to the *ancient six* together with the methanogenesis COGs common to Mka, Mth and Mja add up to 561 LUCA COGs. This 561-COG set gives a low-end estimate for the LUCA proteome. For example, Mka and most likely LUCA contain the SelD (Mk1369) and SelA (Mk0620) genes encoding the pretran synthesis of Cys-tRNA, but this pathway has been abandoned in some of the *ancient six*. Accordingly these two COGs are not included in the 561 COGs. On the other hand, LUCA is located between the *Euryarchaea* and *Crenarchaea* on the tRNA tree. Since the *ancient six* are all primitive euryarchaeons, the COGs common to the *ancient six* might include genes that were originally absent from LUCA but were added to its earliest euryarchaeal offsprings. To counter this possibility, Appendix 15.1 also shows the smaller set of 424 COGs that are common to the *ancient eight* species, which comprise the *ancient six* plus Ape and Pae, the two crenarchaeons with the lowest D_{allo} distances. These 424 COGs define the minimum LUCA proteome. Combining this minimum LUCA proteome with, based on the Mka genome, 39 structural RNA genes yields a minimum *LUCA genome* of 463 genes. It is far smaller than the Mka genome which contains 1731 genes. So either genome size has evolved extensively between LUCA and Mka, or the 463 genes substantially underestimate the LUCA genome, or likely both. An estimate of over 1000 LUCA genes based on ancestral state inference[50] is in agreement with the 463 gene set being only a minimal estimate of the LUCA genome.

Whether LUCA was a rudimentary progenote or an organism with a full-fledged genome has been one of the key unanswered evolutionary questions.[51] The size of the minimal proteome compatible with life is given by models I-VI, which comprise of 150-340 protein encoding genes:

I. Genes in a "limping" life form (Section 2.7): 150 genes
II. The minimalist genome of *Carsonella ruddii*:[52] 182 genes
III. Genes in hypothetical minimal cell:[53] 200 genes
IV. Minimal proteome deduced from a comparison of *Mycoplasma genitalium* and *Haemophilus influenzae* genomes:[54] 256 genes
V. *Bacillus subtilis* essential genes identified by deletions:[55] 271 genes
VI. Core protein genes of bacteria:[56] 340 genes

Since the minimum LUCA proteome of 424 COGs exceeds significantly the minimal proteomes I-VI proposed to be adequate for life, there must be some 'nonminimal' genes in the LUCA genome. For instance, the COG Group A for RNA processing and modification, Group B for chromatin structure and dynamics, Group D for cell cycle control, cell division, chromosome partitioning and Group T for signal transduction mechanisms are part of the minimum LUCA proteome based on either the *ancient six* or *ancient eight* (Appendix 15.1) but are not included in any of the minimal models I-VI. Consequently LUCA was not a progenote or a minimal cell. As the last organism of the prebiotic era and the first modern organism, LUCA evidently had evolved beyond the minimal cell stage of life. Analyses carried out using other methodologies also suggest that LUCA was more complex than a minimal organism.[57,58]

(ii) Invention of DNA

Some viruses employ a DNA genome and some an RNA genome, demonstrating that both DNA and RNA make adequate genetic material. Yet all cellular genomes are based on DNA. Although there are definite advantages of DNA over RNA as genetic material (Section 9.3), the adoption of DNA genes required the development of an intricate machinery. The minimum 424-COG LUCA proteome includes 24 Group L COGs for DNA replication, recombination and repair, 26 Group K COGs for transcription and other DNA-related COGs in Groups B, D, F and R. Thus more than 10% of the minimum LUCA proteome is devoted to the DNA informational machinery. It called for a sustained and powerful evolutionary incentive to develop so many new COGs in order to switch from RNA genes to DNA genes. Forterre proposed that the switch could have been caused by the need of RNA organisms to defend against attack by RNA viruses,[59,60] but this *viral defense* incentive is not without problems:

a. It would take many generations for the cells to develop a multi-COG DNA machinery. There might not be sufficient time with the RNA viruses in hot pursuit.
b. Secondly, even if a switch was finally achieved after many generations, what was there to stop the RNA virus from switching to a DNA genome itself and continuing the attack, knowing that RNA viruses like HIV make the most treacherous enemies on account of their facile mutability?

In view of this, it is necessary to examine also other plausible incentives besides viral defense. One important incentive might be adaptation to hyperthermophilic conditions. As Figure 15.9 shows, the half-life of RNA at hyperthermophilic temperature is so short

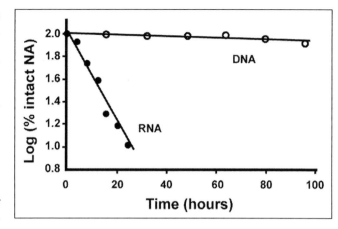

Figure 15.9. Thermal degradation of RNA and DNA at 100°C, pH 7. The RNA sequence employed was gauucaaucugaucucgaugaag and the corresponding DNA sequence was gattcaatctgatctcgatgaag.[48]

that RNA genomes would quickly lose viability. Consequently, as organisms invaded the submarine hydrothermal vents from cooler zones,[61] success versus failure of adaptation could depend on the development of DNA genes. Since the adaptation was a gradual one as the organisms moved closer and closer to the vents, it could be accomplished over many generations, allowing ample time to develop the requisite multi-COG DNA machinery. Such a *hyperthermal adaptation* incentive for the invention of DNA genes could play a pivotal role in the rise of LUCA.

15.7 Hot-Cross Scenario

(i) Thermal Habitats

Present day organisms are divided into four thermal classes according to their optimal growth temperature neighborhoods:[62] psychrophiles 4°C, mesophiles 39°C, thermophiles 60°C and hyperthermophiles 90°C or higher. The formation of the first cell too would occur in one of these four thermal ranges.

Ever since the discovery of living communities at the submarine hydrothermal vents, these vents have been proposed as sites for the origin of life.[63] In the *hyperthermophilic origin* (HYO) scenario, the geothermal energy released at these vents could provide a hospitable environment for prebiotic evolution. The abundant carbon dioxide and hydrogen available at the vents, exergonic carbon fixation at vent temperatures[64] and prebiotic syntheses of a range of organic compounds under vent-like conditions (Section 1.3) supply attractive conditions for an autotrophic HYO. However, there are forbidding drawbacks with elevated temperature at the start of prebiotic evolution: thermal instability of essential biomolecules,[65,66] thermal melting of short base-paired RNA or RNA-like oligomers in the absence of any developed replicases and lack of thermal protection of labile intermediates through metabolite channeling by evolved biocatalysts.

To ameliorate the difficulties of hyperthermophilic temperatures and still retain the biosynthetic advantages of the vents, it has been suggested that life might have originated not at the vents themselves, but close to the vents near 90°C.[67,68] While such a *thermophilic origin* (THO) could bring about a more favorable balance between autotrophic potential and the perils of a hot start, the melting temperature for the binding of a DNA 20-mer to template DNA is only about 50°C. So template-directed incorporation of monomers into RNA or RNA-like oligomers shorter than 20-mer in length would be difficult to initiate under either HYO or THO conditions. It is therefore not surprising that the triple convergence of evidence from genetic code structure, atmospheric amino acid synthesis and meteoritic amino acids relating to prebiotically available amino acids points to a heterotrophic instead of an autotrophic origin of life (Section 1.4).

This leaves a *mesophilic origin* (MEO) or *psychrophilic origin* (PSO) for the first living cell. PSO is more favorable for purine synthesis and RNA polymerization, but all chemical reactions are slower at PSO compared to MEO temperatures. Basically both scenarios are compatible with a heterotrophic origin. The disadvantage of heterotrophy has long been pointed out by Thomas Malthus in An Essay on the Principle of Population, which applies to preLucan life as much as postLucan life: all species including humans reproduce at unsustainably high rates and lacking sufficient food to go around, "Necessity, that imperious all pervading law of nature, restrains them within the prescribed bounds ... among plants and animals its effects are waste of seed, sickness and premature death."[69,70] Pre-Lucan life, like present day life, tended to spread to new ecological niches. It is known that when genetic coding began at the start of the DNA-Protein World about 3.6 billion years ago (Fig. 14.1), the amino acids initially encoded by the genetic code were the Phase 1 amino acids obtainable from the environment (Section 1.4), which suggests that these amino acids were still available from the environment at that stage. However, exponential multiplication of preLucan life sooner or later would threaten to outstrip linear accumulation of environmental organics and any prospect of an impending carbon-fixation crisis would only accelerate the spread to novel niches.

The long term solution to a carbon fixation crisis is autotrophy. To-day, for carbon fixation the living world relies above all on the autotrophic systems of light-dependent photosynthesis and to a much lesser extent hydrogen-dependent methanogenesis/acetogenesis. PreLucan life would be pressed to explore such systems.

(ii) Life at the Vents

Everything organisms do is of great interest to biology, but what they do not do can be just as interesting. The Archaea are experts in autotrophy and carry out carbon fixation through methanogenesis, acetogenesis, reverse citric acid cycle as in green sulfur bacteria, the Calvin cycle, the most widespread autotrophic pathway in Bacteria and Eukarya, using RubisCO and a hydroxypropionate/hydroxybutyrate pathway.[62,71-73] The halophiles have also gone as far as ATP synthesis coupled to photactivated proton pumping with bacteriorhodopsin,[74,75] However, the Archaea apparently have either not or rarely developed full-fledged anoxygenic or oxygenic photosynthesis during the past three billion years. Bacteria use ester-lipids, whereas Archaea use ether-lipids. In photosynthesis, the reducing power of NADH (or NADPH) formed from reduction of NAD^+ (or $NADP^+$) at about -0.32 volts electron potential is needed to reduce CO_2.[62] This negative potential is generated by photosystems in plants and by reversed electron flow in purple bacteria. It would be useful to determine if there might be any inherent difficulty with ether-lipid membranes in handling this level of negative potential, or any other archaeal features not readily compatible with photosynthesis.

Given the aversion of Archaea toward the development of photosynthetic carbon fixation for whatever reasons and the greater competitiveness of methanogenesis over acetogenesis in environments outside the termite hindgut, methanogenesis represents the expected safety exit for the archaea-like preLucans in coping with any approaching carbon fixation crisis. These organisms would at first spread into the vicinity of the hydrothermal vents to feed on the abiotically synthesized organic compounds available there. As competition among the feeders turned fierce, they would have to outdo one another by getting ever closer to the vents. In the process they had to make their proteins more heat-resistant. In addition, as

they neared the vents, the greater heat resistance of DNA relative to RNA became a decisive selective advantage and the preLucans that perfected a DNA genome would claim the vents. Large asteroid impacts on early Earth posed a threat to the primordial biota. It is estimated that the impact of a 440-km diameter (1.3×10^{20} kg) projectile about the size of Vesta and Pallas might bring the oceans to a boil or near boil.[76] Calamities of this kind would further accelerate the DNA-genome powered migration to the vents. At the vents, the DNA-genomed preLucans developed methanogenesis, flourished on the plentiful CO_2 and H_2, produced massive amounts of methane to ward off an ice age even under a faint sun,[77] finalized the 20-amino acid universal genetic code and gave rise to an archaeal hyperthermophilic *Methanopyrus*-like LUCA.

A long standing puzzle pertains to the question of how any single LUCA lineage could possibly impose its genetic code on all life as the universal code, eliminating in the process all other lineages bearing alternative codes.[78] Since there is limited direct competition at any time between organisms inhabiting different temperature zones, how could a hyperthermophile bearing the eventual universal genetic code eliminate any psychrophile bearing an alternative genetic code or vice versa? The development of a DNA-genomed LUCA at the vents suggests a plausible answer. When heterotrophic offsprings of LUCA spread from the hydrothermal vents to the cooler zones, they came out armed with the DNA genome and the universal genetic code. The heterotrophic competitors that crossed their paths in these zones would be equipped with outdated genetic codes and an error-prone RNA genome. The result was total mismatch and elimination of all the heterotrophic competitors. Just as Latin was established over a large usage domain through military conquest by the Roman Empire in the short span of a few centuries, the universality of the 4-letter DNA and 20-letter protein languages of present day life likely arose thanks to the superior biochemical weaponry of LUCA perfected at the vents.

This three-stage *Hot-Cross Origin* (HCO) scenario for the development of modern life—first, crossing of preLucan organisms from cooler zones into the hydrothermal vents; secondly, rise of a hyperthermophilic methanogenic LUCA at the vents; and thirdly, recrossing of LUCA's heterotrophic offsprings back to the cooler zones to establish their dominions throughout all ecological niches on Earth, underlines the importance of thermal adaptations associated with the rise of LUCA. HCO reconciles a psychrophilic-mesophilic hetrotrophic first cell with a hyperthermophilic methanogenic LUCA inhabiting the hydrothermal vents; gives a strong enough evolutionary incentive for the switch from RNA genome to DNA genome; and explains how canonical molecular biological devices might have achieved universality in a DNA-Protein World that has lasted to this day. Migratory birds crisscross continents in search for food, the salmon crosses from ocean to mountain stream to reproduce, *Methanopyrus* spreads out over submarine hydrothermal vents separated oceans apart and LUCA's forebears and offsprings crossed temperature zones in opposite directions—no distance is too great or journey too arduous for organisms in their hunt for livelihood.

15.8 Conclusion

The origin of life began with the first stirrings of membranes, metabolites and replicators and culminated in the emergence of LUCA. The search for LUCA is an almost unattainable goal for biology for two reasons. First, because of the transition from an oxygen-poor to oxygen-rich atmosphere in the history of the Earth, with added environmental fluctuations such as the coming and passing of ice ages, almost all the ecological niches on the planet have undergone profound alterations through the ages. Their inhabitant organisms have no choice but to adapt continuously and in the process lose much of their primordial characteristics. Yet it would be impossible to uncover the nature of LUCA without the persistence of sufficient primordial characteristics in some genomes. Consequently it is fortunate that the submarine hydrothermal vents form a continually well preserved niche on Earth where gaseous composition, temperature and pressure can remain largely constant throughout biological history. This constancy of the vents has allowed ultra-conservative inhabitants like *Methanopyrus* to evolve at an exceptionally slow pace and in so doing retain primordial characters that shed useful light on the nature of LUCA.

Secondly, because of the lack of DNA or ribosomal RNA paralogs inside genomes, these molecules could not be used to search for LUCA. Protein sequences on the other hand are too fast evolving and susceptible to artifacts to be reliable for the task. This leaves only the tRNAs with hopefully adequate information content for the search. Surprisingly, it turns out that tRNA evolution marches to a slow and steady tempo, with some tRNA paralogs diverging minimally over three billion years. This tempo is tailor-made for the search of LUCA and the tRNAs of the ultra-conservative genomes have provided the most reliable guides in this search. Together with evidence from missing genes, ancestral proteins, metabolism and geochemistry, they have made possible the polyphasic identification of a *Methanopyrus*-proximal LUCA.

The developments of the Bacteria and Eukarya domains from their archaeal base were likely to be accompanied right from the start by substantial innovations. For example, the strategic use of multiple anticodon combinations to read different codon boxes by Bacteria and Eukarya may enhance finer adaptations of the translational machinery to changing environments. Likewise, the nuclear membrane of Eukarya enables them to work with multiple chromosomes and large genomes. Since *Methanosarcina acetivorans* (Mac) has the largest genome of 5.75 Mb among all known archaeons and it is also the only archaeon known to enter into an optional multicellular state,[79,80] the correlation suggests that a minimum genome size close to 5.75 Mb might be required for multicellularity. Thus the nuclear membrane invention of Eukarya, making possible genomes far larger than 5.75 Mb, paves the way to the magnificence of multicellular life. The location of LUCA within the Archaea furnishes a starting point for delineating the origins of LBACA and LECA. Here too, the tRNA sequences

make valuable probes. When all the eukaryotic tRNA sequences were compared pairwise with all the bacterial tRNA sequences, only eight out of the 0.98 million cross domain pairs were found to be ultrahi-sim with a genetic distance less than 0.10 and all eight of them paired an Ath tRNA with a cyanobacterial tRNA. These results established not only the migration of some cyanobacteria-derived tRNA genes from plant chloroplast to nucleus, but also the precision of hi-sim and ultrahi-sim tRNA pairs for hunting down primordial kinships between tRNAs from different biological domains. Use of this method to probe Archaea-Bacteria and Eukarya-Archaea relationships has produced hi-sim and ultrahi-sim tRNA pairs that point to a *Thermotoga*-proximal LBACA stemming from an *Aeropyrum*-related crenarchaeon and a *Plasmodium*-proximal LECA stemming from endosymbiosis of a *Ferroplasma*-related archaeal host with a *Rickettsia* and possibly another bacterium.

Because Malthusian population dynamics severely constrain the long term sustainability of heterotrophy in the absence of any autotrophic support much as the plant world supports the animal world to-day, autotrophy was destined to happen sooner or later among the early microbial heterotrophs. Accordingly, given the aversion of archaea-like preLucans to photosynthesis, the rise of methanogenesis and a methanogenic LUCA at the hydrothermal vents was practically a foregone conclusion. In addition, the road leading from psychrophilic-mesophilic first living cells to a hyperthermophilic LUCA and the subsequent spread of LUCA's offsprings back to all thermal niches on Earth must have followed trails that cut across temperature zones, as described by the Hot-Cross Origin or comparable scenarios. The biochemistry of thermal adaptations is therefore a cornerstone of primordial biology at the LUCA stage.

As astrobiological explorations gain momentum targeting at intrasolar and extrasolar planets and moons, prebiological evolution on Earth will continue to serve as a valuable model. In view of the aqueous base of terrestrial life and the identification of a methanogenic LUCA, water and methane will be among the key molecules to look for anywhere in the universe. Undoutedly prebiotic evolution and astrobiology will enrich one another even more in the future. So much awaits investigation and discovery by the prebiotic and astrobiological scientists of tomorrow.

Appendix 15.1. Clusters of Orthologous Groups (COG) in the LUCA Genome.[48] *COG Nos. that are present in all of the ancient eight are shown in bold font; those present in all of the ancient six but not in all of the ancient eight are shown in nonbold font; those not included in any of the minimal gene sets I-VI are shown in italics; the underlined COGs are common to the methanogenesis pathways of Mka, Mth and Mja*

Group	No.	COG
(J) Translation, ribosomal structure and biogenesis	**8**	Glutamyl-tRNA synthetase
	9	Putative translation factor (SUA5)
	12	Predicted GTPase, probable translation factor
	13	Alanyl-tRNA synthetase
	16	Phenylalanyl-tRNA synthetase alpha subunit
	17	Aspartyl-tRNA synthetase
	18	Arginyl-tRNA synthetase
	23	Translation initiation factor 1 (eIF-1/SUI1) and related proteins
	24	Methionine aminopeptidase
	30	Dimethyladenosine transferase (rRNA methylation)
	48	Ribosomal protein S12
	49	Ribosomal protein S7
	51	Ribosomal protein S10
	52	Ribosomal protein S2
	60	Isoleucyl-tRNA synthetase
	72	Phenylalanyl-tRNA synthetase beta subunit
	80	Ribosomal protein L11
	81	Ribosomal protein L1
	87	Ribosomal protein L3
	88	Ribosomal protein L4
	89	Ribosomal protein L23
	90	Ribosomal protein L2
	91	Ribosomal protein L22
	92	Ribosomal protein S3
	93	Ribosomal protein L14
	94	Ribosomal protein L5
	96	Ribosomal protein S8
	97	Ribosomal protein L6P/L9E
	98	Ribosomal protein S5
	99	Ribosomal protein S13
	100	Ribosomal protein S11
	101	Pseudouridylate synthase
	102	Ribosomal protein L13
	103	Ribosomal protein S9
	124	Histidyl-tRNA synthetase
	130	Pseudouridine synthase
	143	Methionyl-tRNA synthetase
	162	Tyrosyl-tRNA synthetase
	172	Seryl-tRNA synthetase
	180	Tryptophanyl-tRNA synthetase
	182	Predicted translation initiation factor 2B subunit, eIF-2B $\alpha/\beta/\delta$ family
	184	Ribosomal protein S15P/S13E
	185	Ribosomal protein S19

continued on next page

Appendix 15.1. Continued

Group	No.	COG
	186	Ribosomal protein S17
	197	Ribosomal protein L16/L10E
	198	Ribosomal protein L24
	199	Ribosomal protein S14
	200	Ribosomal protein L15
	231	Translation elongation factor P (EF-P)/translation initiation factor 5A
	244	Ribosomal protein L10
	252	L-asparaginase/archaeal Glu-tRNAGln amidotransferase subunit D
	255	Ribosomal protein L29
	256	Ribosomal protein L18
	343	Queuine/archaeosine tRNA-ribosyltransferase
	361	Translation initiation factor 1 (IF-1)
	423	Glycyl-tRNA synthetase (class II)
	441	Threonyl-tRNA synthetase
	442	Prolyl-tRNA synthetase
	480	Translation elongation factors (GTPases)
	495	Leucyl-tRNA synthetase
	522	Ribosomal protein S4 and related proteins
	525	Valyl-tRNA synthetase
	532	Translation initiation factor 2 (IF-2; GTPase)
	621	2-Methylthioadenine synthetase
	1093	Translation initiation factor 2, alpha subunit (eIF-2alpha)
	1258	Predicted pseudouridylate synthase
	1325	Predicted exosome subunit
	1358	Ribosomal protein HS6-type (S12/L30/L7a)
	1369	RNase P/RNase MRP subunit POP5
	1383	Ribosomal protein S17E
	1384	Lysyl-tRNA synthetase (class I)
	1471	Ribosomal protein S4E
	1491	Predicted RNA-binding protein
	1498	Protein implicated in ribosomal biogenesis Nop56p homolog
	1499	NMD protein affecting ribosome stability and mRNA decay
	1500	Predicted exosome subunit
	1503	Peptide chain release factor 1 (eRF1)
	1514	2'-5' RNA ligase
	1534	Predicted RNA-binding protein containing KH domain possibly ribosomal protein
	1549	Queuine tRNA-ribosyltransferases, contain PUA domain
	1552	Ribosomal protein L40E
	1588	RNase P/RNase MRP subunit p29
	1601	Translation initiation factor 2, beta subunit (eIF-2beta)/eIF-5 N-terminal domain
	1603	RNase P/RNase MRP subunit p30
	1631	Ribosomal protein L44E
	1632	Ribosomal protein L15E
	1676	tRNA splicing endonuclease
	1717	Ribosomal protein L32E
	1727	Ribosomal protein L18E

continued on next page

Appendix 15.1. Continued

Group	No.	COG
	1736	Diphthamide synthase subunit DPH2
	1746	tRNA nucleotidyltransferase (CCA-adding enzyme)
	1798	Diphthamide biosynthesis methyltransferase
	1841	Ribosomal protein L30/L7E
	1867	Dimethylguanosine tRNA methyltransferase
	1889	Fibrillarin-like rRNA methylase
	1890	Ribosomal protein S3AE
	1911	Ribosomal protein L30E
	1976	Translation initiation factor 6 (eIF-6)
	1997	Ribosomal protein L37AE/L43A
	1998	Ribosomal protein S27AE
	2004	Ribosomal protein S24E
	2007	Ribosomal protein S8E
	2016	Predicted RNA-binding protein (contains PUA domain)
	2023	RNase P subunit RPR2
	2051	Ribosomal protein S27E
	2053	Ribosomal protein S28E/S33
	2058	Ribosomal protein L12E/L44/L45/RPP1/RPP2
	2075	Ribosomal protein L24E
	2092	Translation elongation factor EF-1beta
	2097	Ribosomal protein L31E
	2125	Ribosomal protein S6E (S10)
	2126	Ribosomal protein L37E
	2139	Ribosomal protein L21E
	2147	Ribosomal protein L19E
	2157	Ribosomal protein L20A (L18A)
	2163	Ribosomal protein L14E/L6E/L27E
	2167	Ribosomal protein L39E
	2174	Ribosomal protein L34E
	2238	Ribosomal protein S19E (S16A)
	2260	Predicted Zn-ribbon RNA-binding protein
	2263	Predicted RNA methylase
	2511	Archaeal Glu-tRNAGln amidotransferase subunit E (contains GAD domain)
	2519	tRNA(1-methyladenosine) methyltransferase and related methyltransferases
	2888	Predicted Zn-ribbon RNA-binding protein with a function in translation
	2890	Methylase of polypeptide chain release factors
	3277	RNA-binding protein involved in rRNA processing
	5256	Translation elongation factor EF-1alpha (GTPase)
	5257	Translation initiation factor 2, gamma subunit (eIF-2gamma; GTPase)
(A) RNA processing and modification	*430*	RNA 3'-terminal phosphate cyclase
	2136	Predicted exosome subunit/U3 small nucleolar ribonucleoprotein (snoRNP) component, contains IMP4 domain
(K) Transcription	**85**	DNA-directed RNA polymerase, beta subunit/140 kD subunit
	86	DNA-directed RNA polymerase, beta' subunit/160 kD subunit
	195	Transcription elongation factor
	202	DNA-directed RNA polymerase, alpha subunit/40 kD subunit

continued on next page

Appendix 15.1. Continued

Group	No.	COG
	250	Transcription antiterminator
	640	Predicted transcriptional regulators
	864	Predicted transcriptional regulators containing the CopG/Arc/MetJ DNA-binding domain and a metal-binding domain
	1095	DNA-directed RNA polymerase, subunit E'
	1243	Histone acetyltransferase
	1293	Predicted RNA-binding protein homologous to eukaryotic snRNP
	1308	Transcription factor homologous to NACalpha-BTF3
	1321	Mn-dependent transcriptional regulator
	1378	Predicted transcriptional regulators
	1395	Predicted transcriptional regulator
	1405	Transcription initiation factor TFIIIB, Brf1 subunit
	1522	Transcriptional regulators
	1581	Archaeal DNA-binding protein
	1644	DNA-directed RNA polymerase, subunit N (RpoN/RPB10)
	1675	Transcription initiation factor IIE, alpha subunit
	1758	DNA-directed RNA polymerase, subunit K/omega
	1761	DNA-directed RNA polymerase, subunit L
	1813	Predicted transcription factor, homolog of eukaryotic MBF1
	1846	Transcriptional regulators
	1996	DNA-directed RNA polymerase, subunit RPC10 (contains C4-type Zn-finger)
	2012	DNA-directed RNA polymerase, subunit H, RpoH/RPB5
	2093	DNA-directed RNA polymerase, subunit E''
	2101	TATA-box binding protein (TBP), component of TFIID and TFIIIB
(L) Replication, recombination and repair	**84**	Mg-dependent DNase
	164	Ribonuclease HII
	177	Predicted EndoIII-related endonuclease
	258	5'-3' exonuclease (including N-terminal domain of PolI)
	270	Site-specific DNA methylase
	350	Methylated DNA-protein cysteine methyltransferase
	358	DNA primase (bacterial type)
	417	DNA polymerase elongation subunit (family B)
	419	ATPase involved in DNA repair
	420	DNA repair exonuclease
	468	RecA/RadA recombinase
	470	ATPase involved in DNA replication
	550	Topoisomerase IA
	592	DNA polymerase sliding clamp subunit (PCNA homolog)
	608	Single-stranded DNA-specific exonuclease
	1041	Predicted DNA modification methylase
	1107	Archaea-specific RecJ-like exonuclease, contains DnaJ-type Zn finger domain
	1111	ERCC4-like helicases
	1112	Superfamily I DNA and RNA helicases and helicase subunits
	1241	Predicted ATPase involved in replication control, Cdc46/Mcm family
	1311	Archaeal DNA polymerase II, small subunit/DNA polymerase delta, subunit B
	1389	DNA topoisomerase VI, subunit B

continued on next page

Appendix 15.1. Continued

Group	No.	COG
	1423	ATP-dependent DNA ligase, homolog of eukaryotic ligase III
	1467	Eukaryotic-type DNA primase, catalytic (small) subunit
	1525	Micrococcal nuclease (thermonuclease) homologs
	1591	Holliday junction resolvase—archaeal type
	1599	Single-stranded DNA-binding replication protein A (RPA), large (70 kD) subunit and related ssDNA-binding proteins
	1637	Predicted nuclease of the RecB family
	1697	DNA topoisomerase VI, subunit A
	1793	ATP-dependent DNA ligase
	1933	Archaeal DNA polymerase II, large subunit
	1948	ERCC4-type nuclease
	2219	Eukaryotic-type DNA primase, large subunit
(B) Chromatin structure and dynamics	**123**	Deacetylases, including yeast histone deacetylase and acetoin utilization protein
	2036	Histones H3 and H4
(D) Cell cycle control, cell division, chromosome partitioning	**37**	Predicted ATPase of the PP-loop superfamily implicated in cell cycle control
	206	Cell division GTPase
	455	ATPases involved in chromosome partitioning
	489	ATPases involved in chromosome partitioning
	1192	ATPases involved in chromosome partitioning
	1718	Serine/threonine protein kinase involved in cell cycle control
(T) Signal transduction mechanisms	**467**	RecA-superfamily ATPases implicated in signal transduction
	589	Universal stress protein UspA and related nucleotide-binding proteins
	3642	Mn2+-dependent serine/threonine protein kinase
(M) Cell wall/membrane/envelope biogenesis	**438**	Glycosyltransferase
	449	Glucosamine 6-phosphate synthetase
	451	Nucleoside-diphosphate-sugar epimerase
	463	Glycosyltransferases involved in cell wall biogenesis
	472	UDP-N-acetylmuramyl pentapeptide phosphotransferase/UDP-N-acetylglucosamine-1-phosphate transferase
	668	Small-conductance mechanosensitive channel
	750	Predicted membrane-associated Zn-dependent proteases 1
	794	Predicted sugar phosphate isomerase involved in capsule formation
	1208	Nucleoside-diphosphate-sugar pyrophosphorylase
(U) Intracellular trafficking, secretion and vesicular transport	**201**	Preprotein translocase subunit SecY
	341	Preprotein translocase subunit SecF
	342	Preprotein translocase subunit SecD
	541	Signal recognition particle GTPase
	552	Signal recognition particle GTPase
	681	Signal peptidase I
	1400	Signal recognition particle 19 kDa protein
	1989	Type II secretory pathway, prepilin signal peptidase PulO
	2064	Flp pilus assembly protein TadC
	2443	Preprotein translocase subunit Sss1
	4962	Flp pilus assembly protein, ATPase CpaF

continued on next page

Appendix 15.1. Continued

Group	No.	COG
(O) Posttranslational modification, protein turnover, chaperones	68	Hydrogenase maturation factor
	71	Molecular chaperone (small heat shock protein)
	298	Hydrogenase maturation factor
	309	Hydrogenase maturation factor
	330	Membrane protease subunits, stomatin/prohibitin homologs
	396	ABC-type transport system involved in Fe-S cluster assembly, ATPase component
	409	Hydrogenase maturation factor
	459	Chaperonin GroEL (HSP60 family)
	464	ATPases of the AAA+ class
	492	Thioredoxin reductase
	501	Zn-dependent protease with chaperone function
	533	Metal-dependent proteases with possible chaperone activity
	555	ABC-type sulfate transport system, permease component
	602	Organic radical activating enzymes
	638	20S proteasome, alpha and beta subunits
	1047	FKBP-type peptidyl-prolyl cis-trans isomerases 2
	1067	Predicted ATP-dependent protease
	1180	Pyruvate-formate lyase-activating enzyme
	1222	ATP-dependent 26S proteasome regulatory subunit
	1370	Prefoldin, molecular chaperone implicated in de novo protein folding, alpha subunit
	1382	Prefoldin, chaperonin cofactor
	1730	Predicted prefoldin, molecular chaperone implicated in de novo protein folding
	1899	Deoxyhypusine synthase
	2518	Protein-L-isoaspartate carboxylmethyltransferase
(C) Energy production and conversion	39	Malate/lactate dehydrogenases
	45	Succinyl-CoA synthetase, beta subunit
	74	Succinyl-CoA synthetase, alpha subunit
	243	Anaerobic dehydrogenases, typically selenocysteine-containing
	247	Fe-S oxidoreductase
	371	Glycerol dehydrogenase and related enzymes
	473	Isocitrate/isopropylmalate dehydrogenase
	479	Succinate dehydrogenase/fumarate reductase, Fe-S protein subunit
	543	2-Polyprenylphenol hydroxylase and related flavodoxin oxidoreductases
	636	F0F1-type ATP synthase, subunit c/Archaeal/vacuolar-type H+-ATPase, subunit K
	644	Dehydrogenases (flavoproteins)
	650	Formate hydrogenlyase subunit 4
	674	Pyruvate:ferredoxin oxidoreductase and related 2-oxoacid:ferredoxin oxidoreductases, alpha subunit
	680	Ni,Fe-hydrogenase maturation factor
	716	Flavodoxins
	731	Fe-S oxidoreductases
	778	Nitroreductase
	1012	NAD-dependent aldehyde dehydrogenases
	1013	Pyruvate:ferredoxin oxidoreductase and related 2-oxoacid:ferredoxin oxidoreductases, beta subunit
	1014	Pyruvate:ferredoxin oxidoreductase and related 2-oxoacid:ferredoxin oxidoreductases, gamma subunit

continued on next page

Appendix 15.1. Continued

Group	No.	COG
	1029	Formylmethanofuran dehydrogenase subunit B
	1032	Fe-S oxidoreductase
	1035	Coenzyme F420-reducing hydrogenase, beta subunit
	1036	Archaeal flavoproteins
	1053	Succinate dehydrogenase/fumarate reductase, flavoprotein subunit
	1142	Fe-S-cluster-containing hydrogenase components 2
	1144	Pyruvate:ferredoxin oxidoreductase and related 2-oxoacid:ferredoxin oxidoreductases, delta subunit
	1145	Ferredoxin
	1146	Ferredoxin
	1148	Heterodisulfide reductase, subunit A and related polyferredoxins
	1149	MinD superfamily P-loop ATPase containing an inserted ferredoxin domain
	1150	Heterodisulfide reductase, subunit C
	1151	6Fe-6S prismane cluster-containing protein
	1152	CO dehydrogenase/acetyl-CoA synthase alpha subunit
	1153	Formylmethanofuran dehydrogenase subunit D
	1155	Archaeal/vacuolar-type H+-ATPase subunit A
	1156	Archaeal/vacuolar-type H+-ATPase subunit B
	1229	Formylmethanofuran dehydrogenase subunit A
	1249	Pyruvate/2-oxoglutarate dehydrogenase complex, dihydrolipoamide dehydrogenase (E3) component and related enzymes
	1269	Archaeal/vacuolar-type H+–ATPase subunit I
	1304	L-lactate dehydrogenase (FMN-dependent) and related alpha-hydroxy acid dehydrogenases
	1390	Archaeal/vacuolar-type H+-ATPase subunit E
	1394	Archaeal/vacuolar-type H+-ATPase subunit D
	1436	Archaeal/vacuolar-type H+-ATPase subunit F
	1456	CO dehydrogenase/acetyl-CoA synthase gamma subunit (corrinoid Fe-S protein)
	1527	Archaeal/vacuolar-type H+-ATPase subunit C
	1592	Rubrerythrin
	1614	CO dehydrogenase/acetyl-CoA synthase beta subunit
	1625	Fe-S oxidoreductase, related to NifB/MoaA family
	1819	Glycosyl transferases, related to UDP-glucuronosyltransferase
	1880	CO dehydrogenase/acetyl-CoA synthase epsilon subunit
	1838	Tartrate dehydratase beta subunit/Fumarate hydratase class I, C-terminal domain
	1908	Coenzyme F420-reducing hydrogenase, delta subunit
	1927	Coenzyme F420-dependent N(5), N(10)-methenyltetrahydromethanopterin dehydrogenase
	1941	Coenzyme F420-reducing hydrogenase, gamma subunit
	1951	Tartrate dehydratase alpha subunit/Fumarate hydratase class I, N-terminal domain
	2037	Formylmethanofuran:tetrahydromethanopterin formyltransferase
	2048	Heterodisulfide reductase, subunit B
	2055	Malate/L-lactate dehydrogenases
	2069	CO dehydrogenase/acetyl-CoA synthase delta subunit (corrinoid Fe-S protein)
	2141	Coenzyme F420-dependent N5,N10-methylene tetrahydromethanopterin reductase and related flavin-dependent oxidoreductases

continued on next page

Appendix 15.1. Continued

Group	No.	COG
	2218	Formylmethanofuran dehydrogenase subunit C
	2221	Dissimilatory sulfite reductase (desulfoviridin), alpha and beta subunits
	2710	Nitrogenase molybdenum-iron protein, alpha and beta chains
	3259	Coenzyme F420-reducing hydrogenase, alpha subunit
	3260	Ni,Fe-hydrogenase III small subunit
	3261	Ni,Fe-hydrogenase III large subunit
	4074	H2-forming N5,N10-methylenetetrahydromethanopterin dehydrogenase
(G) Carbohydrate transport and metabolism	57	Glyceraldehyde-3-phosphate dehydrogenase/erythrose-4-phosphate dehydrogenase
	61	Predicted sugar kinase
	63	H2-forming N5,N10-methylenetetrahydromethanopterin dehydrogenase
	120	Ribose 5-phosphate isomerase
	126	3-Phosphoglycerate kinase
	148	Enolase
	149	Triosephosphate isomerase
	235	Ribulose-5-phosphate 4-epimerase and related epimerases and aldolases
	269	3-Hexulose-6-phosphate synthase and related proteins
	483	Archaeal fructose-1,6-bisphosphatase and related enzymes of inositol monophosphatase family
	524	Sugar kinases, ribokinase family
	574	Phosphoenolpyruvate synthase/pyruvate phosphate dikinase
	662	Mannose-6-phosphate isomerase
	1082	Sugar phosphate isomerases/epimerases
	1109	Phosphomannomutase
	1363	Cellulase M and related proteins
	1830	DhnA-type fructose-1,6-bisphosphate aldolase and related enzymes
	1980	Archaeal fructose 1,6-bisphosphatase
	2074	2-phosphoglycerate kinase
	3635	Predicted phosphoglycerate mutase, AP superfamily
	3839	ABC-type sugar transport systems, ATPase components
(E) Amino acid transport and metabolism	2	Acetylglutamate semialdehyde dehydrogenase
	6	Xaa-Pro aminopeptidase
	10	Arginase/agmatinase/formimionoglutamate hydrolase, arginase family
	65	3-Isopropylmalate dehydratase large subunit
	66	3-Isopropylmalate dehydratase small subunit
	75	Serine-pyruvate aminotransferase/archaeal aspartate aminotransferase
	76	Glutamate decarboxylase and related PLP-dependent proteins
	78	Ornithine carbamoyltransferase
	79	Histidinol-phosphate/aromatic aminotransferase and cobyric acid decarboxylase
	111	Phosphoglycerate dehydrogenase and related dehydrogenases
	112	Glycine/serine hydroxymethyltransferase
	119	Isopropylmalate/homocitrate/citramalate synthases
	136	Aspartate-semialdehyde dehydrogenase
	174	Glutamine synthetase
	289	Dihydrodipicolinate reductase
	329	Dihydrodipicolinate synthase/N-acetylneuraminate lyase

continued on next page

Appendix 15.1. Continued

Group	No.	COG
	367	Asparagine synthase (glutamine-hydrolyzing)
	436	Aspartate/tyrosine/aromatic aminotransferase
	440	Acetolactate synthase, small (regulatory) subunit
	460	Homoserine dehydrogenase
	498	Threonine synthase
	527	Aspartokinases
	548	Acetylglutamate kinase
	560	Phosphoserine phosphatase
	620	Methionine synthase II (cobalamin-independent)
	1387	Histidinol phosphatase and related hydrolases of the PHP family
	1812	Archaeal S-adenosylmethionine synthetase
	4992	Ornithine/acetylornithine aminotransferase
(F) Nucleotide transport and metabolism	*5*	Purine nucleoside phosphorylase
	15	Adenylosuccinate lyase
	34	Glutamine phosphoribosyl-pyrophosphate amidotransferase
	41	Phosphoribosyl-carboxy-aminoimidazole mutase
	44	Dihydroorotase and related cyclic amidohydrolases
	46	Phosphoribosylformyl glycinamide synthase, synthetase domain
	47	Phosphoribosylformylglycinamide synthase, glutamine amidotransferase domain
	104	Adenylosuccinate synthase
	105	Nucleoside diphosphate kinase
	125	Thymidylate kinase
	127	Xanthosine triphosphate pyrophosphatase
	150	Phosphoribosylaminoimidazole synthetase
	151	Phosphoribosylamine-glycine ligase
	152	Phosphoribosyl-aminoimidazole-succinocarboxamide synthase
	167	Dihydroorotate dehydrogenase
	284	Orotidine-5′-phosphate decarboxylase
	402	Cytosine deaminase and related metal-dependent hydrolases
	461	Orotate phosphoribosyltransferase
	462	Phosphoribosylpyrophosphate synthetase
	503	Adenine/guanine phosphoribosyltransferases and related PRPP-binding proteins
	504	CTP synthase
	516	IMP dehydrogenase/GMP reductase
	518	GMP synthase—Glutamine amidotransferase domain
	519	GMP synthase, PP-ATPase domain/subunit
	528	Uridylate kinase
	540	Aspartate carbamoyltransferase, catalytic chain
	717	Deoxycytidine deaminase
	1051	ADP-ribose pyrophosphatase
	1102	Cytidylate kinase
	1328	Oxygen-sensitive ribonucleoside-triphosphate reductase
	1437	Adenylate cyclase, class 2 (thermophilic)
	1618	Predicted nucleotide kinase
	1781	Aspartate carbamoyltransferase, regulatory subunit
	1828	Phosphoribosyl-formylglycinamide synthase, PurS component

continued on next page

Appendix 15.1. Continued

Group	No.	COG
	1936	Predicted nucleotide kinase (related to CMP and AMP kinases)
	2019	Archaeal adenylate kinase
(H) Coenzyme transport and metabolism	*43*	3-Polyprenyl-4-hydroxybenzoate decarboxylase and related decarboxylases
	142	Geranylgeranyl pyrophosphate synthase
	157	Nicotinate-nucleotide pyrophosphorylase
	163	3-Polyprenyl-4-hydroxybenzoate decarboxylase
	171	NAD synthase
	214	Pyridoxine biosynthesis enzyme
	237	Dephospho-CoA kinase
	294	Dihydropteroate synthase and related enzymes
	301	Thiamine biosynthesis ATP pyrophosphatase
	303	Molybdopterin biosynthesis enzyme
	311	Predicted glutamine amidotransferase involved in pyridoxine biosynthesis
	315	Molybdenum cofactor biosynthesis enzyme
	351	Hydroxymethylpyrimidine/phosphomethylpyrimidine kinase
	368	Cobalamin-5-phosphate synthase
	379	Quinolinate synthase
	382	4-Hydroxybenzoate polyprenyltransferase and related prenyltransferases
	452	Phosphopantothenoylcysteine synthetase/decarboxylase
	499	S-adenosylhomocysteine hydrolase
	521	Molybdopterin biosynthesis enzymes
	611	Thiamine monophosphate kinase
	720	6-Pyruvoyl-tetrahydropterin synthase
	746	Molybdopterin-guanine dinucleotide biosynthesis protein A
	1270	Cobalamin biosynthesis protein CobD/CbiB
	1339	Transcriptional regulator of a riboflavin/FAD biosynthetic operon
	1635	Flavoprotein involved in thiazole biosynthesis
	1763	Molybdopterin-guanine dinucleotide biosynthesis protein
	1767	Triphosphoribosyl-dephospho-CoA synthetase
	2038	NaMN:DMB phosphoribosyltransferase
	2266	GTP:adenosylcobinamide-phosphate guanylyltransferase
	2896	Molybdenum cofactor biosynthesis enzyme
(I) Lipid transport and metabolism	*20*	Undecaprenyl pyrophosphate synthase
	170	Dolichol kinase
	183	Acetyl-CoA acetyltransferase
	575	CDP-diglyceride synthetase
	615	Cytidylyltransferase
	671	Membrane-associated phospholipid phosphatase
	1257	Hydroxymethylglutaryl-CoA reductase
	1267	Phosphatidylglycerophosphatase A and related proteins
	1577	Mevalonate kinase
	3425	3-Hydroxy-3-methylglutaryl CoA synthase
(P) Inorganic ion transport and metabolism	168	Trk-type K+ transport systems, membrane components
	306	Phosphate/sulphate permeases
	370	Fe2+ transport system protein B
	477	Permeases of the major facilitator superfamily

continued on next page

Appendix 15.1. Continued

Group	No.	COG
	530	Ca2+/Na+ antiporter
	569	K+ transport systems, NAD-binding component
	619	ABC-type cobalt transport system, permease component CbiQ and related transporters
	704	Phosphate uptake regulator
	725	ABC-type molybdate transport system, periplasmic component
	1122	ABC-type cobalt transport system, ATPase component
	1226	3-Hydroxy-3-methylglutaryl CoA synthase
	1918	Fe2+ transport system protein A
(Q) Secondary metabolite biosynthesis, transport and catabolism	**179**	2-Keto-4-pentenoate hydratase/2-oxohepta-3-ene-1,7-dioic acid hydratase(catechol pathway)
	500	SAM-dependent methyltransferases
(R) General function prediction only	**73**	EMAP domain
	312	Predicted Zn-dependent proteases and their inactivated homologs
	375	Zn finger protein HypA/HybF (possibly regulating hydrogenase expression)
	433	Predicted ATPase
	446	Uncharacterized NAD(FAD)-dependent dehydrogenases
	456	Acetyltransferases
	491	Zn-dependent hydrolases, including glyoxylases
	517	FOG: CBS domain
	535	Predicted Fe-S oxidoreductases
	603	Predicted PP-loop superfamily ATPase
	622	Predicted phosphoesterase
	663	Carbonic anhydrases/acetyltransferases, isoleucine patch superfamily
	714	MoxR-like ATPases
	1011	Predicted hydrolase (HAD superfamily)
	1019	Predicted nucleotidyltransferase
	1078	HD superfamily phosphohydrolases
	1084	Predicted GTPase
	1094	Predicted RNA-binding protein (contains KH domains)
	1100	GTPase SAR1 and related small G proteins
	1163	Predicted GTPase
	1201	Lhr-like helicases
	1204	Superfamily II helicase
	1205	Distinct helicase family with a unique C-terminal domain including a metal-binding cysteine cluster
	1234	ATPase components of various ABC-type transport systems, contain duplicated ATPase
	1235	Metal-dependent hydrolases of the beta-lactamase superfamily I
	1237	Metal-dependent hydrolases of the beta-lactamase superfamily II
	1245	Predicted ATPase, RNase L inhibitor (RLI) homolog
	1313	Uncharacterized Fe-S protein PflX, homolog of pyruvate formate lyase activating proteins
	1326	Uncharacterized archaeal Zn-finger protein
	1355	Predicted dioxygenase
	1365	Predicted ATPase (PP-loop superfamily)
	1407	Predicted ICC-like phosphoesterases

continued on next page

Appendix 15.1. Continued

Group	No.	COG
	1412	Uncharacterized proteins of PilT N-term superfamily
	1418	Predicted HD superfamily hydrolase
	1439	Predicted nucleic acid-binding protein, consists of a PIN domain and a Zn-ribbon module
	1458	Predicted DNA-binding protein containing PIN domain
	1537	Predicted RNA-binding proteins
	1545	Predicted nucleic-acid-binding protein containing a Zn-ribbon
	1571	Predicted DNA-binding protein containing a Zn-ribbon domain
	1608	Predicted archaeal kinase
	1646	Predicted phosphate-binding enzymes, TIM-barrel fold
	1759	ATP-utilizing enzymes of ATP-grasp superfamily (probably carboligases)
	1779	C4-type Zn-finger protein
	1782	Predicted metal-dependent RNase, consists of a metallo-beta-lactamase domain and an RNA-binding KH domain
	1818	Predicted RNA-binding protein, contains THUMP domain
	1829	Predicted metal-dependent RNase, consists of a metallo-beta-lactamase domain and an RNA-binding KH domain
	1831	Predicted metal-dependent hydrolase (urease superfamily)
	1855	ATPase (PilT family)
	1907	Predicted archaeal sugar kinases
	1938	Archaeal enzymes of ATP-grasp superfamily
	1964	Predicted Fe-S oxidoreductases
	1988	Predicted membrane-bound metal-dependent hydrolases
	2047	Uncharacterized protein (ATP-grasp superfamily)
	2102	Predicted ATPases of PP-loop superfamily
	2118	DNA-binding protein
	2129	Predicted phosphoesterases, related to the Icc protein
	2151	Predicted metal-sulfur cluster biosynthetic enzyme
	2220	Predicted Zn-dependent hydrolases of the beta-lactamase fold
	2244	Membrane protein involved in the export of O-antigen and teichoic acid
	2520	Predicted methyltransferase
	3269	Predicted RNA-binding protein, contains TRAM domain
(S) Unknown proteins		*11*, **62**, *32, 327, 392,* **432**, *585, 1303,* **1371**, *1379, 1373, 1422, 1430, 1432, 1433,* **1460**, *1469, 1578, 1617, 1627,* **1628**, **1679**, **1690**, *1698,* **1701**, **1720**, *1784,* **1786**, **1814**, *1833, 1844,* **1849**, **1888**, **1909**, *1916,* **1931**, *1945,* **1990**, *1991,* **1992**, *2029, 2034,* **2078**, *2083, 2450, 2454, 2457, 2892*

References

Further Readings

1. Gillis M, Vandamme P, De Vos P et al. Polyphasic taxonomy. In: Boone DR, Castenholz RW, eds. Bergey's Manual of Systematic Bacteriology, 2nd ed. New York:Springer-Verlag, 2001: 43-48.
2. Krane DE, Raymer ML. Fundamental concepts of bioinformatics. Benjamin Cummings, 2003.
3. Salemi M, Vandamme AM, eds. The phylogenetic handbook. Cambrige:Cambridge University Press, 2003.
4. Staley JT, Reysenbach, eds. Biodiversity of microbial life. Wiley-Liss 2002.

Specific References

5. Forterre P, Philippe H. Where is the root of the universal tree of life? BioEssays 1999; 21:871-879.
6. Brinkmann H, Philippe H. Archaea sister group of bacteria? Indications from tree reconstruction artifacts in ancient phylogenies. Mol Biol Evol 1999; 16:817-825.
7. Brown JR, Doolittle WF. Root of the universal tree of life based on ancient aminoacyl-tRNA synthetase gene duplications. Proc Natl Acad Sci USA 1995; 92:2441-2445.
8. Wong JT, Chen J, Mat WK et al. Polyphasic evidence delineating the root of life and roots of biological domains. Gene 2007; 403:39-52.
9. Pennisi E. Is it time to uproot the tree of life? Science 1999; 284:1305-1307.
10. Wolf YI, Rogozin IB, Grishin NV et al. Genome trees and the tree of life. Trends Genet 2002; 18:472-479.
11. Xue H, Tong KL, Marck C et al. Transfer RNA paralogs: evidence for genetic code-amino acid biosynthesis coevolution and an archaeal root of life. Gene 2003; 310:59-66.
12. Xue H, Ng SK, Tong KL et al. Congruence of evidence for a methanopyrus-proximal root of life based on transfer RNA and aminoacyl-tRNA synthetase genes. Gene 2005; 360:120-130.
13. Sun FJ, Caetano-Anolles G. Evolutionary patterns in the sequence and structure of transfer RNA: early origins of archaea and viruses. PLoS Computaional Biol 2008; 4:e1000018.
14. Woese CR, Kandler O, Wheelis ML. Towards a natural system of organisms: Proposal for the domains Archaea, Bacteria and Eucarya. Proc Natl Acad Sci USA 1990; 87:4576-4579.
15. Woese CR. Interpreting the universal phylogenetic tree. Proc Natl Acad Sci USA 2000; 97:8392-8396.
16. Olsen GJ, Woese CR, Overbeek R. The winds of (evolutionary) change: breathing new life into microbiology. J Bacteriol 1994; 176:1-6.
17. Danchin A, Fang G, Noria S. The extant core bacterial proteome is an archive of the origin of life. Proteomics 2007; 7:875-889.
18. Holland M. The unseen extreme: microbiology at deep-sea hydrothermal vents. J Marine Edu 2005; 21:45-48.
19. Eigen M, Winkler-Oswatitsch R. Transfer-RNA, an early gene? Naturwissen 1981; 68:282-292.
20. Florentz C, Giege R. tRNA-like structures in plant viral RNAs. In: Soll D, RajBhandary UL, eds. tRNA Structure, Biosynthesis and Function: Washington DC:ASM Press, 1995:141-163.
21. Maizels N, Weiner AM. The genomic-tag hypothesis: what molecular fossils tell us about the evolution of tRNA. In: Gesteland RF, Cech TR, Atkins JF, eds. The RNA World. Woodbury:Cold Spring Harbor Laboratory Press, 1999:79-111.
22. Tong KL, Wong JT. Anticodon and wobble evolution. Gene 2004; 333:169-177.
23. Simonson AB, Servin JA, Skophammer RG et al. Decoding the genome tree of life. Proc Natl Acad Sci USA 2005; 102:6608-6613.
24. Qin Y, Polacek N, Vesper O et al. The highly conserved LepA is a ribosomal elongation factor that back-translocates the ribosome. Cell 2006; 127:721-733.
25. Bocchetta M, Gribaldo S, Sanangelantoni A et al. Phylogenetic depth of the bacterial Genera Aquifex and Thermotoga inferred from analysis of ribosomal protein, elongation factor and RNA polymerase subunit sequences. J Mol Evol 2000; 50:366-380.
26. Brown JR, Douady CJ, Italia MJ et al. Universal trees based on large combined protein sequence data sets. Nature Genet 2001; 28:281-285.
27. Altschul SF et al. Gapped BLAST and PSI-BLAST: a new generation of protein database search programs. Nulc Acid Res 1997; 25:3389-3402.
28. Battistuzzi FU, Feijao A, Hedges SB. 2004. A genomic timescale of prokaryote evolution: insight into the origin of methanogenesis, phototrophy and the colonization of land. BMC Evol Biol 2004; 4:44-57.
29. Raymond J, Segre D. The effect of oxygen on biochemical networks and the evolution of complex life. Science 2006; 311:1764-1767.
30. Archetti M, Di Giulio M. The evolution of the genetic code took place in an anaerobic environment. J Theoret Biol 2007; 245:169-174.
31. Di Giulio M. The universal ancestor and the ancestor of bacteria were hyperthermophiles. J Mol Evol 2003; 57:721-730.
32. Di Giulio M. A comparison of proteins of Pyrococcus furiosus and Pyrococcus abyssi: barophily in the physicochemical properties of amino acids and in the genetic code. Gene 2005; 346:1-6.
33. Di Giulio M. Structuring of the genetic code took place at acidic pH. J Theoret Biol 2005; 237:219-226.
34. Stetter KO. Hyperthermophilic prokaryotes. FEMS Microbiol Rev 1996; 18:149-158.
35. Gaucher EA, Govindarajan S, Ganesh OK. Palaeotemperature trend for Precambrian life inferred from resurrected proteins. Nature 2008; 451:704-708.
36. Gouy M, Chaussidon M. Evolutionary biology: ancient bacteria liked it hot. Nature 2008; 451:635-636.
37. Leigh JA. Chapter 4. Evolution of energy metabolism. In: Biodiversity of Microbial Life, eds. Staley JT and Reysenbach A-L. Wiley-Liss 2002; 103-120.
38. Falkowski PG. Tracing oxygen's imprint on Earth's metabolic evolution. Science 2006; 311:1724-1725.
39. Slesarev AI et al. The complete genome of hyperthermophilic Methanopyrus kandleri AV19 and monophyly of archaeal methanogens. Proc Natl Acad Sci USA 2002; 99:4644-4649.
40. Dopson M, Baker-Austin C, Hind A et al. Characterization of Ferroplasma isolates and Ferroplasma acidarmanus sp., extreme acidophiles from acid mine drainage and industrial bioleaching environments, Appl Environ Microbiol 2004; 70:2079-2088.
41. Golyshina OV, Timmis KN. Ferroplasma and relatives, recently discovered cell wall-lacking archaea making a living in exteremely acid, heavy metal-rich environments. Environ Microbiol 2005; 7:1277-1288.
42. Searcy DG, Hixon WG. Cytoskeletal origins in sulfur-metabolizing archaebacteria, BioSystems 1991; 25:1-11.
43. Margulis L, Chapman M, Guerrero R et al. The last eukaryotic common ancestor (LECA): acquisition of cytoskeletal motility from aerotolerant spirochaetes in the Proterozoic Eon. Proc Natl Acad Sci USA 2006; 103:13080-13085.
44. Pisani D, Cotton JA, McInerney JO. Supertrees disentangle the chimeric origin of eukaryotic genomes, Mol Biol Evol 2007; 24:1752-1760.
45. Kurland CG, Collins LJ, Penny D. Genomics and the irreducible nature of eukaryotic cells. Science 2006; 312:1011-1014.
46. Fitzpatrick DA, Creevy CJ, McInerney JO. Genome phylogenies indicate a meaningful proteobacterial phylogeny and support a grouping of the mitochnodria with the Rikettsiales. Mol Biol Evol 2006; 23:74-85.
47. Poole AM, Penny D. Evaluating hypotheses for the origin of eukaryotes. BioEssays 2006; 29:74-84.
48. Mat WK, Xue H, Wong JT. The genomics of LUCA. Frontiers in Biosc 2008; 13:5605-5613.
49. Tatusov RL, Koonin EV, Lipman DJ. A genomic perspective on protein families. Science 1997; 278:631-637.
50. Ouzounis CA, Kunin V, Darzentas N et al. A minimum estimate for the gene content of the last universal common ancestor—exobiology from a terrestrial perspective. Res Micobiol 2006; 157:57-68.
51. Woese CR. Evolutionary questions: the "Progenote". Science 1990; 247:789.
52. Nakabachi A, Yamashita A, Toh H et al. The 160-kilobase genome of the bacterial endosymbiont Carsonella. Science 2006; 314:267.
53. Islas S, Becerra A, Luisi PL et al. Comparative genomics and the gene complement of a minimal cell. Orig Life Evol Biosph 2004; 34:243-56.
54. Glass JI, Assad-Garcia N, Alperovich N et al. Essential genes of a minimal bacterium. Proc Natl Acad Sci USA 2006; 103:425-430.
55. Kobayashi K. et al. Essential Bacillus subtilis genes. Proc Natl Acad Sci USA 2003; 100:4678-4683.
56. Danchin A, Fang G, Noria S. The extant core bacterial proteome is an archive of the origin of life. Proteomics 2007; 7:875-889.
57. Ranea JA, Sillero A, Thornton JM et al. Protein superfamily evolution and the Last Universal Common Ancestor (LUCA). J Mol Evol 2006; 63:513-525.
58. Wang M, Yafremava LS, Caetano-Anolles D et al. Reductive evolution of architectural repertoires in proteomes and the birth of the tripartite world. Genome Res 2007; 17:1572-1585.
59. Forterre P. The two ages of the RNA world and the transition to the DNA world: a story of viruses and cells. Biochimie 2005; 87:793-803.
60. Forterre P. Three RNA cells for ribosomal lineages and three DNA viruses to replicate their genomes: a hypothesis for the origin of cellular domains. Proc Natl Acad Sci USA 2006; 103:3669-3674.
61. Islas S, Velasco AM, Becerra A et al. Hyperthermophily and the origin and earliest evolution of life. Int Microbiol 2003; 6:87-94.

62. Madigan MT, Martinko JM, Parker J. Brock Biology of Microorganisms. 10th ed. Prentice Hall 2002; 152:448-559.

63. Corliss JB, Baross JA, Hoffman SE. A hypothesis concerning the relationship between submarine hot springs and the origin of life on Earth. Oceanol Acta 1981; 4:59-69.

64. Shock EL. Chemical environment in submarine hydrothermal systems. Orig Life Evol Biosphere (suppl) 1992; 22:67-107.

65. Miller SL, Bada JL. Submarine hot springs and the origin of life. Nature 1988; 334:609-611.

66. Bada JL, Lazcano A. Origin of life. Some like it hot, but not the first biomolecules. Science 2002; 296:1982-1983.

67. Cody GD, Boctor NZ, Filley TR et al. Primordial carbonylated iron-sulfur compounds and the synthesis of pyruvate. Science 2000; 289:1337-1340.

68. Koonin EV, Martin W. On the origin of genomes and cells within inorganic compartments. Trends in Genet 2005; 21:647-654.

69. Malthus TR. An Essay on the Principle of Population. New York:Norton, 1976 rpt.

70. Larson EJ. On the Origins of Darwinsism Evolution. Modern Library Chronicles, 2004:68.

71. Henstra A, Dijkema C, Stams AJM. Archaeoglobus fulgidus couples CO oxidation to sulfate reduction and acetogenesis with transient formate accumulation. Environ Microbiol 2007; 9:1836-1841.

72. Berg IA, Kockelkorn D, Buckel W et al. A 3-hydroxypropionate/ 4-hydroxybutyrate autotrophic carbon dioxide assimilation pathway in Archaea. Science 2007; 318:1782-1786.

73. Thauer RK. A fifth pathway of carbon fixation. Science 2007; 318:1732-1733.

74. Danon A, Stoeckenius W. Photophosphorylation in Halobacterium halobium. Proc Natl Acad Sci USA 1974; 71:1234-1238.

75. Schafer G, Engelhard M, Muller V. Bioenergetics of the Archaea. Microbiol Mol Biol Rev 1999; 63:570-620.

76. Sleep NH, Zahnle KJ, Kasting JF et al. Annihilation of ecosystems by large asteroid impacts on the early Earth. Nature 1989; 342:139-142.

77. Kasting JF, Siefert JL. Life and the evolution of Earth's atmosphere. Science 2002; 296:1066-1068.

78. Wong JT. The evolution of a universal genetic code. Proc Natl Acad Sci USA 1976; 73:2336-2340.

79. Macario AJL, Conway de Macario E. The molecular chaperon system and other anti-stress mechanisms in archaea. Frontiers Biosc 2001; 6:d262-283.

80. Galagan JE et al. The genome of M. acetivorans reveals extensive metabolic and physiological diversity. Genome Res 2002; 12:532-542.

INDEX